Engineering Fundamentals & Problem Solving

Arvid R. Eide
Iowa State University

Roland D. Jenison
Iowa State University

Larry L. Northup
Iowa State University

Steven K. Mickelson
Iowa State University

Sixth Edition

Connect
Learn
Succeed™

The McGraw-Hill Companies

ENGINEERING FUNDAMENTALS AND PROBLEM SOLVING, SIXTH EDITION

ISBN: 978-0-07-131507-4
MHID: 0-07-131507-1

Vice President & Editor-in-Chief: *Marty Lange*
Vice President EDP/Central Publishing Services: *Kimberly Meriwether David*
Publisher: *Raghu Srinivasan*
Sponsoring Editor: *Peter Massar*
Marketing Manager: *Curt Reynolds*
Developmental Editor: *Lora Neyens*
Project Manager: *Melissa Leick*
Design Coordinator: *Brenda A. Rolwes*
Cover Design: *Studio Montage, St. Louis, Missouri*
Cover Images: *Aerial View of Wind Generators:* © Thinkstock/Masterfile; *Solar Power:* © Royalty-Free/CORBIS; *Hydroelectric Dam in Imatra, Finland:* © Kai Honkanen/PhotoAlto; *Aerial View of Rice Fields:* © Ingram Publishing/SuperStock
Buyer: *Susan K. Culbertson*
Media Project Manager: *Balaji Sundararaman*
Compositor: *Lachina Publishing Services*
Typeface: *10/12 Palatino*
Printer: *R. R. Donnelley*

Library of Congress Cataloging-in-Publication Data

Engineering fundamentals & problem solving / Arvid R. Eide . . . [et al.]. — 6th ed.
    p. cm.
    ISBN-13: 978-0-07-353491-6
    ISBN-10: 0-07-353491-9
    1. Engineering. 2. Problem solving. I. Eide, Arvid R. II. Title: Engineering fundamentals and problem solving.

TA147.E52 2011
620—dc22

2010048327

www.mhhe.com

Contents

Preface

To the Student

As you begin the study of engineering no doubt you are filled with enthusiasm, curiosity, and a desire to succeed. Your first year will be spent primarily in establishing a solid foundation in mathematics, basic sciences, and communications. Also, you will be introduced to selected engineering topics that will demonstrate how engineers approach problem solving, arrive at correct solutions, and interface with other engineering professionals and the general public to implement the solutions. You will see how mathematics, science, and communications provide the means to solve problems and convey the solutions in a manner that can be clearly understood and quickly verified by the appropriate persons. You will quickly discover the need for more in-depth study in many engineering subjects in order to solve increasingly complex problems. As authors, we believe the material presented in this book will provide you with a fundamental understanding of how engineers function in today's technological world. After your study of topics in this text, we believe you will be eager to enter the advanced engineering subjects in your chosen discipline, confident that you will successfully achieve your educational goals.

To the Instructor

Engineering courses for first-year students cover a wide range of topics from an overview of the engineering profession to discipline-specific subjects. A broad set of course goals, including coverage of prerequisite material, motivation, and retention, have spawned a variety of first-year activity. Courses in introductory engineering and problem solving routinely utilize spreadsheets and mathematical solvers in addition to teaching the rudiments of a computer language. The Internet has become a major instructional tool, providing a wealth of data to supplement your class notes and textbooks. This sixth edition continues the authors' intent to introduce the profession of engineering and to provide students with many of the tools and techniques needed to succeed.

The sixth edition of this text draws on the experiences the authors have encountered with the first five editions and incorporates many excellent suggestions from faculty and students using the text. Over the past 30 years the fundamentals of problem solving have remained nearly the same, but the numerical tools and presentation techniques have improved tremendously. Therefore our general objectives remain the same for this sixth edition, and we have concentrated on new problems and improvements in the textual material.

The objectives are (1) to motivate engineering students during their first year when exposure to the subject matter of engineering is limited, (2) to provide students with experience in solving problems in both SI and customary units while presenting solutions in a logical manner, (3) to introduce students to subject areas common to most engineering disciplines that require the application of fundamental engineering concepts, and (4) to develop students' skills in solving open-ended problems.

The material in this book is presented in a manner that allows each of you to emphasize certain aspects more than others without loss of continuity. In the sixth edition, five chapters from the fifth edition have been subdivided to clarify the material coverage and aid you in planning your course. The problems that follow most chapters vary in difficulty so that students can experience success rather quickly and still be challenged as problems become more complex. A large number of the problems are new or revised.

There is sufficient material in the seventeen chapters for a three-credit semester course. By omitting some chapters and/or by varying coverage from term to term, you can present a sound introductory problem-solving course in two to four quarter credits or two semester credits.

The book may be visualized as having three major sections. The first, encompassing the first three chapters, is an introduction to the engineering profession. Chapter 1 provides information on engineering disciplines and functions. If a formal orientation course is given separately, Chapter 1 can be simply a reading assignment and the basis for students to investigate disciplines of interest. Chapter 2 outlines the course of study and preparation for an engineering work environment. Interdisciplinary projects, teaming, and ethics are discussed. Chapter 3 is an introduction to the design process. If time permits, this material can be supplemented with case studies and your personal experiences to provide an interesting and motivating look at engineering.

The second major section, Processing Engineering Data, includes materials we believe that all engineering students require in preparation for success in the engineering profession. Chapters 4 and 5 provide procedures for approaching an engineering problem, determining the necessary data and method of solution, and presenting the results. The authors have found that emphasis in this area will reap benefits when the material and problems become more difficult later.

Chapters 6 and 7 include engineering estimations and dimensions and units (including both customary and SI units). Throughout the book discussions and example problems tend to emphasize SI metric. However, other dimension systems are used extensively today, so a number of our examples and problems contain nonmetric units to ensure that students are exposed to conversions and other units that are commonly used.

Chapters 8 and 9, Engineering Economy, demonstrate the importance of understanding the time value of money in making engineering decisions. Chapter 8 emphasizes basic calculations using everyday information such as credit card debt, savings accounts, and current interest rates. Chapter 9 follows with applications to engineering decision making for equipment selection, depreciation, and taxes. Chapters 10 and 11, Statistics, provide an introduction to a subject that is assuming a greater role in engineering decision making. Chapter 10 introduces basic descriptive statistics, linear regression and coefficient of correlation. Chapter 11 includes normal distributions but adds several new distributions including *Student's t, F* and *Chi-Square*. It also adds new material on the use of inferential statistics and a general introduction to randomized sampling and experimental design. The ability to take large amounts of test or field data, perform statistical analyses, and draw correct conclusions is crucial in establishing performance parameters. Engineering Economy and Statistics are subdivided, permitting you to choose the first chapter for an introduction to the fundamentals and, if time permits, applications to specific engineering activities can be covered.

The third major section provides engineering content that you can use to reinforce fundamentals from the previous section. Chapters 12 through 17 allow you as an instructor a great deal of flexibility. Chapters 12 and 13 on engineering mechanics provide an introduction to statics and strength of materials. Force vectors, two-dimensional force systems, and the conditions of equilibrium are emphasized in Chapter 12. Chapter 13 emphasizes stresses and strains and requires Chapter 12 as a prerequisite. Chapters 15 and 16, covering energy topics, have undergone significant updating. Chapter 15 discusses energy forms and sources. The authors believe that engineering students need to become aware of the world's current dependence on fossil fuels very early in their studies so they may apply this knowledge to the use and development of alternative sources of energy.

Chapter 16 follows with an introduction to thermodynamics and applications of the First and Second Laws of Thermodynamics. The study of Chapter 16 should be preceded by coverage of Chapter 15.

Chapters 14, Material Balance, and Chapter 17, Electrical Theory, complete the third major section and contain upgraded example problems.

Certain problems suggest the use of a computer or spreadsheet for solution. These are open-ended or "what-if" problems. Depending on the students' prior work with programming or spreadsheets, additional instruction may be required before attempting these problems.

The appendixes are provided as a ready reference on selected areas that will enable students to review topics from algebra and trigonometry. The National Society of Professional Engineers' Code of Ethics for Engineers is included and is highly recommended for reading and class discussion. A brief section on flow-charting for computer programming is included. Other appendixes include tables, unit conversions, formulas, and selected answers to chapter problems.

Because the text was written for first-year engineering students, mathematical expertise beyond algebra, trigonometry, and analytical geometry is not required for any material in the book. The authors have found, however, that additional experience in pre-calculus mathematics is very helpful as a prerequisite for this text.

Online Resources

The sixth edition features an accompanying website at www.mhhe.com/eide6e. The site features numerous instructor resources, including solutions to the problems, an image library, and new lecture PowerPoints for each chapter. New to this edition is also a test bank for each chapter that allows instructors to assign online homework and quizzes. This online homework is gradable and can be edited by instructors.

McGraw-Hill Create™

Craft your teaching resources to match the way you teach! With McGraw-Hill Create™, www.mcgrawhillcreate.com, you can easily rearrange chapters, combine material from other content sources, and quickly upload content you have written like your course syllabus or teaching notes. Find the content you need in Create by searching through thousands of leading McGraw-Hill textbooks. Arrange your book to fit your teaching style. Create even allows you to personalize your book's appearance by selecting the cover and adding your name, school, and course information. Order a Create book and you'll receive a complimentary print review copy in 3–5 business days or a complimentary electronic review copy (eComp) via email in minutes. Go to www.mcgrawhillcreate.com today and register to experience how McGraw-Hill Create™ empowers you to teach your students your way.

McGraw-Hill Higher Education and Blackboard have teamed up.

Blackboard, the Web-based course-management system, has partnered with McGraw-Hill to better allow students and faculty to use online materials and activities to complement face-to-face teaching. Blackboard features exciting social learning and teaching tools that foster more logical, visually impactful and active learning opportunities for students. You'll transform your closed-door classrooms into communities where students remain connected to their educational experience 24 hours a day.

This partnership allows you and your students access to McGraw-Hill's Create™ right from within your Blackboard course—all with one single sign-on. McGraw-Hill and Blackboard can now offer you easy access to industry leading technology and content, whether your campus hosts it, or we do. Be sure to ask your local McGraw-Hill representative for details.

Acknowledgments

The authors are indebted to many who assisted in the development of this edition of the textbook. First, we would like to thank the faculty of the former Division of Engineering Fundamentals and Multidisciplinary Design at Iowa State University, who have taught the engineering computations courses over the past 30 years. They, with the support of engineering faculty from other departments, have made the courses a success by their efforts. Several thousands of students have taken the courses, and we want to thank them for their comments and ideas, which have influenced this edition. The many suggestions of faculty and students alike have provided us with much information that was necessary to improve the previous editions. Special thanks go to the reviewers for this edition whose suggestions were extremely valuable. These suggestions greatly shaped the manuscript in preparation of the sixth edition.

The authors are also indebted to Dr. Cheryl Eide for her contributions to several of the book chapters and assistance in the acquisition of student profiles. Special thanks go to all those who provided biographical information for Chapter 1: Dean Hawkinson at Boeing Company, Julie Friend at DuPont, Donna Faust at ConAgra Foods, David Shallbert at Black and Veatch, Mark Land at Synder & Associates, Morgan Halverson at John Deere, Sheela Rajendran at Caterpillar's Global Engine Development–North America Division, Amy Dee Schlechte at General Mills, and Nick Mohr at Methodist Medical Center and Wishard Memorial Hospital in Indianapolis.

Finally, we thank our families for their continuing support of our efforts.

Arvid R. Eide
Roland D. Jenison
Larry L. Northup
Steven K. Mickelson

About the Authors

Arvid R. Eide received his baccalaureate degree in mechanical engineering from Iowa State University. Upon graduation he spent two years in the U.S. Army as a commissioned officer and then returned to Iowa State as an instructor while completing a master's degree in mechanical engineering. Professor Eide has worked for Western Electric, John Deere, and the Trane Company. He received his Ph.D. in 1974 and was appointed professor and Chair of Freshman Engineering, a position he held from 1974 to 1989, at which time Dr. Eide was appointed Associate Dean of Academic Affairs. In 1996, he returned to teaching as a professor of mechanical engineering. In January 2000 he retired from Iowa State University as professor emeritus of mechanical engineering.

Roland D. (Rollie) Jenison taught for 35 years in aerospace engineering and lower-division general engineering. He taught courses in engineering problem solving, engineering design graphics, aircraft performance, and aircraft stability and control, in addition to serving as academic adviser to many engineering students. He was a member of the American Society for Engineering Education (ASEE) and the American Institute of Aeronautics and Astronautics (AIAA), and published numerous papers on engineering education. He served as chair of the Engineering Design Graphics Division of ASEE in 1986–1987. He was active in the development of improved teaching methodologies through the application of team learning, hands-on projects, and open-ended problem solving. He retired in June 2000 as professor emeritus in the Department of Aerospace Engineering and Engineering Mechanics at Iowa State University.

Larry L. Northup is a professor emeritus of civil, construction, and environmental engineering at Iowa State University. He has 40 years of teaching experience, with 25 years devoted to lower-division engineering courses in problem solving, graphics, and design. He has two years of industrial experience and is a registered engineer in Iowa. He has been active in ASEE (Engineering Design Graphics Division), having served as chair of the Freshman Year Committee and Director of Technical and Professional Committees (1981–1984). He also served as chair of the Freshman Programs Constituent Committee of ASEE in 1983–1984.

Steven K. Mickelson is a professor in the Department of Agricultural and Biosystems Engineering (ABE) at Iowa State University. Dr. Mickelson is also the Director of the Center for Excellence in Learning and Teaching, the Co-director of Learning Communities, and the Associate Chair for the ABE department. His teaching specialties include computer-aided graphics, engineering problem solving, engineering design, and soil and water conservation engineering. His research areas include evaluation of best management practices for reducing surface and groundwater contamination, manure management evaluation for environmental protection of water resources, and the scholarship of teaching and learning. Dr. Mickelson has been very active in the American Society for Engineering Education and the American Society of Agricultural and Biosystems Engineers for the past 25 years. He received his agricultural engineering degrees from Iowa State University in 1982, 1984, and 1991.

CHAPTER 1

The Engineering Profession

Chapter Objectives

When you complete your study of this chapter, you will be able to:

- Understand the role of engineering in the world
- Understand how to prepare for a meaningful engineering career
- Understand the role of an engineer in the engineering workplace
- Describe the responsibilities and roles of the most common engineering disciplines
- Gain academic career advice from past engineering graduates from various engineering disciplines

1.1 An Engineering Career

The rapidly expanding and developing sphere of science and technology may seem overwhelming to the individual exploring a career in a technological field. A technical specialist today may be called engineer, scientist, technologist, or technician, depending on education, industrial affiliation, and specific work. For example, about 380 colleges and universities offer close to 1 900 engineering programs accredited by the Accreditation Board for Engineering and Technology (ABET) or the Canadian Engineering Accreditation Board (CEAB). Included in these programs are such traditional specialties as aerospace, agricultural, architectural, chemical, civil, computer, construction, electrical, industrial, manufacturing, materials, mechanical, and software engineering—as well as expanding bioengineering, biomedical, biological, electromechanical, environmental and telecommunications. Programs in engineering, mechanics, mining, nuclear, ceramic, software, and petroleum engineering add to a lengthy list of career options in engineering alone. Coupled with thousands of programs in science and technical training offered at hundreds of universities, colleges, and technical schools, the task of choosing the right field no doubt seems formidable (Fig. 1.1).

Since you are reading this book, we assume that you are interested in studying engineering or at least are trying to decide whether to do so. Up to this point in your academic life you probably have had little experience with engineering as a career and have gathered your impressions from advertising materials, counselors, educators, and perhaps a practicing engineer or two. Now you must investigate as many careers as you can as soon as possible to be sure of making the right choice.

Figure 1.1

Imagine the number of engineers who were involved in the design of the windmill related to construction, material choices, electrical systems, and mechanical systems.

The study of engineering requires a strong background in mathematics and the physical sciences. Section 1.5 discusses typical areas of study within an engineering program that lead to the bachelor's degree. You also should consult with your academic counselor about specific course requirements. If you are enrolled in an engineering program but have not chosen a specific discipline, consult with an adviser or someone on the engineering faculty about particular course requirements in your areas of interest.

When considering a career in engineering or any closely related fields, you should explore the answers to several questions:

- What is engineering?
- What are the career opportunities for engineers?
- What are the engineering disciplines?
- Where does the engineer fit into the technical spectrum?
- How are engineers educated?
- What is meant by professionalism and engineering ethics?
- What have engineers done in the past?
- What are engineers doing now? What will engineers do in the future?
- What are the workplace competencies needed to be a successful engineer?

Finding answers to such questions is difficult and time consuming, but essential to determining the proper path for you as an individual. To assist you in assessing your educational goals we have included a number of student profiles. These are students that have recently graduated from an accredited engineering program and selected different career paths. Each student background is unique and each career path is different. We hope you find these helpful.

In 1876, 15 men led by Thomas Alva Edison gathered in Menlo Park, New Jersey, to work on "inventions." By 1887, the group had secured over 400 patents, including ones for the electric lightbulb and the phonograph. Edison's approach typified that used for early engineering developments. Usually one person possessed nearly all the knowledge in one field and directed the research, development, design, and manufacture of new products in this field.

Today, however, technology has become so advanced and sophisticated that one person cannot possibly be aware of all the intricacies of a single device or process. The concept of systems engineering thus has evolved; that is, technological problems are studied and solved by a technology team.

Scientists, engineers, technologists, technicians, and craftspersons form the *technology team*. The functions of the team range across what often is called the *technical spectrum*. At one end of the spectrum are functions that involve work with scientific and engineering principles. At the other end of this technical spectrum are functions that bring designs into reality. Successful technology teams use the unique abilities of all team members to bring about a successful solution to a human need.

Each of the technology team members has a specific function in the technical spectrum, and it is of utmost importance that each specialist understands the role of all team members. It is not difficult to find instances where the education and tasks of team members overlap. For any engineering accomplishment, successful team performance requires cooperation that can be realized only through an understanding of the functions of the technology team. The technology team is one part of a larger team that has the overall responsibility for bringing a device, process, or system into reality. This team, frequently called a project or design team, may include managers, sales representatives, field service persons, financial representatives, and purchasing personnel in addition to the technology team members. These project teams meet frequently from the beginning of the project to ensure that schedules and design specifications are met, and that potential problems are diagnosed early. We will now investigate each of the team specialists in more detail.

1.2.1 Scientist

Scientists have as their prime objective increased knowledge of nature (see Fig. 1.2). In the quest for new knowledge, the scientist conducts research in a systematic manner. The research steps, referred to as the *scientific method*, are often summarized as follows:

1. Formulate a hypothesis to explain a natural phenomenon.
2. Conceive and execute experiments to test the hypothesis.
3. Analyze test results and state conclusions.
4. Generalize the hypothesis into the form of a law or theory if experimental results are in harmony with the hypothesis.
5. Publish the new knowledge.

An open and inquisitive mind is an obvious characteristic of a scientist. Although the scientist's primary objective is that of obtaining an increased knowledge of nature, many scientists are also engaged in the development

Figure 1.2

Scientists use the laboratory for discovery of new knowledge.

of their ideas into new and useful creations. But to differentiate quite simply between the scientist and engineer, we might say that the true scientist seeks to understand more about natural phenomena, whereas the engineer primarily engages in applying new knowledge.

1.2.2 Engineer

The profession of engineering takes the knowledge of mathematics and natural sciences gained through study, experience, and practice and applies this knowledge with judgment to develop ways to utilize the materials and forces of nature for the benefit of all humans.

An engineer is a person who possesses this knowledge of mathematics and natural sciences, and through the principles of analysis and design applies this knowledge to the solution of problems and the development of devices, processes, structures, and systems. Both the engineer and scientist are thoroughly educated in the mathematical and physical sciences, but the scientist primarily uses this knowledge to acquire new knowledge, whereas the engineer applies the knowledge to design and develops usable devices, structures, and processes. In other words, the scientist seeks to know, the engineer aims to do.

You might conclude that the engineer is totally dependent on the scientist for the knowledge to develop ideas for human benefit. Such is not always the case. Scientists learn a great deal from the work of engineers. For example, the science of thermodynamics was developed by a physicist from studies of practical steam engines built by engineers who had no science to guide them. On the other hand, engineers have applied the principles of nuclear fission discovered by scientists to develop nuclear power plants and numerous other devices and systems requiring nuclear reactions for their operation. The scientist's and engineer's functions frequently overlap, leading at times to a somewhat blurred image of the engineer. What distinguishes the engineer from the scientist in broad terms, however, is that the engineer often conducts research but does so for the purpose of solving a problem.

The end result of an engineering effort—generally referred to as *design*—is a device, structure, system, or process that satisfies a need. A successful design is achieved when a logical procedure is followed to meet a specific need. The procedure, called the *design process*, is similar to the scientific method with respect to a step-by-step routine, but it differs in objectives and end results. The design process encompasses the following activities, all of which must be completed.

1. Define the problem to be solved.
2. Acquire and assemble pertinent data.
3. Identify solution constraints and criteria.
4. Develop alternative solutions.
5. Select a solution based on analysis of alternatives.
6. Communicate the results.

As the designer proceeds through each step, new information may be discovered and new objectives may be specified for the design. If so, the designer must backtrack and repeat steps. For example, if none of the alternatives appears to be economically feasible when the final solution is to be selected, the designer must redefine the problem or possibly relax some of the constraints to admit less expensive alternatives. Thus, because decisions must frequently be made at each step as a result of new developments or unexpected outcomes, the design process becomes iterative.

As you progress through your engineering education, you will solve problems and learn the design process using the techniques of analysis and synthesis. Analysis is the act of separating a system into its constituent parts, whereas synthesis is the act of combining parts into a useful system. In the design process you will observe how analysis and synthesis are utilized to generate a solution to a human need.

1.2.3 Technologist and Technician

Much of the actual work of converting the ideas of scientists and engineers into tangible results is performed by technologists and technicians (see Fig. 1.3). A technologist generally possesses a bachelor's degree and a technician an associate's degree. Technologists are involved in the direct application of their education and experience to make appropriate modifications in designs as the need arises. Technicians primarily perform computations and experiments and prepare design drawings as requested by engineers and scientists. Thus technicians (typically) are educated in mathematics and science but not to the depth required of scientists and engineers. Technologists and technicians obtain a basic knowledge of engineering and scientific principles in a specific field and develop certain manual skills that enable them to communicate technically with all members of the technology team. Some tasks commonly performed by technologists and technicians include drafting, estimating, model building,

Figure 1.3

A technician makes sound measurements in an acoustics laboratory.

data recording and reduction, troubleshooting, servicing, and specification. Often they are the vital link between the idea on paper and the idea in practice.

1.2.4 Skilled Tradespersons/Craftspersons

Members of the skilled trades possess the skills necessary to produce parts specified by scientists, engineers, technologists, and technicians. Craftspersons do not need to have an in-depth knowledge of the principles of science and engineering incorporated in a design (see Fig. 1.4). They often are trained on the job, serving an apprenticeship during which the skills and abilities to build and operate specialized equipment are developed. Specialized positions include welder, machinist, electrician, carpenter, plumber, and mason.

1.3 The Engineering Profession

Engineering is an exciting profession. Engineers don't just sit in a cubicle and solve mathematical equations; they work in teams to solve challenging engineering problems to make life safer, easier, and more efficient for the world we live in. Engineers must demonstrate competence in initiative, professionalism, engineering knowledge, teamwork, innovation, communication, cultural adaptability, safety awareness, customer focus, general knowledge, continuous learning, planning, analysis and judgment, quality orientation, and integrity. Engineers help to shape government policies, international development, and education at all levels. Engineering is fun and challenging, and it provides for a meaningful career.

Figure 1.4

7

The Engineering
Functions

Skilled craftspersons are key elements in a manufacturing process.

1.4 The Engineering Functions

As we alluded to in Sec. 1.2, engineering feats dating from earliest recorded history up to the Industrial Revolution could best be described as individual accomplishments. The various pyramids of Egypt were usually designed by one individual, who directed tens of thousands of laborers during construction. The person in charge called every move, made every decision, and took the credit if the project was successful or accepted the consequences if the project failed.

The Industrial Revolution brought a rapid increase in scientific findings and technological advances. One-person engineering teams were no longer practical or desirable. Today, no single aerospace engineer is responsible for a jumbo jet and no one civil engineer completely designs a bridge. Automobile manufacturers assign several thousand engineers to the design of a new model. So we not only have the technology team as described earlier, but we have engineers from many disciplines who are working together on single projects.

One approach to explaining an engineer's role in the technology spectrum is to describe the different types of work that engineers do. For example, agricultural, biological, civil, electrical, mechanical, and other engineers become involved in design, which is an engineering function. The *engineering functions*, which are discussed briefly in this section, are research, development, design, production, testing, construction, operations, sales, management, consulting, and teaching. Several of the *engineering disciplines* will be discussed later in the chapter.

To avoid confusion between "engineering disciplines" and "engineering functions," let us consider the following. Normally a student selects a curriculum (e.g., aerospace, chemical, mechanical) either before or soon after admission to an engineering program. When and how the choice is made varies with each school. The point is, the student does not choose a function but rather a discipline. To illustrate further, consider a student who has chosen mechanical engineering. This student will, during an undergraduate education, learn how mechanical engineers are involved in the engineering functions of research, development, design, and so on. Some program options allow a student to pursue an interest in a specific subdivision within the curriculum, such as energy conversion in a mechanical engineering program. Most other curricula have similar options.

Upon graduation, when you accept a job with a company, you will be assigned to a functional team performing in a specific area such as research, design, or sales. Within some companies, particularly smaller ones, you may become involved in more than one function—design *and* testing, for example. It is important to realize that regardless of your choice of discipline, you may become involved in one or more of the functions discussed in the following paragraphs.

An Engineer in Industry—On the Leading Edge of Commercial Aircraft Development

Dean A. Hawkinson

After receiving his BS degree in aerospace engineering in 2000, Dean Hawkinson accepted a job with The Boeing Company's commercial airplane division in the Seattle, Washington, area. In 2009, he earned his MS degree in aeronautical engineering from the University of Washington by attending classes virtually while continuing to work at Boeing.

One of Hawkinson's interesting experiences at Boeing was conducting 787 wind tunnel tests. "We use these expensive tests to verify the aerodynamic efficiency and stability and control characteristics predicted by computational modeling and simulations," he explains.

Hawkinson believes that gathering a variety of skills and working with a mentor can help engineers remain flexible and excel at all opportunities afforded them. To achieve this, he took advantage of a job rotation program, which is common in many large companies. Hawkinson gained experience in several facets of the airplane design process with three one-year rotations in a variety of aerodynamics disciplines.

Furthermore, through consultation with a mentor, Hawkinson stretched beyond aerodynamic properties to examine 787 airframe structural capabilities. Recently, he was responsible for coordinating the engineering effort in full-scale structural testing of the 787-8. "The major static test represents the capstone of proof tests that demonstrate structural response of the 787 airframe to external loads prior to first flight and ulitmately certification," he explains.

During his undergraduate education, Hawkinson participated in many extracurricular activities, including the student chapter of the American Institute of Aeronautics and Astronautics (AIAA). Later, after joining Boeing, he was appointed to the Board of Directors for AIAA as the Young Professional Liaison. He also participated in an aerospace engineering internship and a co-op program. Hawkinson credits these activities with strengthening his ability to adapt as an engineer, which has proven to be valuable in his work. As he says, "To succeed in a high technology industry, one must be prepared to apply good engineering judgment to an ever-changing set of challenges."

1.4.1 Research

Successful research is one catalyst for starting the activities of a technology team or, in many cases, the activities of an entire industry. The research engineer seeks new findings, as does the scientist; but keep in mind that the research engineer also seeks a way to use the discovery.

Key qualities of a successful research engineer are perceptiveness, patience, and self-confidence. Most students interested in research will pursue the master's and doctor's degrees in order to develop their intellectual abilities and the necessary research skills. An alert and perceptive mind is needed to recognize nature's truths when they are encountered. When attempting to reproduce natural phenomena in the laboratory, cleverness and patience are prime attributes. Research often involves tests, failures, retests, and so on for long periods of time (see Fig. 1.5). Research engineers therefore are often discouraged and frustrated and must strain their abilities and rely on their self-confidence in order to sustain their efforts to a successful conclusion.

Billions of dollars are spent each year on research at colleges and universities, industrial research laboratories, government installations, and independent research institutes. The team approach to research is predominant today primarily because of the need to incorporate a vast amount of technical information into the research effort. Individual research also is carried out but not to the extent it was several years ago. A large share of research monies are channeled into the areas of energy, environment, health, defense, and space exploration. A fast growing research area is nanotechnology. The Royal Academy of Engineering describes it this way: "Nanotechnology is

Figure 1.5

Research requires high-cost, sophisticated equipment.

the application of nanoscale science, engineering and technology to produce novel materials and devices, including biological and medical applications." Research funding from federal agencies is very sensitive to national and international priorities. During a career as a research engineer you might expect to work in many diverse, seemingly unrelated areas, but your qualifications will allow you to adapt to many different research efforts.

1.4.2 Development

Using existing knowledge and new discoveries from research, the development engineer attempts to produce a functional device, structure, or process (see Fig. 1.6). Building and testing scale or pilot models is the primary means by which the development engineer evaluates ideas. This has been made easier to accomplish with 3-D modeling software, rapid prototyping equipment, and virtual reality software and visualization tools. A major portion of development work requires use of well-known devices and processes in conjunction with established theories. Thus reading available literature and having a solid background in the sciences and in engineering principles are necessary for the engineer's success.

Many people who suffer from heart irregularities are able to function normally today because of the pacemaker, an electronic device that maintains

Conducting Engineering Research
Julie Friend

Julie Friend received her undergraduate degrees in chemical engineering and French and her PhD in chemical engineering. She is currently a senior research engineer with DuPont. DuPont, headquartered in Wilmington, Delaware, with over 60 000 employees in 70 countries worldwide, offers a wide range of innovative products and services for markets including agriculture, nutrition, electronics, communications, safety and protection, home and construction, transportation, and apparel.

Dr. Friend works in research and development for DuPont and focuses on early-stage work, meaning that she determines the appropriate combination and quantity of chemicals as well as the steps required to make a desired compound. Her work can involve lab experiments, modeling, and economic evaluations. Her goal is to come up with a method to take through development and scale up to a safe and cost-effective commercial process. To do this, Dr. Friend works in teams with other chemists, biologists, and engineers.

Dr. Friend was in marching band and participated in various campus groups while pursuing her undergraduate degree, but her other main "activity" was studying abroad in Switzerland her junior year. She felt it was very important while in school to have something to do that was completely different than engineering to take her mind off the engineering coursework.

Dr. Friend has found that a degree in engineering can take a person in many different directions, something she didn't fully appreciate as an undergrad. She knows engineers who work in the more traditional areas of manufacturing, consulting, and research, but also engineers who work in law, medicine, marketing, regulatory affairs, and human resources. As for her future, who knows! The last five years have been nothing like she had expected, she says, and she doesn't expect the next five to be any different in that respect.

Figure 1.6

11

The Engineering
Functions

Development engineers take an idea and produce a concept of a functional product or system. The result of this activity is passed on to the design engineers for completing necessary details for production.

a regular heartbeat. The pacemaker is an excellent example of the work of development engineers. The first pacemakers developed in the 1950s were externally AC powered units that sent pulses of energy through an implanted lead wire to the heart. However, the power requirement for heart stimulus was so great that patients suffered severe burns on their chests. As improvements were studied, research in surgery and electronics enabled development engineers to devise a battery-powered external pacemaker that eliminated concerns about chest burns and allowed the patients more mobility. Although more efficient from the standpoint of power requirements, the devices were uncomfortable, and patients frequently suffered infection where the wires entered the chest. Finally, two independent teams developed the first internal pacemaker, eight years after the original pacemaker had been tested. Their experience and research with tiny pulse generators for spacecraft led to this achievement. But the very fine wire used in these early models proved to be inadequate and quite often failed, forcing patients to have the entire pacemaker replaced. A team of engineers at General Electric developed a pacemaker that incorporated a new wire, called a *helicable*. The helicable consisted of 49 strands of wire coiled together and then wound into a spring. The spring diameter was about 46 μm, one-half the diameter of a human hair. Thus, with doctors and development engineers working together, an effective, comfortable device was perfected that has enabled many heart patients to enjoy a more active life.

To illustrate how important technological changes can arise from the work of development engineers, here's what's become of the pacemaker. Today pacemakers have been developed by engineers and doctors to operate at more than one speed, enabling the patient to speed up or slow down heart rate depending on physical activity. Motion sensors are used to detect breathing rate and can be programmed by the doctor for individual needs. The new pager-sized pacemakers are also now programmed to act as sensors to detect problems with intermittent atrial fibrillation and switch the pacemaker to a different mode that paces the lower heart chamber only. For those with damage to the heart muscles and the heart's electrical systems, the new biventricular pacemaker can pace both of the lower heart chambers so that they beat at the same rate. In addition to the advances in pacemakers, medical research has evolved many other procedures for correcting heart deficiencies. The field of electrophysiology, combining cardiology with electrical and computer engineering, is enabling thousands of persons with heart irregularities to live productive and happy lives.

We have discussed the pacemaker in detail to point out that important changes in technology can arise from the work of development engineers. That it took only 13 years to develop an efficient, dependable pacemaker, five years to develop the transistor, and 25 years to develop the digital computer indicates that modern engineering methods generate and improve products nearly as fast as research generates new knowledge.

Successful development engineers are ingenious and creative. Astute judgment often is required in devising models that can be used to determine whether a project will be successful in performance and economical in production. Obtaining an advanced degree is helpful, but not as important as it is for an engineer who will be working in research. Practical experience more than anything else produces the qualities necessary for a career as a development engineer.

Development engineers frequently are asked to demonstrate that an idea will work. Within certain limits they do not work out the exact specifications that a final product should possess. Such matters are usually left to the design engineer if the idea is deemed feasible.

1.4.3 Design

The development engineer produces a concept or model that is passed on to the design engineer for converting into a device, process, or structure (see Fig. 1.7). The designer relies on education and experience to evaluate many possible design options, keeping in mind the cost of manufacture, ease of production, availability of materials, and performance requirements. Usually several designs and redesigns will be undertaken before the product is brought before the general public.

To illustrate the role that the design engineer plays, we will discuss the development of the side air bags for added safety in automobiles. Side air bags created something of a design problem, as designers had to decide where and how the air bags would be fastened to the car body. They had to determine what standard parts could be used and what parts had to be designed from scratch. Consideration was given to how to hide the air bags so not to

Figure 1.7

13

*The Engineering
Functions*

A team of design engineers review a proposed design solution.

take away from the aesthetics of the interior of the car while still providing maximum impact safety. Materials to be used for attaching the air bags to the frame and for the air bag itself had to be selected. An inflation device had to be designed that would give flawless performance.

The 12 000 or so other parts that form the modern automobile also demand numerous considerations: Optimum placement of engine accessories, comfortable design of seats, maximization of trunk space, and aesthetically pleasing body design all require thousands of engineering hours if the model is to be successful in a highly competitive industry.

Like the development engineer, the designer is creative. But where the development engineer is usually concerned only with a prototype or model, the designer is restricted by the state of the art in engineering materials, production facilities, and, perhaps most important, economic considerations. An excellent design from the standpoint of performance may be completely impractical from a monetary point of view. To make the necessary decisions, the designer must have a fundamental knowledge of many engineering specialty subjects as well as an understanding of economics and people.

1.4.4 Production and Testing

When research, development, and design have created a device for use by the public, the production and testing facilities are geared for mass production (see Figs. 1.8 and 1.9). The first step in production is to devise a schedule that will efficiently coordinate materials and personnel. The production engineer is responsible for such tasks as ordering raw materials at the optimum times,

Figure 1.8

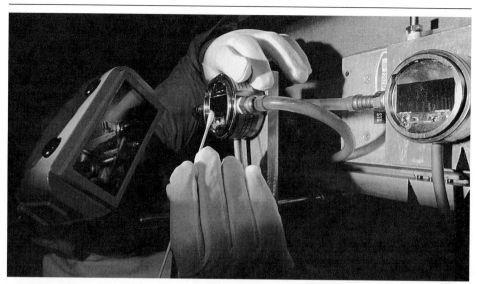

Test engineering is a major component in the development of new products.

Figure 1.9

Modern production facilities are often controlled from a central computer system.

setting up the assembly line, and handling and shipping the finished product. The individual who chooses this field must be able to visualize the overall operation of a particular project as well as know each step of the production effort. Knowledge of design, economics, and psychology is of particular importance for production engineers.

Test engineers work with a product from the time it is conceived by the development engineer until such time as it may no longer be manufactured. Some testing can be conducted prior to the creation of an actual physical model through the use of 3D modeling and analysis software. Finite element analyses allow the test engineer to evaluate load changes, temperature changes, pressure change, and many other physical variations prior to building a prototype for testing. In the automobile industry, for example, test engineers evaluate new devices and materials that may not appear in automobiles for several years. At the same time they test component parts and completed cars currently coming off the assembly line. Test engineers are usually responsible for quality control of the manufacturing process. In addition to the education requirements of the design and production engineers, a fundamental knowledge of statistics is beneficial to the test engineer.

1.4.5 Construction

The counterpart of the production engineer in manufacturing is the construction engineer in the building industry (see Fig. 1.10). When an organization bids on a competitive construction project, the construction engineer begins the process by estimating material, labor, and overhead costs. If the bid is successful, a construction engineer assumes the responsibility of coordinating the project. On large projects, a team of construction engineers may supervise the individual segments of construction such as mechanical (plumbing), electrical (lighting), and civil (building). In addition to a strong background in engineering fundamentals, the construction engineer needs on-the-job experience and an understanding of labor relations. With the increased regulation of energy and sustainability, the construction engineer must also be familiar with LEED, an internationally "green" building certification system. LEED certification

Figure 1.10

Numerous engineers from many disciplines are involved in the design and construction of massive structures such as this hydroelectric plant.

Figure 1.11

Operations engineers help to lay out manufacturing facilities for optimum efficiency.

assesses building design for energy savings, improved indoor environmental quality, CO2 emissions reduction, water efficiency, and stewardship of resources.

1.4.6 Operations

Up to this point, discussion has centered on the results of engineering efforts to discover, develop, design, and produce products that are of benefit to humans. For such work engineers obviously must have offices, laboratories, and production facilities in which to accomplish it. The major responsibility for supplying such facilities falls on the operations engineer (see Fig. 1.11). Sometimes called a plant engineer, this individual selects sites for facilities, specifies the layout for all facets of the operation, and selects the fixed equipment for climate control, lighting, and communication. Once the facility is in operation, the plant engineer is responsible for maintenance and modifications as requirements demand. Because this phase of engineering comes under the economic category of overhead, the operations engineer must be very conscious of cost and keep up with new developments in equipment so that overhead is maintained at the lowest possible level. Knowledge of basic engineering, industrial engineering principles, economics, and law are prime educational requirements of the operations engineer. Operation engineers also are trained in total quality improvement methods and lean manufacturing production products that help to provide "value" to the process and final product.

1.4.7 Sales

In many respects all engineers are involved in selling. To the research, development, design, production, construction, and operations engineers, selling means convincing management that money should be allocated for development of particular concepts or expansion of facilities. This is, in essence, selling one's own ideas. Sales engineering, however, means finding or creating a market for a product. The complexity of today's products requires an individual who is thoroughly familiar with materials in and operational procedures for consumer products to demonstrate to the consumer in layperson's terms how the products can be of benefit. The sales engineer is thus the liaison between the company and the consumer, a very important part of influencing a company's reputation. Therefore, excellent communication and teamwork skills are important to becoming a successful sales engineer. Some engineering schools are now providing a sales engineering major or minor.

An engineering background plus a sincere interest in people and a desire to be helpful are the primary attributes of a sales engineer. The sales engineer usually spends a great deal of time in the plant learning about the product to be sold. After a customer purchases a product, the sales engineer is responsible for coordinating service and maintaining customer satisfaction. As important as sales engineering is to a company, it still has not received the interest from new engineering graduates that other engineering functions have. (See Fig. 1.12.)

Figure 1.12

Sales engineers interact with people around the world using many forms of communications media.

1.4.8 Management

Traditionally management has consisted of individuals who are trained in business and groomed to assume positions leading to the top of the corporate ladder (Fig. 1.13). However, with the influx of scientific and technological data being used in business plans and decisions, and hence the increasing need for managers with knowledge and experience in engineering and science, a growing percentage of management positions are being assumed by engineers and scientists. Inasmuch as one of the principal functions of management is to use company facilities to produce an economically feasible product, and decisions often must be made that may affect thousands of people and involve millions of dollars over periods of several years, a balanced education of engineering or science and business seems to produce the best managerial potential. Engineering programs are now partnering with business programs to offer undergraduates the opportunity to obtain an undergraduate engineering degree and a master's of business degree at the same time, with as little as one year added onto the undergraduate degree program time.

1.4.9 Consulting

For someone interested in self-employment, a consulting position may be an attractive one (see Fig. 1.14). Consulting engineers operate alone or in partnership furnishing specialized help to clients who request it. Of course, as in any business, risks must be taken. Moreover, as in all engineering disciplines, a sense of integrity and a knack for correct engineering judgment are primary necessities in consulting.

Figure 1.13

Managers often must balance design projects, staff, and finances.

Figure 1.14

Consulting engineers often partner with clients in a team setting to provide engineering expertise and guidance.

A consulting engineer must possess a professional engineer's license before beginning practice. Consultants usually spend many years in the private, corporate or government world gaining experience in a specific area before going on their own. A successful consulting engineer maintains a business primarily by being able to solve unique problems for which other companies have neither the time nor capacity. In many cases large consulting firms maintain a staff of engineers of diverse backgrounds so that a wide range of engineering problems can be contracted.

1.4.10 Teaching

Individuals interested in helping others to become engineers will find teaching very rewarding (see Fig. 1.15). The engineering teacher must possess an ability to communicate abstract principles and engineering experiences in a manner that students can understand and appreciate. Teachers must follow general guidelines but are usually free to develop their own method of teaching and means of evaluating its effectiveness. In addition to teaching, the engineering educator can become involved in student advising, extension, and research. At research universities, faculty members rarely have a 100% teaching load. They usually split their time between teaching, research, and extension (off-campus instruction often for engineering professionals). These faculty members are rewarded and promoted based on peer-reviewed scholarly work related to their appointments, the ability to obtain external funds for research, and national recognition for their contribution to their profession.

Figure 1.15

Teaching provides opportunities for student coaching, mentoring, and research collaboration.

Engineering teachers today must have a mastery of fundamental engineering and science principles and knowledge of applications. Customarily, they must obtain an advanced degree in order to improve their understanding of basic principles, to perform research in a specialized area, and perhaps to gain teaching experience on a part-time basis.

The emphasis in the classrooms today is moving from teaching to learning. Methods of presenting material and involving students in the learning process to meet designed outcomes follow the sound educational principles developed by our education colleagues. You as a student will benefit greatly by these learning processes, which empower you to take control of your education through teamwork, active participation, and hands-on learning.

If you are interested in a teaching career in engineering or engineering technology, you should observe your teachers carefully as you pursue your degrees. Note how they approach the teaching process, the methodologies they use to stimulate learning, and their evaluation methods. Your initial teaching methods likely will be based on the best methods you observe as a student.

1.5 The Engineering Disciplines

There are over 25 specific disciplines of engineering that can be pursued for the baccalaureate degree. The opportunities to work in any of these areas are numerous. Most engineering colleges offer some combination of the disciplines, primarily as four-year programs. In some schools two or more disciplines, such as industrial, management, and manufacturing engineering, are combined within one department that may offer separate degrees, or include one discipline as a specialty within another discipline. In this case a degree in the area of specialty is not offered. Other common combinations of engineering disciplines include civil/construction/environmental, mechanical/aerospace, and electrical/computer.

Figure 1.16 gives a breakdown of the number of engineering degrees in six categories for 2008. Note that each category represents combined disciplines and does not provide information about a specific discipline within that category. The "other" category includes, among others, agricultural, biomedical, ceramic, materials, metallurgical, mining, nuclear, safety, and ocean engineering. The percentage of bachelor's degrees awarded to women was 18 percent of the total number of engineering graduates.

Seven of the individual disciplines will be discussed in this section. Engineering disciplines which pique your interest may be investigated in more detail by contacting the appropriate department, checking the library at your institution, and searching the Internet.

Managing Engineering Innovation

Donna Faust

Donna Faust received her BS in chemical engineering and her MBA in marketing, and is now a senior marketing manager with ConAgra Foods in Omaha, Nebraska. ConAgra Foods is one of the largest packaged foods companies in North America with major presence in retail outlets, foodservice and restaurants. Popular ConAgra Foods consumer brands include Banquet, Chef Boyardee, Healthy Choice, Hunt's, Marie Callender's, and many others,

Faust leads cross-functional development teams to identify new product opportunities that support ConAgra's growth strategy. And although her job title is marketing, she is still solving problems like an engineer. Faust's team talks with consumers about their needs, generates ideas of products that might solve the problems, works with engineers and product developers to create a prototype, sends the prototype home for feedback from consumers, works to optimize the products, manages total production financials to ensure profitability, then develops the marketing strategy and business plan for bringing new product into the marketplace.

As an undergraduate student, Faust's involvement in the Society of Women Engineers exposed her to many different engineering career opportunities and leadership experiences that were a valuable introduction to working with cross-functional teams and a good foundation for a career in management. She was also involved with the Salt Company Fellowship Leadership Team, which she credits with helping to develop strong interpersonal skills and self-awareness that serves her well in the business world.

Faust's advice for engineering students is to take advantage of every opportunity to work on projects with students in other majors. "Don't just hang out with engineers—when you get into the workforce you'll find that we really do have a slightly different language and way of approaching problems than our cross-functional partners," she advises. "Learn how to bridge those differences in perspectives and show understanding of your teammates' motivations and needs now so that you can effectively sell your ideas and succeed together after you graduate."

Figure 1.16

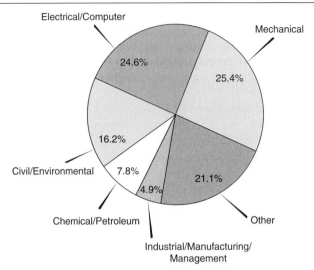

Engineering degrees by discipline. Total engineering degrees awarded in 2008 were 68 206. (ASEE Profiles of Engineering & Technology Colleges, 2008 Edition.)

1.5.1 Aerospace Engineering

Aerospace engineers study the motion and control of all types of vehicles, including aircraft weighing upwards of one million pounds and spacecraft flying at 17 000 miles per hour. Aerospace engineers must understand many flight environments, from turbulent air in Earth's atmosphere to the winds and sandstorms on Mars. They design, develop, and test aircraft, missiles, space vehicles, helicopters, hydrofoils, ships, and submerging ocean vehicles (see Fig. 1.17). Areas of specialty include aerodynamics, propulsion, orbital mechanics, stability and control, structures, design, testing, and supervision of the manufacturing of aerospace vehicles. Aerospace engineers work for private aviation companies, government defense contractors, and research organizations.

Aerodynamics is the study of the effects of moving a vehicle through the Earth's atmosphere. The air produces forces that have both a positive effect on a properly designed vehicle (lift) and a negative effect (drag). In addition, at very high speeds (e.g., reentry velocities of spacecraft), the air generates heat on the vehicle that must be dissipated to protect crews, passengers, and cargo. Aerospace engineering students learn to determine such things as optimum wing and body shapes, vehicle performance, and environmental impact.

The operation and construction of turboprops, turbo and fan jets, rockets, ram and pulse jets, and nuclear and ion propulsion are part of the aerospace engineering student's study of propulsion. Such constraints as efficiency, noise levels, and flight distance enter into the selection of a propulsion system for a flight vehicle.

The aerospace engineer develops plans for interplanetary missions based on knowledge of orbital mechanics. The problems encountered include determination of trajectories, stabilization, rendezvous with other vehicles, changes in orbit, and interception.

Figure 1.17

23
*The Engineering
Disciplines*

Many aerospace engineers work in avionics and design for new aircraft.

Stability and control involves the study of techniques for maintaining stability and establishing guidance and control of vehicles operating in the atmosphere or in space. Automatic control systems for autopilots and unmanned vehicles are part of the study of stability and control.

The study of structures is primarily involved with thin-shelled, flexible structures that can withstand high stresses and extreme temperature ranges. The structural engineer works closely with the aerodynamics engineer to determine the geometry of wings, fuselages, and control surfaces. The study of structures also involves thick-shelled structures that must withstand extreme pressures at ocean depths and lightweight composite structural materials for high-performance vehicles.

The aerospace design engineer combines all the aspects of aerodynamics, propulsion, orbital mechanics, stability and control, and structures into the optimum vehicle. Design engineers work in a team and must learn to compromise in order to determine the best design satisfying all criteria and constraints.

The final proofing of a design involves the physical testing of a prototype. Aerospace test engineers use testing devices such as wind tunnels, lasers, and data acquisition systems. They may work on the flight line for aircraft testing, at a launch facility for spacecraft, or for organizations that build and test component parts such as automatic control systems, engines, landing gear, and control surface operating mechanisms.

New and exciting areas for aerospace development evolve continually. The construction and utilization of the international space station is stimulating new developments in service vehicles making round-trips from Earth. Perhaps you may someday participate in a commercial venture to add the space station or the Moon as an exotic vacation destination. Other areas of development include the unmanned drone aircraft for defense activity, hypersonic vehicles for suborbital flights between points on the Earth, innovative methods to improve the safety of aircraft through nondestructive structural testing, new engine designs that significantly reduce noise levels, and turbulence detection.

1.5.2 Chemical Engineering

Chemical engineers deal with the chemical and physical principles that allow us to maintain a suitable environment. They create, design, and operate processes that produce useful materials, including fuels, plastics, structural materials, food products, health products, fibers, and fertilizers (Fig. 1.18). As our natural resources become scarce, chemical engineers are finding ways to extend them or creating substitutes.

Graduates of chemical engineering programs have a strong background in basic chemistry and advanced chemistry (such as organic, inorganic, physical, analytical, materials chemistry, and biochemistry). They also need working knowledge of occupational safety, material and energy balance,

Figure 1.18

Chemical engineers design processing plants for many of the products on which we depend in our daily lives.

thermodynamics, heat transfer, chemical reaction engineering, separation operations, process design and modern experimental and computing techniques.

The chemical engineer, in the development of new products, in designing processes, and in operating plants, may work in a laboratory, pilot plant, or full-scale plant. In the laboratory the chemical engineer searches for new products and materials that benefit humankind and the environment. This laboratory work would be classified as research engineering.

In a pilot plant the chemical engineer is trying to determine the feasibility of carrying on a process on a large scale. There is a great deal of difference between a process working in a test tube in the laboratory and a process working in a production facility. The pilot plant is constructed to develop the necessary unit operations to carry out the process. Unit operations are fundamental chemical and physical processes that are uniquely combined by the chemical engineer to produce the desired product. A unit operation may involve separation of components by mechanical means, such as filtering, settling, and floating. Separation also may take place by changing the form of a component—for example, through evaporation, absorption, or crystallization. Unit operations also involve chemical reactions such as oxidation and reduction. Certain chemical processes require the addition or removal of heat or the transfer of mass. The chemical engineer thus works with heat exchanges, furnaces, evaporators, condensers, and refrigeration units in developing large-scale processes.

In a full-scale plant the chemical engineer will continue to fine-tune the unit operations to produce the optimum process based on the lowest cost. The day-by-day operations problems in a chemical plant, such as piping, storage, and material handling, are the responsibility of chemical engineers.

Many chemical engineering degree programs are also active in biological engineering education and research. This might include working in the areas of biomedical engineering, bioinformatics, biomaterial engineering, biobased products, metabolic and tissue engineering, or biocatalyst and separation engineering.

1.5.3 Civil Engineering

Civil engineering is the oldest branch of the engineering profession. The term "civil" was used to distinguish this field from military engineers. Military engineers originated in Napoleon's army. The first engineers trained in this country were military engineers at West Point. Civil engineering involves application of the laws, forces, and materials of nature to the design, construction, operation, and maintenance of facilities that serve our needs in an efficient, economical manner. Civil engineers work for consulting firms engaged in private practice, for manufacturing firms, and for federal, state, and local governments. Civil engineers, more than any other specialists, work outdoors at least some of the time. Thus persons who enjoy working outside may find civil engineering attractive.

Because of the nature of their work, civil engineers assume a great deal of responsibility, which means that professional registration is a vital goal for the civil engineer who is beginning to practice. Typical specialties within civil engineering include structures, transportation, environmental (including water/wastewater treatment and water resources), geotechnical, surveying, and construction.

Structural engineers design bridges, buildings, dams, tunnels, and supporting structures (building structures, for example). The designs include consideration of loads, winds, temperature extremes, and other natural phenomena such as earthquakes and hurricanes. Advanced technology such as mathematical modeling and computer simulation is used extensively in this specialty. Civil engineers with a strong structural background often are found in aerospace and manufacturing firms, playing an integral role in the design of vehicular structures.

Civil engineers in transportation plan, design, construct, operate, and maintain facilities that move people and goods throughout the world, whether by land, sea or air. For example, civil engineers decide where a freeway system should be located and describe the economic impact of the system on the affected public (Fig. 1.19). They plan for growth of residential and industrial sectors of the nation. The modern rapid transit systems are another example of a public need satisfied by transportation engineers. Transportation specialists often work for public agencies or consulting firms that support public agencies.

Chemical Engineering Consultant/ Project Manager
David Shallberg

David Shallberg received his BS degree in chemical engineering and his MBA in finance. He is currently a senior consultant/project manager for Black & Veatch. Black & Veatch has headquarters in Kansas City, Missouri, and employs over 8 000 people at 90 sites. It is one of the largest and most diversified engineering and construction firms in the world, having completed more than 35 000 projects for over 6 500 clients worldwide.

Shallberg is a registered professional engineer and member of the Enterprise Management Solutions division of Black & Veatch. Currently, he is responsible for performing independent engineering assessments, strategic planning, and economic analysis for owners, developers, and project lenders of electric power generation assets. These reviews provide technical, financial, and economic analysis of projects across the spectrum of the power industry. Prior to his current assignment, Shallberg served in a variety of roles including power project development, business development, and detailed engineering assignments.

Shallberg estimates that at least 90% of his work involves interaction with coworkers, clients, suppliers, and others. Involvement in extracurricular activities and organizations (including Tau Beta Phi and Omega Chi Epsilon honor societies, residence hall governance, clubs, and intramural sports) while he was an undergraduate student provided opportunities for personal and professional development in areas such as interpersonal communication and leadership skills. Participation also provided a convenient way to demonstrate these skills to potential employers at graduation.

Analysis and problem-solving skills are highly valued in most endeavors, and in Shallberg's experience, an engineering degree, while being very specific in its studies, serves as a springboard to a wide variety of career opportunities. The photo shows Shallberg in Albania while he was doing some work related to power-plant siting.

Figure 1.19

27
*The Engineering
Disciplines*

Civil engineers had a major role in the design of the route of this new highway. The terrain had to be prepared to support the highway and care had to be taken to protect the local environment.

Environmental and water/wastewater treatment engineers are concerned with maintaining a healthful environment by proper treatment and distribution of drinking water, treatment of wastewater, and control of all forms of pollution. The water resources engineer specializes in the evaluation of potential sources of new water for increasing or shifting populations, irrigation, and industrial needs and in the prevention and control of flooding.

Before any structure can be erected, geotechnical engineers carefully study the soil, rock, and groundwater conditions to ensure stability of pavements and structures. Foundations must be designed to support all structures. The geotechnical engineer also plans portland cement and asphaltic concrete mixes for all types of construction.

Surveying engineers develop maps for any type of engineering project. For example, if a road is to be built through a mountain range, the surveyors will determine the exact route and develop the topographical survey, which is then used by the transportation engineer to lay out the roadway. Global positioning systems (GPS) and geographic information systems (GIS) make surveying projects easier to complete and more accurate and are helpful in making complex decisions.

Construction engineering is a significant portion of civil engineering, and many engineering colleges offer a separate degree in this area. Often construction engineers will work outside at the actual construction site. They are involved with the initial estimating of construction costs for surveying, excavation, and construction. They supervise the construction, start-up, and initial operation of the facility until the client is ready to assume operational responsibility. Construction engineers work around the world on many construction projects such as highways, skyscrapers, and power plants.

1.5.4 Electrical/Computer Engineering

Electrical/computer engineering is the largest branch of engineering, representing about 31% of the graduates entering the engineering profession. Because of the rapid advances in technology associated with electronics and computers, this branch of engineering also is the fastest growing. Areas of specialty include communications, power systems, analog/digital electronics, controls, signals and systems, power systems, computing and networking systems, security and reliable computing, semiconductor devices, linear systems, software systems, and electromagnetic fields, antennas, and propagation (Fig. 1.20).

Almost every minute of our lives, we depend on communication equipment developed by electrical engineers. Telephones, television, radio, and radar are common communications devices that we often take for granted. Our national defense system depends heavily on the communications engineer and on the hardware used for our early warning and detection systems.

The power engineer is responsible for producing and distributing the electricity demanded by residential, business, and industrial users throughout the world. The production of electricity requires a generating source such as fossil fuels, nuclear reactions, or hydroelectric dams. The power engineer may be involved with research and development of alternative generation sources, such as sun, wind, and fuel cells. Transmission of electricity involves conductors and insulating materials. On the receiving end, appliances are designed by power engineers to be highly efficient in order to reduce both electrical demand and costs.

The area of analog/digital electronics is the fastest-growing specialty in electrical engineering. The development of solid-state circuits (functional electronic circuits manufactured as one part rather than wired together) has produced high reliability in electronic devices. Microelectronics has revolutionized the computer industry and electronic controls. Circuit components that are much smaller than one micrometer in width enable reduced costs and higher electronic speeds to be attained in circuitry. The microprocessor, the principal component of a digital computer, is a major result of solid-state

Figure 1.20

29
*The Engineering
Disciplines*

Electrical engineers work with new electronics and control systems.

circuitry and microelectronics technology. Home computers, cellular telephones, automobile control systems, and a multitude of electrical application devices conceived, designed, and produced by electronic engineers have greatly improved our standard of living.

Great strides have been made in the control and measurement of phenomena that occur in all types of processes. Physical quantities such as temperature, flow rate, stress, voltage, and acceleration are detected and displayed rapidly and accurately for optimal control of processes. In some cases the data must be sensed at a remote location and accurately transmitted long distances to receiving stations. The determination of radiation levels is an example of the electrical process called *telemetry*.

The impact of microelectronics on the computer industry has created a multibillion-dollar annual business that in turn has enhanced all other industries. The design, construction, and operation of computer systems are the tasks of computer engineers. This specialty within electrical engineering in many schools has become a separate degree program. Computer engineers deal with both hardware and software problems in the design and application of computer systems. The areas of application include research, education, design engineering, scheduling, accounting, control of manufacturing operations, process control, and home computing needs. No single development in history has had as great an impact on our lives in such a short time span as has the computer.

1.5.5 Environmental Engineering

Environmental engineering deals with the appropriate use of our natural resources and the protection of our environment (Fig. 1.21). For the most part environmental engineering curricula are relatively new and in many instances reside as a specialty within other disciplines, such as civil, chemical, and agricultural engineering. Environmental engineers focus on at least one of the following environmental issues areas: air, land, water, or environmental health.

The construction, operation, and maintenance of the facilities in which we live and work have a significant impact on the environment. Environmental engineers with a civil engineering background are instrumental in the design of water and wastewater treatment plants, facilities that resist natural disasters such as earthquakes and floods, and facilities that use no hazardous or toxic materials. The design and layout of large cities and urban areas must include protective measures for the disturbed environment.

Environmental engineers with a chemical engineering background are interested in air and water quality, which is affected by many by-products of chemical and biological processes. Products that are slow to biodegrade are studied for recycling possibilities. Other products that may contaminate or be hazardous are being studied to develop either better storage or replacement products that are less dangerous to the environment. With the concerns of energy conservation, environmental engineers consider the use of energy and

Figure 1.21

When engineers design a new product or system, such as an offshore drilling rig, the design must minimize the impact on the environment.

resources such that the rate of use does not compromise the environment. This is often called sustainable engineering. Sustainable engineering minimizes waste in the design process and also considers the material used in the design process to ensure that the product will not be detrimental to the environment or can be recycled after its useful life.

With an agricultural engineering background, environmental engineers study air and water quality that is affected by animal production facilities, chemical runoff from agricultural fertilizers, and weed control chemicals. As

Civil Engineer Emphasizes Leadership and Internship Experiences
Mark Land

Mark Land earned his BS in civil engineering in 1992 and immediately joined Snyder & Associates, a 200-employee firm that provides municipalities, schools, public agencies, and private developers with comprehensive engineering and planning services. Now a Certified Professional in Erosion and Sediment Control, Land serves as head of the transportation and civil engineering business units for the Ankeny, Iowa, office and as the stormwater practice leader. He is a member of the board of directors of the firm and oversees approximately 60 employees in disciplines such as traffic and transportation, storm water, water, wastewater, and geographic information systems (GIS). The accompanying photo is of the Ada Hayden Heritage Park project near Ames, Iowa, completed under Land's direction.

As an undergraduate, Land was active in the American Society of Civil Engineers, Steel Bridge Team, and Civil Engineering Curriculum Committee and served in dormitory leadership positions including floor vice president. He believes that these activities helped him to develop his communication and "people" skills, which in turned helped him to attain his current leadership position in his company. Land's current outside activities include membership on the boards of directors of the Iowa Engineering Society and the Ankeny Chamber of Commerce. He is considering running for elective office.

Land had four years of summer job experience with Snyder & Associates. "This helped me understand what being a consulting engineer was about so that I really understood what I was getting into when I went to work there full time," he says. "My father worked as an engineer in the public sector so I had a good idea of what that was about. Consulting engineering is much different so I was glad to have the opportunity to gain experience in the field. It convinced me to be a consultant."

Land offers the following advice to engineering students: "There are numerous opportunities in engineering for those who want to work hard to achieve their goals. In order to make the most of your opportunities, understand that in the early stages of your engineering education, you will question the importance of all the general education classes like math, physics, etc. I just took one semester at a time and always made sure to have some fun along the way. Your grade point average will help you get some interviews early in your career, but the social aspects you learn doing things other than study will carry over into all areas of your life. Even though I have not used much of the math and physics learned in early classes, it did help me gain critical thinking skills. Also, take advantage of opportunities to volunteer. You will gain great experience in dealing with all types of people as well as learn better communication skills. As a person who hires numerous people, those three things, critical thinking skills, communication, and dealing with people, are the most important to me."
(Photo courtesy of Snyder & Associates Engineers and Planners)

we become more environmentally conscious, the demand for designs, processes, and structures that protect the environment will create an increasing demand for environmental engineers. They will provide the leadership for protecting our resources and environment for generations to come.

1.5.6 Industrial Engineering

Industrial engineering covers a broad spectrum of activities in organizations of all sizes. A primary objective of this engineering specialty is to improve the competitiveness and vitality of industry, government, and nonprofit institutions through the application of theory to human endeavors. Industrial engineering education requires a balanced understanding of mathematics, physical, and engineering sciences, as well as laboratory and industrial experiences. The principal efforts of industrial engineers are directed to the design of production systems for goods and services (Fig. 1.22). Most departments allow for specialization in at least one of three areas:

> *Manufacturing:* Industrial engineers must understand the fundamentals of modern and economic manufacturing; use product specifications as the keystone of part interchangeability; verify a product's conformance to its specifications; apply manufacturing principles to a manufacturing process; program flexible manufacturing equipment and system controllers; design logical manufacturing layouts; and implement contemporary systems issues such as lean manufacturing.

Figure 1.22

Industrial engineers design the assembly lines for production of products such as new electronics units.

Human Factors: Industrial engineers analyze and design both the job and the worksite in a cost-effective manner using time studies, as well as measure the resulting output; they also design, implement, and evaluate human-computer interfaces according to principles outlined in foundational human-computer interaction readings.

Management and Information Systems: Industrial engineers apply time value of money to make financial decisions; use probability concepts to solve engineering problems; estimate parameters; conduct tests of hypotheses and create regression models; apply statistical quality control methods such as process capability, control charts, and tolerance allocation; design experiments; optimize and solve mathematical models of real problems using linear programming, dynamic programming, networking, Markov chains, queuing, and inventory models; and create simulation models of manufacturing and service systems and analyze simulation output; understand object-oriented programming foundations; and develop applications of information technology in industrial engineering.

1.5.7 Mechanical Engineering

Mechanical engineers are involved with all forms of energy generation, distribution, utilization and conversion, the design and development of machines, the control of automated systems, manufacturing and processing of materials, and the creative solutions to environmental problems (Fig. 1.23). Practicing

Figure 1.23

Here a mechanical engineer is involved with the design of alternative energy solar panels.

mechanical engineers are typically associated with research, manufacturing, operations, testing, marketing, administration and teaching.

A typical undergraduate curriculum includes required and elective courses in the following areas:

Design

Fluid mechanics and
    propulsion

Heat transfer

Solid mechanics

Noise and vibration control

Modeling and simulation

Acoustics

Robotics

HVAC (heating, ventilation,
    thermodynamics
        and air-conditioning)

Industrial Engineering Manufacturing Engineer

Morgan Halverson

Morgan Halverson received her BS in industrial engineering and minor in business and is currently pursuing her MBA degree. Halverson is employed as an industrial engineer at John Deere in Ankeny, Iowa, where the company manufactures cotton harvesting equipment, self-propelled sprayers, and tillage equipment. John Deere is a major industrial corporation that does business around the world and employs approximately 47,000 people. John Deere has four manufacturing divisions: agricultural equipment, construction and forestry equipment, commercial and consumer equipment, and power systems, as well as the support areas of credit and parts.

As an industrial engineer at Deere, Halverson works on projects such as designing the manufacturing sequence of events for new and upcoming products. She also writes and maintains simulations of the assembly lines to analyze many different manufacturing scenarios. Halverson is involved in improving ergonomics and manufacturing efficiency while considering variables such as manpower rates, demand rates, work balance, and time needed for upcoming vehicle design changes.

While a student, Halverson was on the central committee for Engineering Week and was responsible for its Web site, advertising, fundraising, and more. She was president of the Engineering Spring Career Expo, a career fair that attracts over 100 companies from across the country looking for students to fill internships, co-ops, and summer job positions. She was also involved in the student chapter of the Institute of Industrial Engineers. All of these experiences aided her not only in her job search but also in broadening her network of friends and contacts, experience with teamwork and project management, and knowledge about key issues in manufacturing.

Halverson worked for a year as a co-op student at Lennox Manufacturing in Marshalltown, Iowa, where she gained broad exposure to the manufacturing environment such as material and inventory management, purchasing, ergonomics, process improvements, and Kaizen events. She found this to be a very important experience. Halverson's advice to students: Work hard and don't give up. "An engineering degree is challenging," she says, "but it prepares you to accomplish a wide variety of things in life . . . not just in your career!"

Combustion and energy
 utilization
Manufacturing and materials
 processing

Mechatronics
Measurements and
 instrumentation
Automatic controls

The energy crisis and environmental challenges in the current decade have reprioritized a need for new sources of energy as well as new and improved methods of energy conversion. Mechanical engineers are involved in the research and development of solar, geothermal, biomass and wind energy sources, along with research to increase the efficiency of producing electricity from fossil fuel, hydroelectric, and nuclear sources.

Machines and mechanisms that are used daily in all forms of manufacturing and transportation have been designed and developed by mechanical engineers. Automobiles, airplanes, and trains combine a source of power with an aerodynamically designed enclosure to provide modern transportation. Tractors and other farm implements aid the agricultural community. Automated machinery and robotics are rapidly growing areas for mechanical engineers. Machine design requires a strong mechanical engineering background and a vivid imagination.

In order to drive modern machinery, a source of power is needed. The mechanical engineer is involved with the generation of electricity by converting chemical energy in fuels to thermal energy in the form of steam, then to mechanical energy through a turbine to drive the electric generator. Internal combustion devices such as gasoline, turbine, and diesel engines are designed for use in most areas of transportation. The mechanical engineer studies engine

Mechanical Engineer Performance Analyst

Sheela Rajendran

Sheela Rajendran received her BS degree in mechanical engineering in 2004. She is currently a senior associate engineer with Caterpillar's Global Engine Development—North America Division. Rajendran is a performance analyst for engine development in Caterpillar's Engine Systems Technology & Solutions division, which provides comprehensive engine testing and development services to designers, manufacturers, and other customers. The division applies a variety of analytical and developmental techniques to ensure that engines meet high standards for performance, durability, and emission standards. Rajendran's responsibilities require working in a lab environment that focuses on the testing and development of diesel engines. The goal is to ensure that Caterpillar's diesel engines comply with increasingly stringent emissions standards, while maintaining high performance.

Rajendran's five-year plan is to continue developing a sound technical engineering foundation and eventually transition into management. She also plans to pursue a master's degree in engineering or business.

Rajendran honed many necessary job skills (including communication, networking, and coordinating/planning) through her involvement in the Society of Women Engineers and through working at the Industrial Assessment Center while pursuing her undergraduate studies. She uses many of these skills in her job on a daily basis.

cycles, fuel requirements, ignition performance, power output, cooling systems, engine geometry, and lubrication in order to develop high-performance, low-energy-consuming engines.

Mechanical engineers work on engineering teams that are responsible for design and development of products and systems. Virtually any machine or process you can imagine has benefited from the influence of a mechanical engineer. More than 20% of the nearly 3 million engineers in the United States are mechanical engineers.

1.6 Conclusion

We have touched only briefly on the possibilities for exciting and rewarding work in all engineering areas. The first step is to obtain the knowledge during your college education that is necessary for your first technical position. After that you must continue your education, either formally by seeking an advanced degree or degrees or informally through continuing education courses or appropriate reading to maintain pace with the technology, an absolute necessity for a professional. Many challenges await you. Prepare to meet them well.

Problems

1.1 Describe which engineering roles best fit your long-term career goals.

1.2 Compare the definitions of an engineer and a technologist from at least two sources.

1.3 When considering the engineering functions of research, development, design, production, testing, construction, operations, sales, management, consulting, and teaching, which two best fit your personality and long-term goals? Explain why.

1.4 Find the name of a prominent engineer in the field of your choice and write a brief paper on the accomplishments of this individual.

1.5 Select a specific discipline of engineering and list at least 10 different companies and/or government agencies that utilize engineers from this field.

1.6 For the discipline selected in Exercise 1.5, choose a function such as design. Select one of the 10 industrial organizations utilizing engineers in this discipline and write a brief report on some typical activities that are undertaken by the engineers who perform that function in this organization. Be specific in your discussion of the activities. For example: A mechanical engineer for Company X designs the steering linkage for a garden tractor, specifies all the parts, and conducts prototype tests.

1.7 Read through each of the personal profiles of engineering graduates from this chapter. List and describe five lessons you learned from their advice.

1.8 Locate a full-time job description for an engineering major that interests you. You can find this on the Internet or on your campus's career services job posting Web site. What knowledge, skills, and abilities does the job require? Include the job description along with your answer.

1.9 After learning about the different types of engineering programs, select a program that is offered at your school and read the catalog descriptions for courses offered in this major. Find two courses that best fit your personal interest in this major and describe why they interest you.

1.10 Prepare a brief paper on the engineering degree that you are most interested in. Explain why this degree program best fits your academic and professional goals.

1.11 Prepare a five-minute talk to present to your class describing one of the engineering disciplines that most interests you.

1.12 Choose one of the following topics (or one suggested by your instructor) and write a paper that discusses technological changes that have occurred in this area in the past 15 years. Include commentary on the social and environmental impact of the changes and on new problems that may have arisen because of the changes.

(*a*) passenger automobiles

(*b*) electric power-generating plants

(*c*) computer graphics

(*d*) heart surgery

(*e*) heating systems (furnaces)

(*f*) microprocessors

(*g*) water treatment

(*h*) road paving (both concrete and asphalt)

(*i*) composite materials

(*j*) robotics

(*k*) air-conditioning

CHAPTER 2

Education for Engineering

Chapter Objectives

When you complete your study of this chapter, you will be able to:

- Understand the skills and abilities needed to pursue an engineering degree
- Understand how to prepare for a meaningful engineering career workplace
- Understand the importance of obtaining an internship/cooperative education experience
- Realize the importance of the engineering profession and the steps to becoming a professional engineer

2.1 Education for Engineering

The amount of information coming from the academic and business world is increasing exponentially, and at the current rate it will double in less than 20 years. More than any other group, engineers are using this knowledge to shape civilization. To keep pace with a changing world, engineers must be educated to solve problems that are as yet unheard of. A large share of the responsibility for this mammoth education task falls on the engineering colleges and universities. But the completion of an engineering program is only the first step toward a lifetime of education. The engineer, with the assistance of the employer and the university, must continue to study. (See Fig. 2.1.)

Logically, then, an engineering education should provide a broad base in scientific and engineering principles, some study in the humanities and social sciences, and specialized studies in a chosen engineering curriculum. But specific questions concerning engineering education still arise. We will deal here with the questions that are frequently asked by students. What are the desirable characteristics for success in an engineering program? What knowledge and skills should be acquired in college? What is meant by continuing education with respect to an engineering career?

2.1.1 Desirable Characteristics

Years of experience have enabled engineering educators to analyze the performance of students in relation to the abilities and desires they possess when entering college. The most important characteristics for an engineering student can be summarized as follows:

Figure 2.1

The Internet is an important tool for engineering design, research, and education.

1. A strong interest in and ability to work with mathematics and science taken in high school.
2. An ability to think through a problem in a logical manner.
3. A knack for organizing and carrying through to conclusion the solution to a problem.
4. An unusual curiosity about how and why things work.
5. A passion for helping solve modern-day challenges (e.g., renewable, non-polluting energy; abundant clean water; modern health care; sustainable agriculture and manufacturing; safe roads and bridges; designs for natural and man-made disasters).

Although such attributes are desirable, having them is no guarantee of success in an engineering program. Simply a strong desire for the job has made successful engineers of some individuals who did not possess any of these characteristics; and, conversely, many who possessed them did not complete an engineering degree. Moreover, an engineering education is not easy, but it can offer a rewarding career to anyone who accepts the challenge. To succeed, it is important to choose an engineering field of study that best matches your personal passions and interests.

Developing into a professional engineer takes more than just taking the classes required for an engineering degree. Undergraduate engineers also need to develop their knowledge, skills, and abilities outside the classroom. There are many opportunities for students to become leaders in professional and nonprofessional student clubs, for internship or cooperative education experiences with industries closely associated with the student's chosen engineering field, for community or service learning project, and participation in many other academic and nonacademic related activities.

Engineering Is More Than Just Taking Courses

Amy Dee Schlechte

Amy Dee Schlechte graduated with a degree in agricultural engineering and is currently a systems engineer at General Mills, dedicated to the Fruit Roll-Ups and Fruit by the Foot packaging systems. Her responsibilities include identifying and driving system improvements, optimizing systems performance and product quality, and driving strategic productivity and capacity solutions.

During college Schlechte held three internships. "These work experiences allowed me to apply my learning from the classroom and extracurricular activities in order to develop into a well-rounded engineer," she says. "With the assistance of my summer work experiences, I was able to quickly figure out what my strengths and interests were. Upon the conclusion of my final internship, I evaluated my experiences and chose to be an engineer for a food manufacturing company."

Schlechte was also very involved in extracurricular activities. Professional leadership included region executive in the Society of Women Engineers, vice president of the American Society of Agricultural Engineers, and general co-chair and event chair for Engineering Week. Schlechte was also a peer mentor for a team of freshmen students in her college's Agricultural Engineering Learning Community. Membership in other committees included Engineering Student Council, Cardinal Key Honor Society, and Student Leadership. "Getting involved and participating in clubs/committees at the college level was a great way for me to begin to grow and develop into a leader," Schlechte remembers. "Being involved and active not only in the classroom but through extracurricular activities taught me valuable lessons early on that I could apply directly to the workplace and life in general. The most important lessons that I learned firsthand through participating in various clubs was the importance of goal setting, prioritization, good communication, team work, leadership, and delegation. Pairing those tools with the technical knowledge that I learned through the classroom and on the job training makes me a more effective engineer."

Schlechte sees herself returning to school part-time to obtain a master's degree in business administration. She would like to increase her responsibilities at General Mills and become a plant technical manager. Within her community, she'd like to become more active with charitable/philanthropic organizations such as Junior Achievement and Big Brothers/Big Sisters.

Schlechte has three areas of advice for students beginning their engineering studies:

1. "Acquire internships, co-op positions, and summer job experiences in order to evaluate different engineering positions on a short-term basis. I encourage students studying engineering to participate in as many different work experiences as they can in order to best identify their ideal professional role."

2. "Become actively involved in some type of organization, club, or committee. This is very valuable and will provide overall character growth and leadership development. This is also a great way to become better connected with fellow peers and future colleagues."

3. "Remember that a person with an engineering degree can travel down many different professional avenues—find one that you enjoy! One of the things I like most about my engineering job is that every day brings on new and different challenges. By understanding and leveraging technology, anything is possible."

2.1.2 Course of Study

The quality control of engineering programs is effected through the accreditation process. The engineering profession, through ABET and CEAB, has developed standards and criteria for the education of engineers entering the profession. Through visitations, evaluations, and reports, the written criteria and standards are compared with the engineering curricula at a university. For ABET, individual degree-granting programs complete a self-study report that must address several criteria in areas related to students, program educational objectives, program outcomes and assessment, the professional or technical component of the curriculum, faculty, facilities, institutional support and financial resources, and program criteria. Each program, if operating according to the standards and criteria, may receive up to six years of accreditation. If some discrepancies appear, accreditations may be granted for a shorter time period or may not be granted at all until appropriate improvements are made.

It is safe to say that for any given engineering discipline, no two schools will have identical offerings. However, close scrutiny will show a framework within which most courses can be placed, with differences occurring only in textbooks used, topics emphasized, and sequences followed. Figure 2.2 depicts

Figure 2.2

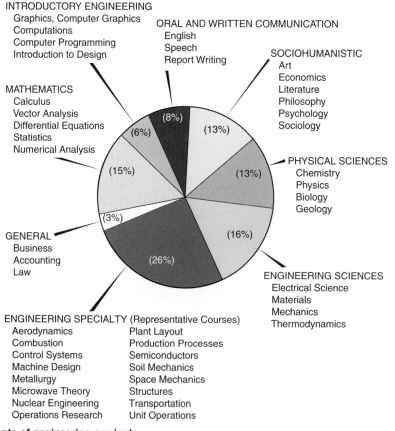

INTRODUCTORY ENGINEERING
  Graphics, Computer Graphics
  Computations
  Computer Programming
  Introduction to Design

ORAL AND WRITTEN COMMUNICATION
  English
  Speech
  Report Writing

SOCIOHUMANISTIC
  Art
  Economics
  Literature
  Philosophy
  Psychology
  Sociology

MATHEMATICS
  Calculus
  Vector Analysis
  Differential Equations
  Statistics
  Numerical Analysis

PHYSICAL SCIENCES
  Chemistry
  Physics
  Biology
  Geology

GENERAL
  Business
  Accounting
  Law

ENGINEERING SCIENCES
  Electrical Science
  Materials
  Mechanics
  Thermodynamics

(8%) (6%) (13%) (13%) (15%) (16%) (3%) (26%)

ENGINEERING SPECIALTY (Representative Courses)
  Aerodynamics   Plant Layout
  Combustion   Production Processes
  Control Systems   Semiconductors
  Machine Design   Soil Mechanics
  Metallurgy   Space Mechanics
  Microwave Theory   Structures
  Nuclear Engineering   Transportation
  Operations Research   Unit Operations

Elements of engineering curricula.

this framework and some of the courses that fall within each of the areas. The approximate percentage of time spent on each course grouping is indicated.

The general education block is a small portion of most engineering curricula, but it is important because it helps the engineering student to understand and develop an appreciation for the potential impact of engineering to undertakings on the environment and general society. When the location of an ethanol plant is being considered, the engineers involved in this decision must respect the concerns and feelings of all individuals who might be affected by the location. Discussions of the interaction between engineers and the general public take place in few engineering courses; general education courses (sociology, economics, history, management, etc.) thus are needed to furnish engineering students with an insight into the needs and aspirations of society.

Chemistry and physics are almost universally required in engineering. They are fundamental to the study of engineering science. The mathematics normally required for college chemistry and physics is more advanced than that for the corresponding high school courses. Higher-level chemistry and physics also may be required, depending on departmental structure. Finally, other physical science, biological, or business courses may be required in some programs or taken as electives.

An engineer cannot be successful without the ability to communicate ideas and the results of work efforts. The research engineer writes reports and orally presents ideas to management. The production engineer must be able to converse with craftspersons in understandable terms. And all engineers have dealings with the public and must be able to communicate on a nontechnical level. Engineers must become proficient in written, oral, visual, and electronic communication. For some, this may mean taking communication courses above and beyond the required communication courses, becoming involved in extracurricular activities, or obtaining an internship that will develop communication competence.

Mathematics is the engineer's most powerful problem-solving tool. The amount of class time spent in this area is indicative of its importance. Courses in calculus, vector analysis, and differential equations are common to all degree programs. Statistics, numerical analysis, and other mathematics courses support some engineering specialty areas. Students desiring an advanced degree may want to take mathematics courses beyond the baccalaureate-level requirements.

In the early stages of an engineering education, introductory courses in graphical communication, computational techniques, design, and computer programming are often required. Engineering schools vary somewhat in their emphasis on these areas, but the general intent is to develop skills in the application of theory to practical problem solving and familiarity with engineering terminology. Design is presented from a conceptual point of view to aid the student in creative thinking. Graphics develops the visualization capability and assists the student in transferring mental thoughts into well-defined concepts on paper. The tremendous potential of the computer to assist the engineer has led to the requirement of computer programming in almost all curricula. Use of the computer to perform many tedious calculations has increased the efficiency of the engineer and has allowed more time for creative thinking. Computer graphics is also an important part for many engineering curricula. Its ability to enhance the visualization of geometry and to depict engineering quantities graphically has increased productivity in the design process.

With a strong background in mathematics and physical sciences, you can begin to study engineering sciences, courses that are fundamental to all engineering specialties. Electrical science includes study of charges, fields, circuits, and electronics. Materials science courses involve study of the properties and chemical compositions of metallic and nonmetallic substances (see Fig. 2.3). Mechanics includes study of statics, dynamics, fluids, and mechanics of materials. Thermodynamics is the science of heat and is the basis for study of all types of energy and energy transfer. A sound understanding of the engineering sciences is most important for anyone interested in pursuing postgraduate work and research.

Figure 2.2 shows only a few examples of the many specialized engineering courses given. Scanning course descriptions in a college general bulletin or catalog will provide a more detailed insight into the specialized courses required in the various engineering disciplines.

Most curricula allow students flexibility in selecting a few outside courses. For example, a student interested in management may take some courses in business and accounting. Another may desire some background in law or medicine, with the intent of entering a professional school in one of these areas upon graduation from engineering. Many new curricula are integrating biological sciences into their programs to meet industry demand for those who can design systems or products that have a biological twist. Other new programs are integrating entrepreneurship and leadership into their curricula.

Figure 2.3

Material engineers select appropriate materials for designing solar panels that will endure extreme conditions.

2.1.3 Preparation for an Engineering Work Environment

As an engineering student, you must prepare yourself for the engineering workplace by gaining key knowledge and by developing core skills and abilities/competencies. The Accreditation Board for Engineering and Technology (ABET) requires each accredited engineering program to assess that their engineering students have demonstrated, by the time of graduation, eleven student learning outcomes. In ABET's 2009–2010 Criteria for Accrediting Engineering Programs document, they state that:

> *Engineering programs must demonstrate that their students attain the following outcomes:*
> (a) *an ability to apply knowledge of mathematics, science, and engineering*
> (b) *an ability to design and conduct experiments, as well as to analyze and interpret data*
> (c) *an ability to design a system, component, or process to meet desired needs within realistic constraints such as economic, environmental, social, political, ethical, health and safety, manufacturability, and sustainability*
> (d) *an ability to function on multidisciplinary teams*
> (e) *an ability to identify, formulate, and solve engineering problems*
> (f) *an understanding of professional and ethical responsibility*
> (g) *an ability to communicate effectively*
> (h) *the broad education necessary to understand the impact of engineering solutions in a global, economic, environmental, and societal context*
> (i) *a recognition of the need for, and an ability to engage in life-long learning*
> (j) *a knowledge of contemporary issues*
> (k) *an ability to use the techniques, skills, and modern engineering tools necessary for engineering practice.*

Note that seven of these outcomes start with "an ability to . . . .", one requires "an understanding of. . .", one requires "a knowledge of . . .", and another calls for "the broad education necessary to . . . ." All of these are difficult to measure. So where are the best settings and what are the best activities to participate in to develop these "a–k" student learning outcomes? In the past, it was expected that most if not all of these should be addressed in the college classroom during the pursuit of an undergraduate degree. Today, faculty and industry representatives realize that these outcomes can only be met through active participation in a combination of curricular, co-curricular, and extracurricular activities. The curricular activities would include the traditional classroom experiences (e.g., lecture, laboratory, and capstone design courses) and participation in an academic engineering cooperative education/internship program. We will talk about cooperative education/internship programs in the next section. Co-curricular activities might include participation in an engineering study abroad program, an engineering service learning project, or an engineering student club. Extracurricular activities could include participation in a non-engineering club, intramural sports, college sports, community service, or volunteer programs.

Industries are interested in what you will contribute to their company to help make them successful. They determine future behavior of potential employees by evaluating past behavior. When interviewing engineering undergraduates for cooperative education/internship programs or full-time employment, many companies today use behavioral-based interviewing. Instead of asking you an interview question like, "How well do you think you would work in a team within our company?" they would instead ask, "Can you tell us of a time when you worked on an engineering design project, where you contributed significantly to the success of the final design?" Did you notice the difference between these two questions? The first question allows you to make up a story about how you would work on a team, whereas with the second question you have to reflect on a specific time when you contributed to a team's success either in a class project or during your internship.

To best determine which workplace competencies a company is looking for, conduct a key word or phrase analysis of a job posting from an engineering company that you are interested in working for. The most common competencies that a company is looking for include communication (oral, written, visual, and electronic), teamwork, initiative, engineering knowledge, and analysis and judgment. Other competencies might include innovation, safety awareness, general knowledge, cultural adaptability, planning, professional impact, quality orientation, customer focus, and integrity.

As you plan out your engineering academic program, make sure that you have a well-balanced program that includes key curricular, co-curricular, and extra-curricular activities. When companies evaluate your resume, they will typically rank your engineering work experience as the most important, followed by your grade point average (GPA), leadership experiences, and useful skills learned in class, on a job, or through other activities. When you first start your degree program, take a course load that helps you to be successful in the classroom. Establishing a good grade point average is critical to obtaining a cooperative education/internship experience. As you establish good study and time management skills and a strong GPA, start to integrate co-curricular and extra-curricular activities that will help to strengthen your resume. Don't forget to have some fun along the way.

2.1.4 Cooperative Education/Internship Programs

Almost all engineering degree programs offer their students the opportunity to obtain meaningful industry work experience through some sort of engineering experiential education. The three basic forms of engineering experiential education are cooperative education, an engineering internship, and an engineering summer work experience. Although the definitions given below are fairly common across engineering programs, there are variations. Check the Web site information from your local engineering career services office for the definitions at your institution. Also note that for some engineering programs, an engineering experiential education experience may be required, while for other programs it may be voluntary.

Cooperative Education

Cooperative education is typically defined as alternating periods of full-time academic college training with periods of full-time work experience of approximately equal length. At least one calendar year equivalent of institutionally supervised work experience is typically required (e.g., a minimum of two semesters and one summer).

An internship is most often defined as a single work period of institutionally supervised full-time employment of at least one semester. This experience would typically not include two semesters in the same academic year, but instead one semester, or one semester plus one summer.

Summer

A summer internship is defined as a single work period of institutionally supervised, engineering-related, full-time employment of typically 8-10 weeks. If you are interested in pursuing one of these experiences, it is important to establish a strong GPA, obtain meaningful non-engineering or technical work experience, and develop competencies required for the engineering field you are interested in. Some campuses hold career fairs that you can attend in order to interact with industrial representatives interested in hiring coops, interns, or summer hires. If this is not the case at your institution, you may have to pursue these opportunities by sending your resume to human resources personnel at the company you are interested in. Also check for career fairs that you can attend at your professional society meetings or at another school within driving distance. Your academic advisor is a great resource for determining the best options for you.

2.1.5 Continuing Education

Once you have received a bachelor of science degree in engineering it would be a mistake to think the learning process has been completed. On the contrary, it has just begun. Continued education can take many forms. It may involve preparation for a professional license, and it certainly will include seminars, short courses and professional conferences to maintain an up-to-date understanding of your selected area of expertise. Many students elect to continue their undergraduate studies with a master of science (MS) or even a PhD in their discipline.

Another common area for engineering grads to consider is a master of business administration (MBA). A few students will pursue law and occasionally medicine as a path toward continued professional development.

2.2 The Engineer as a Professional

Engineering is a learned vocation, demanding an individual with high standards of ethics and sound moral character. When making judgments, which may create controversy and affect many people, the engineer must keep foremost in mind a dedication to the betterment of humanity.

2.2.1 Professionalism

Professionalism is a way of life. A professional person is one who engages in an activity that requires a specialized and comprehensive education and is motivated by a strong desire to serve humanity. A professional thinks and acts in a manner that brings favor upon the individual and the entire profession. Developing a professional frame of mind begins with your engineering education.

The professional engineer can be said to have the following:

1. Specialized knowledge and skills used for the benefit of humanity.
2. Honesty and impartiality in engineering service.
3. Constant interest in improving the profession.
4. Support of professional and technical societies that represent the professional engineer.

It is clear that these characteristics include not only technical competence but also a positive attitude toward life that is continually reinforced by educational accomplishments and professional service.

2.2.2 Professional Registration

The power to license engineers rests with each of the 50 states. Since the first registration law in Wyoming in 1907 all states have developed legislation specifying requirements for engineering practice. The purpose of registration laws is to protect the public. Just as one would expect a physician to provide competent medical service, an engineer can be expected to provide competent technical service. However, the laws of registration for engineers are quite different from those for lawyers or physicians. An engineer does not have to be registered to practice engineering. Legally, only the chief engineer of a firm needs to be registered for that firm to perform engineering services. Individuals testifying as expert engineering witnesses in court and those offering engineering consulting services need to be registered. In some instances the practice of engineering is allowed as long as the individual does not advertise as an engineer.

The legal process for becoming a licensed professional engineer consists of four parts, two of which entail examinations. The parts include

1. An engineering degree from an acceptable institution as defined by the state board for registration. Graduation from an ABET-accredited institution satisfies the degree requirement automatically.
2. Successful completion of the Fundamentals of Engineering Examination entitles one to the title "engineer-in-training" (EIT). This eight-hour examination may be taken during the last term of an undergraduate program that is ABET-accredited. The first half of the exam covers fundamentals in the areas of mathematics, chemistry, physics, engineering mechanics, electrical science, thermal science, economics, and ethics. For the second half of the exam, students can chose a general engineering focus or a specific discipline focus such as civil engineering. The passing grade is determined by the state board.
3. Completion of work experience as an EIT under the supervision of an engineer who is already licensed. Four years is a typical requirement.
4. Successful completion of the Principles and Practice Examination completes the licensing process. This eight-hour examination covers problems normally encountered in the area of specialty such as mechanical or chemical engineering.

It should be noted that once the license is received, it is permanent although there is an annual renewal fee. In addition, the trend is toward specific requirements in continuing education each year in order to maintain the license. Licensed engineers in some states may attend professional meetings in their specialty, take classes, and write professional papers or books to accumulate sufficient professional development activities beyond their job responsibilities to maintain their licenses. This trend is a reflection of the rapidly changing technology and the need for engineers to remain current in their area.

Registration does have many advantages. Most public employment positions, all expert witness roles in court cases, and some high-level company positions require the professional engineer's license. However, less than one-half of the eligible candidates are currently registered. You should give serious consideration to becoming registered as soon as you qualify. Satisfying the requirements for registration can be started even before graduation from an ABET-accredited curriculum.

2.2.3 Professional Ethics

Ethics is the guide to personal conduct of a professional. Most technical societies have a written code of ethics for their members. The preamble for the NSPE Code of Ethics for Engineers is shown in Figure 2.4. Figure 2.5 is the "Engineer's Creed" as published by the NSPE.

2.2.4 Professional Societies

Over 550 colleges and universities offer programs in engineering that are accredited by ABET or CEAB. These boards have as their purpose the quality control of engineering and technology programs offered in the United States and Canada. The basis of the boards is the engineering profession, which is represented through the participating professional groups.

Figure 2.4

Preamble to the NSPE Code of Ethics for Engineers *(National Society of Professional Engineers).*

NSPE Code of Ethics for Engineers

PREAMBLE

Engineering is an important and learned profession. As members of this profession, engineers are expected to exhibit the highest standards of honesty and integrity. Engineering has a direct and vital impact on the quality of life for all people. Accordingly, the services provided by engineers require honesty, impartiality, fairness, and equity, and must be dedicated to the protection of the public health, safety, and welfare. Engineers must perform under a standard of professional behavior that requires adherence to the highest principles of ethical conduct.

Figure 2.5

Engineer's Creed *(National Society of Professional Engineers)*

Engineers' Creed

As a Professional Engineer, I dedicate my professional knowledge and skill to the advancement and betterment of human welfare.
I pledge:

- To give the utmost of performance;
- To participate in none but honest enterprise;
- To live and work according to the laws of man and the highest standards of professional conduct;
- To place service before profit, the honor and standing of the profession before personal advantage, and the public welfare above all other considerations.

In humility and with need for Divine Guidance, I make this pledge.

Adopted by National Society of Professional Engineers, June 1954.

Table 2.1 is a partial listing of the numerous engineering societies that support the engineering disciplines and functions. These technical societies are linked because of their support of the accreditation process. Over 60 other societies exist for the purpose of supporting the professional status of engineers. Among these are the Society of Women Engineers (SWE), the National Society of Black Engineers (NSBE), the Society of Hispanic Professional Engineers (SHPE), the Acoustical Society of America (ASA), the Society of Plastics Engineers (SPE), and the American Society for Quality Control (ASQC).

A primary reason for the rapid development in science and engineering is the work of technical societies. The fundamental service provided by a society is the sharing of ideas, which means that technical specialists can publicize their efforts and assist others in promoting excellence in the profession. When information is distributed to other society members, new ideas evolve and duplicated efforts are minimized. The societies conduct meetings on international, national, and local bases. Students of engineering will find a technical society in their specialty that may operate as a branch of the regular society or as a student chapter on campus. The student organization is an important link with professional workers, providing motivation and the opportunity to make acquaintances that will help students to formulate career objectives.

2.3 Conclusion

Pursuing an engineering degree will be both challenging and rewarding as you make your way through your degree program. Your undergraduate program will provide many challenges, regardless of whether your studies are in mathematics, physical sciences, engineering sciences, engineering specialties,

Table 2.1 Participating Bodies in the Accreditation Process

Organization	Abbreviation
American Academy of Environmental Engineers	AAEE
American Institute of Aeronautics and Astronautics, Inc.	AIAA
American Institute of Chemical Engineers	AICHE
American Nuclear Society	ANS
American Society for Engineers Education	ASEE
American Society of Agricultural and Biological Engineers	ASABE
American Society of Civil Engineers	ASCE
American Society of Heating, Refrigerating, and Air-Conditioning Engineers, Inc.	ASHRAE
The American Society of Mechanical Engineers	ASME
American Society of Safety Engineers	ASSE
American Society for Testing and Materials	ASTM
Biomedical Engineering Society	BMES
Institute of Biological Engineering	IBE
The Institute of Electrical and Electronics Engineers, Inc.	IEEE
Institute of Industrial Engineers, Inc.	IIE
The Minerals, Metals & Materials Society	TMS
National Council of Examiners for Engineering and Surveying	NCEES
National Society of Professional Engineers	NSPE
Society of Automotive Engineers	SAE
Society of Manufacturing Engineers	SME
Society of Petroleum Engineers	SPE

Source: Compiled from Lichtenberger Engineering Library, University of Iowa, 2005.

communication, or social and human sciences. It is important that you develop the overall knowledge, skills, and abilities needed for a successful engineering career through a well-designed academics program, including experiential education and co-curricular and extra-curricular experiences. Engage with others in your chosen profession of study by joining your local professional student club. This is a great way to meet graduates from your degree program who are currently working in the industry. Finally, start to build strong relationships with other engineering students who are serious about their academic success.

Problems

2.1 For a particular discipline of engineering, such as electrical engineering, find the program of study for the first two years and compare it with the program offered at your school approximately 20 years ago. Comment on the major differences.

2.2 Compare the knowledge, skills, abilities, and competencies of a job description for a cooperative education/internship position versus that of a full-time position for a company that you would like to work with. What differences do you see?

2.3 List five of your own personal characteristics and compare that list with the one in Sec. 2.1.1.

2.4 Prepare a brief paper on the requirements for professional registration in your state. Include the type and content of the required examinations.

2.5 Prepare a five-minute talk to present to your class describing one of the technical societies listed in Table 2.1 and how it can benefit you as a student.

2.6 Go to your engineering career service Web site and locate the description for cooperative education, internships, and summer work experiences. What are the requirements for each of these programs for your degree program? What are the key differences between a cooperative education experience and an internship experience?

CHAPTER 3

Introduction to Engineering Design

Chapter Objectives

When you complete your study of this chapter, you will be able to:

- Identify and explain the key steps in the design process
- Explain the importance of the customer's role in the design process
- Apply the design process to solving an open-ended problem
- Understand the importance of the engineering design process in development of engineering solutions to society's needs

3.1 An Introduction to Engineering Design

What do you say when asked why you are planning to be an engineer? One possible response is, "I want to become an engineer to design . . . ." It might be to design a water-quality system for a developing country, a new spacecraft for NASA, the tallest building in the world, an auto-guidance device for automobiles, new and improved sports equipment, or even synthetic blood. The key is that engineers design devices, systems, or processes to help humankind.

So what is engineering design? Engineering design is a systematic process by which solutions to the needs of humankind are obtained. Design is the essence of engineering. The design process is applied to problems (needs) of varying complexity. For example, mechanical engineers will apply the design process to develop an effective, efficient vehicle suspension system; electrical engineers will apply the process to design lightweight, compact wireless communication devices; and materials engineers will apply the process to design strong, lightweight composites for aircraft structures.

The vast majority of complex problems in today's high technology society do not depend for solutions on a single engineering discipline; rather, they depend on teams of engineers, scientists, environmentalists, economists, sociologists, legal personnel, and others. Solutions are dependent not only on the appropriate applications of technology but also on public sentiment as executed through government regulations and political influence. As engineers we are empowered with the technical expertise to develop new and improved products and systems; however, at the same time we must be increasingly aware of the impact of our actions on society and the environment in general and work conscientiously toward the best solution in view of all relevant factors.

The systematic design process can be conveniently represented by the six steps introduced in Sec. 3.2.

1. Define the problem to be solved.
2. Acquire and assemble pertinent data.
3. Identify solution constraints and criteria.
4. Develop alternative solutions.
5. Select a solution based on analysis of alternatives.
6. Communicate the results.

Building on an Engineering Degree

Nick Mohr

Nick Mohr received his BS in mechanical engineering and then went on to obtain his medical degree. He is currently a resident physician in emergency medicine, caring for patients in the emergency departments of two trauma centers and on a helicopter transport service in Indianapolis, Indiana.

Once he finishes his residency, he plans then either to look for a faculty appointment at a university or to pursue further training, perhaps in a postgraduate aerospace medicine program offered by Johnson Space Center (JSC) in conjunction with the University of Texas. Recently, he spent some time at JSC working with the flight surgeons in space medicine, which he found to be "incredible."

Dr. Mohr was involved with numerous student organizations and activities while pursuing his engineering degree, including Team PrISUm (Iowa State University's solar car team), the Cosmic Ray Observation Project, and the Ames (Iowa) Free Clinic. He feels that his involvement outside the classroom was one of the most important aspects of his undergraduate training. It provided him experience in learning how (1) to solve novel problems and make decisions without knowing the right answers; (2) to succeed and fail when the stakes for failure are high; and (3) to identify what consequences are worth fearing, and using those consequences to choose risks worth taking.

Dr. Mohr recalls that as a student, he had no idea how many doors an engineering degree could open and offers the following advice to students just beginning their engineering education:

"There is no harm in being uncertain about what path your career may take, and in fact, many people change their direction along the way—that's not bad. The important part is to have dreams, and to follow them wholeheartedly until they change. If we do that, we will solve some very interesting problems in our lives and can improve the world in which we live. A strong and diverse educational foundation in engineering has opened doors that I never could have imagined when I was in college."

A formal definition of engineering design is found in the curriculum guidelines of the Accreditation Board for Engineering and Technology (ABET). ABET accredits curricula in engineering schools and derives its membership from the various engineering professional societies. Each accredited curriculum has a well-defined design component that falls within the ABET guidelines. The ABET statement on design reads as follows:

Engineering design is the process of devising a system, component, or process to meet desired needs (Fig. 3.1). It is a decision-making process (often iterative), in which the basic sciences, mathematics, and engineering sciences are applied to convert resources optimally to meet a stated objective. Among the fundamental elements of the design process are the establishment of objectives and criteria, synthesis, analysis, construction, testing, and evaluation. The engineering design component of a curriculum must include most of the following features: development of student creativity, use of open-ended problems, development and use of modern design theory and methodology, formulation of design problem statements and specifications, consideration of alternative solutions, feasibility considerations, production processes, concurrent engineering design, and detailed system descriptions. Further, it is essential to include a variety of realistic constraints such as economic factors, safety, reliability, aesthetics, ethics, and social impact.

Figure 3.1

The engineering design process was very critical in the design of the international space station.

3.2 The Design Process

A simple definition of design is "a structured problem-solving activity." A process, on the other hand, is a phenomenon identified through step-by-step changes that lead toward a required result. Both these definitions suggest the idea of an orderly, systematic approach to a desired end. The design process, however, is not linear. That is, one does not necessarily achieve the best solution by simply proceeding from one step in the process to the next. New discoveries, additional data, and previous experience with similar problems generally will result in several iterations through some or all the steps of the process (Fig. 3.2).

It is important to recognize that any project will have time constraints. Normally before a project is approved a time schedule and a budget will be approved by management.

For your initial introduction to the design process, we will explain in more detail what is involved at each of the six steps above. Simply memorizing the steps will not give you the needed understanding of design. We suggest that you take one or more of the suggested design problems at the end of the chapter, organize a team of two to four students, and develop a workable solution for each problem selected. By working as a team you will generate more and better solution ideas and develop a deeper understanding of the process.

The process begins with a definition of the problem (Step 1) to be solved. In many cases the engineering design team does not identify or define the

Figure 3.2

Environmental engineers used the engineering design process to maximize the efficiency of this recycling system without causing damage to the environment.

problem. Instead customers, field representatives for the company, and management will provide the initial request. The team must be careful not to define a solution at this step. If it does it has not satisfied the design process. For example, assume company management asks a team to design a cart to transport ingots of metal from one building to another, the buildings approximately 200 meters apart. The solution to this problem is already known: a cart. It is a matter of seeing what is available on the market for handling the required load. It may be to the company's benefit to find a "new system" that effectively and efficiently moves heavy loads over a short distance. This would open up the possibility for a rail system, conveyor, or other creative solution. Usually a simple problem definition allows the most flexibility for the design team. For example, the initial problem could be defined as simply "Currently there are ingots of metal stored in building one. We need a way to get the ingots from that building to building two."

The team next acquires and assembles all pertinent information on the problem (Step 2). Internal company documents, available systems, Internet searches, and other engineers are all possible sources of information. Once all team members are up to speed on the available information, the solution constraints and criteria are identified (Step 3). A constraint is a physical or practical limitation on possible solutions; for example, the system must operate with 220-volt electricity. Criteria are desirable characteristics of a solution; for example, the solution must be reliable, must be easy to operate, must have an acceptable cost, and must be durable. You might think of a constraint as a requirement—all possible solutions must meet it—while a criterion is a relative consideration, in that one solution is better than another ("durable" is a criterion, for example).

Now the team is ready for the creative part of the process, developing alternative solutions (Step 4). This is where experience and knowledge, combined with group activities such as brainstorming, yield a variety of possible solutions (Fig. 3.3). Each of the alternatives is now analyzed using the constraints and comparing each to the specified criteria. In many cases prototypes are built and tested to see if they meet constraints and criteria. Computer modeling

Figure 3.3

The brainstorming of new ideas for solving an engineering problem is important in the design process.

and analysis are used heavily during this step. Then, using a device such as a decision matrix, a solution is selected (Step 5).

The last step of the design process often involves the most time and requires resources outside the original design team. Communicating the results (Step 6) involves developing all the details and reports necessary for the design to be built or manufactured as well as presentations for management and customers.

Although the systematic design process appears to end at Step 6, it really remains open throughout the product life cycle. Field testing, customer feedback, and new developments in materials, manufacturing processes, and so on may require redesign any time during the life cycle. Today many products are required to have a disposal plan prior to marketing. In these cases the original design needs to include disposal as a constraint on the solution. Although a six (6) step design process is outlined above, other more expanded steps are in common use. For example, Figure 3.4 illustrates a nine-step design process.

To further illustrate the iterative nature of the design process, study Figure 3.5 for a typical industrial activity. The process begins with a conceptual design and proceeds to preliminary design, detailed design, prototype design, and the final design. Note that design evaluation is conducted frequently during the process. Also note that the design is optimized at the detailed design stage. Optimization is beyond the scope of this introduction, but suffice it to say that it occurs after the solution is determined and is based on the analysis of alternatives (Step 5).

Figure 3.4

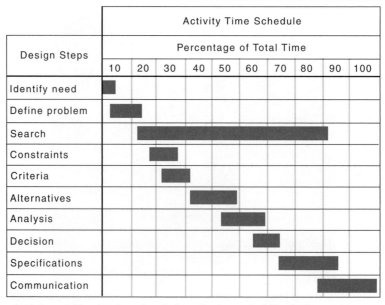

Design Steps	Activity Time Schedule									
	Percentage of Total Time									
	10	20	30	40	50	60	70	80	90	100
Identify need	▮									
Define problem	▮▮									
Search		▮▮▮▮▮▮▮▮▮▮▮▮▮								
Constraints		▮▮								
Criteria			▮▮							
Alternatives				▮▮						
Analysis					▮▮					
Decision						▮▮				
Specifications							▮▮▮			
Communication									▮▮▮	

A time schedule must be developed early in order to control the design process.

Figure 3.5

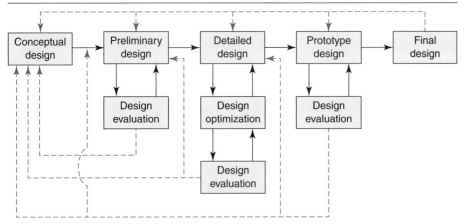

A flow diagram for the categories of engineering design.

3.3   Design and the Customer

Often customer requirements are not well defined. The design team must determine, in consultation with the customer, the expectations of the solution. The customer therefore must be kept informed of the design status at all times during the process. It is likely that compromises will have to be made. Both the design team and customer may have to modify their requirements in order to meet deadlines, cost limits, manufacturing constraints, and performance requirements. Figure 3.6 is a simple illustration of the Kano model showing the relationship between degree of achievement (horizontal axis) and customer satisfaction (vertical axis). Customer requirements are categorized in three areas: basic, performance related, and exciting.

Figure 3.6

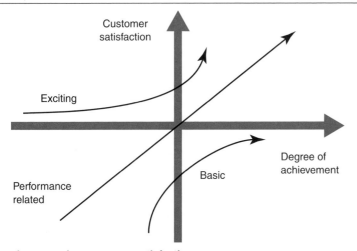

Factors in generating customer satisfaction.

Basic customer requirements are simply expected by the customer and assumed to be available. For example, if the customer desires a new electric-powered barbeque grill, the customer assumes that the design team and the company have proven their ability, with existing successful products, to design and manufacture electric-powered barbeque grills.

Performance-related customer requirements are the basis for requesting the new product. In the example of the barbecue grill, cooking time, cooking effectiveness, ease of setting the controls, and ease of cleaning are among the many possible performance-related items that a customer may specify. As time goes by and more electric powered grills reach the marketplace, these requirements may become basic.

Exciting customer requirements are generally suggested by the design team. The customer is unlikely to request these features because they are often outside the range of customer knowledge or vision. The exciting requirements are often a strong selling point in the design because they give the customer an unexpected bonus in the solution. Perhaps the capability of programming a cooking cycle to vary the temperature during the cooking process would be a unique (but perhaps costly) addition to the solution.

Figure 3.6 indicates that the basic requirements are a must for customer satisfaction. The customer will be satisfied once a significant level of performance-related requirements is met. The exciting requirements always add to customer satisfaction, so the more of these features that can be added, the greater is the satisfaction.

3.4 The Nature of Engineering Design

In the first half of the 20th century, engineering design was considered by many to be a creative, ad hoc process that did not have a significant scientific basis. Design was considered an art, with successful designs emanating from a few talented individuals in the same manner as great artwork is produced by talented artists. However, there are now a wealth of convincing arguments that engineering design is a cognitive process that requires a broad knowledge base, intelligent use of information, and logical thinking. Today successful designs are generated by design teams, comprised of engineers, marketing personnel, economists, management, customers, and so on, working in a structured environment and following a systematic strategy. Utilizing tools such as the Internet, company design documentation, brainstorming, and the synergy of the design team, information is gathered, analyzed, and synthesized with the design process yielding a final solution that meets the design criteria.

What do we mean by a cognitive process? In the 1950s Benjamin Bloom developed a classification scheme for cognitive ability that is called Bloom's taxonomy. Figure 3.7 shows the six levels of complexity of cognitive thinking and provides an insight into how the design process is an effective method of producing successful products, processes, and systems. The least complex level, knowledge, is simply the ability to recall information, facts, or theories. [What was the date of the Columbia space shuttle accident?]

The next level is comprehension, which describes the ability to make sense of (understand) the material. [Explain the cause of the Columbia accident.] The

Figure 3.7

61

*The Nature of
Engineering Design*

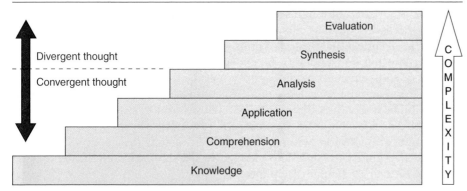

Bloom's taxonomy on learning aligns with the engineering design process.

third level is application, which is the ability to use knowledge and comprehension in a new situation and to generalize the knowledge. [What would you have done to prevent the Columbia accident?]

The fourth level is analysis, which is the ability to break learned material into its component parts so that the overall structure may be understood. It includes part identification, relationships of the parts to each other and to the whole, and recognition of the organizational principles involved. The individual must understand both the content and structure of the material. Figure 3.7 shows that analysis is the highest level of convergent thinking, whereby the individual recalls and focuses on what is known and comprehended to solve a problem through application and analysis. [What lessons did we learn about the space program from the Columbia accident?]

Levels 5 and 6 on Bloom's taxonomy represent divergent thinking, in which the individual processes information and produces new insights and discoveries that were not part of the original information (thinking outside the box). Synthesis refers to the ability to put parts together to form a new plan or idea. Everyone synthesizes in a different manner. Some accomplish synthesis by quiet mental musing; others must use pencil and paper to doodle, sketch, outline ideas, and so on. [Propose an alternative to the Columbia fuel tank insulation design that would perform the required functions.]

Evaluation is the highest level of thinking. It is the ability to judge the value of material based on specific criteria. Usually the individual is responsible for formulating the criteria to be used in the evaluation. [Assess the impact of the Columbia accident on the U.S. space program.]

To help your understanding of the levels of cognitive thinking, review several exams you have taken in college in mathematics, chemistry, physics, and general education courses (e.g., economics, sociology, history, etc). For each question, decide which level of thinking was required to obtain a successful result. You will find while moving along in your engineering curriculum that exam questions, homework problems, and projects will reflect higher and higher levels of thinking.

3.5   Experiencing the Design Process in Education

The design process, although structured, is an iterative process with flexibility to make necessary adjustments as the design progresses. The emphasis in this chapter is on conceptual design. At this stage of your engineering education it is important that you undergo the experience of applying the design process to a need with which you can identify based on your personal experiences. As you approach the baccalaureate degree you will have acquired the technical capability to conduct the necessary analyses and to make the appropriate technical decisions required for complex products, systems, and processes. Most engineering seniors will participate in a capstone design experience that will test their ability to apply knowledge toward solving a complicated design problem in their particular discipline.

3.6   Design Opportunities and Challenges of the Future

The world continues to undergo rapid and sometimes tumultuous change. As a practicing engineer, you will occupy center stage in many of these changes in the near future and will become even more involved in the more distant future. The National academy of Engineering has identified 14 "Engineering Grand Challenges." These include: 1) make solar energy economical; 2) provide energy from fusion; 3) develop carbon sequestration methods; 4) manage the nitrogen cycle; 5) provide clear water; 6) restore and improve urban infrastructure; 7) advance health informatics; 8) engineer better medicines; 9) reverse-engineer the brain; 10) prevent nuclear terror; 11) secure cyberspace; 12) enhance virtual reality; 13) advance personalized learning; and 14) engineer tools of scientific discovery. (Source: National Academy of Engineering of the National Academies, "Grand Challenges for Engineering," www.engineeringchallenges.org, viewed 8/16/2010.) The huge tasks of providing solutions to these problems will challenge the technical community beyond anyone's imagination.

Engineers of today have nearly instantaneous access to a wealth of information from technical, economic, social, and political sources. A key to the success of engineers in the future will be the ability to study and absorb the appropriate information in the time allotted for producing a design or solution to a problem. A degree in engineering is only the beginning of a lifelong period of study in order to remain informed and competent in the field.

Engineers of tomorrow will have even greater access to information and will use increasingly powerful computer systems to digest this information. They will work with colleagues around the world solving problems and creating new products. They will assume greater roles in making decisions that affect the use of energy, water, and other natural resources. Engineering design solution considerations for energy, the environment, infrastructure, and global competitiveness are addressed in the following sections.

3.6.1   Energy

In order to develop technologically, nations of the world require vast amounts of energy. With a finite supply of our greatest energy source, fossil fuels, alternate supplies must be developed and existing sources must be controlled with a worldwide usage plan. A key factor in the design of products must be minimum use of energy.

As demand increases and supplies become scarcer, the cost of obtaining the energy increases and places additional burdens on already financially strapped regions and individuals. Engineers with great vision are needed to develop alternative sources of energy from the sun, radioactive materials, wind, biomaterials, and ocean and to improve the efficiency of existing energy consumption devices (Fig. 3.8). Ethanol and biodiesel are two fuels that are produced in the United States from renewable resources that can assist in reducing America's dependence on foreign sources of energy. Waste-to-energy and biomass resources are also recognized by the U.S. Department of Energy as renewable energy source and are included in the department's tracking of progress toward achieving the federal government's renewable energy goal.

Along with the production and consumption of energy come the secondary problems of pollution and global warming. Such pollutants as smog, acid rain, heavy metals, nutrients, and carbon dioxide must receive attention in order to maintain the balance of nature. Also, increasing concentrations of greenhouse gases are likely to accelerate the rate of climate change, thus causing global warming. According to the National Academy of Sciences, the Earth's surface temperature has risen by about 1 degree Fahrenheit in the past century, with accelerated warming during the past two decades.

3.6.2 Environment

Our insatiable demand for energy, water, and other national resources creates imbalances in nature that only time and serious conservation efforts can

Figure 3.8

Windmill farms are an increasingly significant factor in the electrical infrastructure.

keep under control (Fig. 3.9). The concern for environmental quality is focused on four areas: cleanup, compliance, conservation, and pollution prevention. Partnerships among industry, government, and consumers are working to establish guidelines and regulations in the gathering of raw materials, the manufacturing of consumer products, and the disposal of material at the end of its designed use.

The American Plastics Council publishes a guide titled *Designing for the Environment,* which describes environmental issues and initiatives affecting product design. All engineers need to be aware of these initiatives and how they apply in their particular industries:

> *Design for the Environment (DFE): Incorporate environmental consider-ations into product designs to minimize impacts on the environment.*
>
> *Environmentally Conscious Manufacturing (ECM) or Green Manufacturing: Incorporating pollution prevention and toxics use reduction into product manufacturing.*
>
> *Extended Product or Producer Responsibility (Manufacturer's Responsibility or Responsible Entity): Product manufacturers are responsible for taking back their products at the product's end of life and managing them accord-ing to defined environmental criteria.*

Figure 3.9

Hydroelectric generating stations produce electricity important for industry and residence areas.

Life Cycle Assessment (LCA): Quantified assessment of the environmental impacts associated with all phases of a product's life, often from the extraction of base minerals through the product's end of life.

Pollution Prevention: Prevent pollution by reducing pollution sources (e.g., through design) as opposed to addressing pollution after it is generated.

Product Life Cycle Management (PLCM): Managing the environmental impacts associated with all phases of a product's life, from inception to disposal.

Product Takeback: The collection of products by manufacturer at the product's end of life.

Toxic Use Reduction: Reduce the amount, toxicity, and number of toxic chemicals used in manufacturing.

As you can see from these initiatives, all engineers regardless of discipline must be environmentally conscious in their work. In the next few decades we will face tough decisions regarding our environment. Engineers will play a major role in making the correct decisions for our small, delicate world.

The basic water cycle—from evaporation to cloud formation, then to rain, runoff, and evaporation again—is taken for granted by most people. (See Fig. 3.10.) However, if the rain or the runoff is polluted, then the cycle is interrupted

Figure 3.10

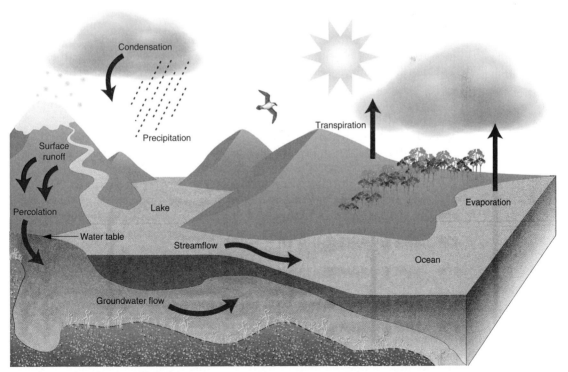

Understanding the water cycle (hydrologic cycle) is necessary in order to be able to engineer systems to control water pollution to our water resources.

and our water supply becomes a crucial problem. In addition, some highly populated areas have a limited water supply and must rely on water distribution systems from other areas of the country. Many formerly undeveloped agricultural regions are now productive because of irrigation systems. However, the irrigation systems deplete the underground streams of water that are needed downstream.

These problems must be solved in order for life to continue to exist as we know it. Because of the regional water distribution patterns, the federal government must be a part of the decision-making process for water distribution. One of the concerns that must be eased is the amount of time required to bring a water distribution plan into effect. Government agencies and the private sector are strapped by regulations that cause delays of several years in planning and construction. Greater cooperation and a better informed public are goals that public works engineers must strive to achieve. Developing nations around the world need additional water supplies because of increasing population growth. Many of these nations do not have the necessary freshwater and must rely on desalination, a costly process. The continued need for water is a concern for leaders of the world, and engineers will be asked to create additional sources of this life-sustaining resource.

3.6.3 Infrastructure

All societies depend on an infrastructure of transportation, waste disposal, and water distribution systems for the benefit of the population (Fig. 3.11). In the United States much of the infrastructure is in a state of deterioration without sound plans for upgrading. For example:

1. Commercial jet fleets include aircraft that are 35 to 40 years old. Major programs are now underway to extend safely the service life of these jets. In order to survive economically, airlines must balance new replacement jets with a program to keep older planes flying safely.
2. One-half of the sewage treatment plants cannot satisfactorily handle the demand.
3. The interstate highway system, over 50 years old in many areas, needs major repairs throughout. Over-the-road trucking has increased wear and tear on a system designed primarily for the automobile. Local paved roads are deteriorating because of a lack of infrastructure funds.
4. Many bridges are potentially dangerous to the traffic loads on them.
5. Railroads continue to struggle with maintenance of railbeds and rolling stock in the face of stiff commercial competition from the air freight and truck transportation industries.
6. Municipal water and wastewater systems require billions of dollars in repairs and upgrades to meet public demands and stricter water-quality requirements.

It is estimated that the total value of the public works facilities is over $2 trillion. To protect this investment, innovative thinking and creative funding must be fostered. Some of this is already occurring in road design and repair. For example, a new method of recycling asphalt pavement actually produces a stronger product. Engineering research is producing extended-life pavement

Figure 3.11

Development of new and improved infrastructure, such as this new rail system, are important to the economy of our nation.

with new additives and structural designs. New, relatively inexpensive methods of strengthening old bridges have been used successfully.

3.6.4 A Competitive Edge in the World Marketplace

We have all purchased or used products that were manufactured outside the country. Many of these products incorporate technology that was developed in the United States. In order to maintain our strong industrial base, we must develop practices and processes that enable us to compete not just with other U.S. industries but with international industries (Fig. 3.12). Engineers must also be able to design products that will be accepted by other cultures and work in their environments. It is therefore most important that engineers develop their global awareness and cultural adaptability competence.

The goal of any industry is to generate a profit. In today's marketplace this means creating the best product in the shortest time at a lower price than the competition. A modern design process incorporating sophisticated analysis procedures and supported by high-speed computers with graphical displays increases the capability for developing the "best" product. The concept of integrating the design and manufacturing functions shortens the design-to-market time for new products and for upgraded versions of existing products. The development of the automated factory is an exciting concept that is receiving a great deal of attention from manufacturing engineers today. Remaining

Figure 3.12

A global design team at work on an engineering design problem.

competitive by producing at a lesser price requires a national effort involving labor, government, and distribution factors. In any case engineers are going to have a significant role in the future of our industrial sector.

Problems

3.1 Complete the statement "I want to become an engineer to design . . ." in as much detail as you can.

3.2 Choose an engineering company that you would someday consider working for. From the information found on their Web site, write a one-page paper on the engineering problems that they are solving and how they are addressing their customer's needs.

3.3 Find three textbooks that introduce the engineering design process. Copy the steps in the process from each textbook. Compare with the six steps in Sec.

3.1. Note similarities and differences and write a paragraph describing your conclusions.

3.4 Interview an engineer working in your chosen field of study, describe in a one-page paper what steps of the design process he/she is engaged with in their job.

3.5 Select a specific discipline of engineering and list at least 20 different companies and/or government agencies that utilize engineers from this field.

3.6 Choose a product that was most likely designed by an engineer in your chosen field of study. Identify what problem this product solved, what constraints were applied to its design, and what criteria were most likely used to evaluate this design.

3.7 Choose one of the National Academy of Engineering's Greatest Challenges found in Sec. 3.6 and write a one-page paper on how you as an engineer could be involved in helping to solve this challenge.

3.8 Choose one of the products from the list below and note key features and functions for the product as produced today. Then, go back one generation (18–25 years) to family, relatives, or friends and ask them to describe the key features and functions of the same product as produced at that time. Note changes and improvements and prepare a brief report.

(*a*) toaster

(*b*) electric coffeemaker

(*c*) color television

(*d*) landline telephone

(*e*) cookware (pots, pans, skillets)

(*f*) vacuum cleaner

(*g*) microwave oven

3.9 Choose a product that you use every day and evaluate how effective the company that designed it was in meeting your customer needs and the needs of others. What suggestions would you make to help the designers improve the project? How could this product be used for another application?

3.10 Choose a device or product that you believe can be improved upon. Answer the following questions: 1) what do you already know about this device/product; 2) what do you think you know about this device/product; and (3) what do you need to know about this device/product? Based on your responses to these questions conduct research to confirm or reveal what you know and don't know. Write a short report to summarize your findings. Use proper citation methods to list your research sources.

3.11 Choose one of the following topics (or one suggested by your instructor) and write a paper that discusses technological changes that have occurred in this area in the past 15 years. Include commentary on the social and environmental impact of the changes and on new problems that may have arisen because of the changes.

(*a*) passenger automobiles

(*b*) electric power-generating plants

(*c*) computer graphics

(*d*) heart surgery

(*e*) heating systems (furnaces)

(*f*) microprocessors

(*g*) water treatment

(*h*) road paving (both concrete and asphalt)

(*i*) composite materials

(*j*) robotics

(*k*) air-conditioning

3.12 Investigate current designs for one or more of the items listed below. If you do not have the items in your possession, purchase them or borrow from friends. Conduct the following "reverse engineering" procedures on each of the items:

(*a*) Write down the need that the design satisfies.

(*b*) Disassemble the item and list all the parts by name.

(*c*) Write down the function of each of the parts in the item.

(*d*) Reassemble the item.

(*e*) Write down answers to the following questions:

- Does the item satisfactorily solve the need you stated in part (*a*)?
- What are the strengths of the design?
- What are the weaknesses of the design?
- Can this design be easily modified to solve other needs? If so, what needs and what modifications should be made?
- What other designs can solve the stated need?

The items for your study are the following:

- Mechanical pencil
- Safety razors from three vendors; include one disposable razor
- Flashlight
- Battery-powered slide viewer
- Battery-powered fabric shaver

3.13 The following list of potential design projects can be addressed by following the six-step design process discussed in the chapter. A team approach to a proposed solution, with three or four members on each team, is recommended. Develop a report and oral presentation as directed by your instructor.

- A device to prevent the theft of helmets left on motorcycles
- An improved rack for carrying packages or books on a motorcycle or bicycle
- A child's seat for a motorcycle or bicycle
- A device to permit easier draining of the oil pan by weekend mechanics
- A heated steering wheel for cold weather
- A sun shield for an automobile
- An SOS sign for cars stalled on freeways
- A storage system for a cell phone in a car (including charger)
- An improved wall outlet
- A beverage holder for a card table
- A better rural mailbox
- An improved automobile traffic pattern on campus
- An alert for drowsy or sleeping drivers
- Improved bicycle brakes
- A campus transit system
- Improved pedestrian crossings at busy intersections
- Improved parking facilities in and around campus
- A device to attach to a paint can for pouring
- An improved soap dispenser
- A better method of locking weights to a barbell shaft
- A shoestring fastener to replace the knot
- A better jar opener
- A system or device to improve efficiency of limited closet space

- A shoe transporter and storer
- A pen and pencil holder for college students
- A rack for mounting electric fans in dormitory windows
- A device to pit fruit without damage
- An automatic device for selectively admitting and releasing pets through an auxiliary door
- A device to permit a person loaded with packages to open a door
- A more efficient toothpaste tube
- A fingernail catcher for fingernail clippings
- A more effective alarm clock for reluctant students
- A clock with a display showing that the alarm has been set to go off
- A device to help a parent monitor small children's presence and activity in and around the house
- A simple pocket alarm that is difficult to shut off, used for discouraging muggers
- An improved storage system for luggage, books, and so on in dormitories
- A lampshade designed to permit one roommate to study while the other is asleep
- A device that would permit blind people to vote in an otherwise conventional voting booth
- A one-cup coffeemaker
- A silent wake-up alarm
- Home aids for the blind (or deaf)
- A safer, more efficient, and quieter air mover for room use
- A can crusher
- A rain-sensitive house window that would close automatically when it rains
- A better grass catcher for a riding lawn mower
- A built-in auto refrigerator
- A better camp cooler
- A dormitory cooler
- An impact-hammer adapter for electric drills
- An improved method of detecting and controlling the level position of the bucket on a bucket loader
- An automatic tractor-trailer-hitch aligning device
- A jack designed expressly for motorcycle use (special problems involved)
- Improved road signs for speed limits, curves, deer crossings, and so on
- Automatic light switches for rooms
- A device for dealing with oil slicks
- An egg container (light, strong, compact) for camping and canoeing
- Ramps or other facilities for handicapped students

CHAPTER 4

Engineering Solutions

Chapter Objectives

When you complete your study of this chapter, you will be able to:

- Recognize the importance of engineering problem analysis
- Recall and explain the engineering method
- Apply general guidelines for problem-solving presentation and solution documentation
- Develop an ability to solve and present simple or complex problems in an orderly, logical, and systematic way

4.1 Introduction

The practice of engineering involves the application of accumulated knowledge and experience to a wide variety of technical situations. Two areas, in particular, that are fundamental to all of engineering are design and problem solving. The professional engineer is expected to approach, analyze, and solve a range of technical problems intelligently and efficiently. These problems can vary from single-solution, reasonably simple problems to extremely complex, open-ended problems that require a multidisciplinary team of engineers.

Problem solving is a combination of experience, knowledge, process, and art. Most engineers through either training or experience solve many problems by a process. The design process, for example, is a series of logical steps that when followed produce an optimal solution given time and resources as two constraints. The total quality (TQ) method is another example of a process. This concept suggests a series of steps leading to desired results while exceeding customer expectations.

This chapter provides a basic guide to problem analysis, organization, and presentation. Early in your education, you must develop an ability to solve and present simple or complex problems in an orderly, logical, and systematic way.

4.2 Problem Analysis

A distinguishing characteristic of a qualified engineer is the ability to solve technical problems. Mastery of problem solving involves a combination of art and science. By *science* we mean the knowledge of the principles of mathematics, chemistry, physics, mechanics, and other technical subjects that must be learned so that they can be applied correctly. By *art* we mean the proper

73

judgment, experience, common sense, and know-how that must be used to reduce a real-life problem to such a form that science can be applied to its solution. To know when and how rigorously science should be applied and whether the resulting answer reasonably satisfies the original problem is an art.

Much of the science of successful problem solving comes from formal education in school or from continuing education after graduation. But most of the art of problem solving cannot be learned in a formal course; rather, it is a result of experience and common sense. Its application can be more effective, however, if problem solving is approached in a logical and organized method—that is, if it follows a process.

To clarify the distinction, let us suppose that a manufacturing engineer and a logistics specialist working for a large electronics company are given the task of recommending whether the introduction of a new computer that will focus on the computer-aided-design (CAD) market can be profitably produced. At the time this task is assigned, the competitive selling price has already been estimated by the marketing division. Also, the design group has developed working models of the computer with specifications of all components, which means that the approximate cost of these components is known. The question of profit thus rests on the costs of assembly and distribution. The theory of engineering economy (the science portion of problem solving) is well known and applicable to the cost factors and time frame involved. Once the production and distribution methods have been established, these costs can be computed using standard techniques. Selection of production and distribution methods (the art portion of problem solving) depends largely on the experience of the engineer and logistics specialist. Knowing what will or will not work in each part of these processes is a must in the cost estimate; however, these data cannot be found in handbooks, but, rather, they are found in the minds of the logistics specialist and the engineer. It is an art originating from experience, common sense, and good judgment.

Before the solution to any problem is undertaken, whether by a student or a practicing professional engineer, a number of important ideas must be considered. Think about the following questions: How important is the answer to a given problem? Would a rough, preliminary estimate be satisfactory, or is a high degree of accuracy demanded? How much time do you have and what resources are at your disposal? In an actual situation, your answers may depend on the amount of data available or the amount that must be collected, the sophistication of equipment that must be used, the accuracy of the data, the number of people available to assist, and many other factors. Most complex problems require some level of computer support such as a spreadsheet or a math analysis program. What about the theory you intend to use? Is it state of the art? Is it valid for this particular application? Do you currently understand the theory, or must time be allocated for review and learning? Can you make assumptions that simplify without sacrificing needed accuracy? Are other assumptions valid and applicable?

The art of problem solving is a skill developed with practice. It is the ability to arrive at a proper balance between the time and resources expended on a problem and the accuracy and validity obtained in the solution. When you can optimize time and resources versus reliability, problem-solving skills will serve you well.

The *engineering method* is an example of process. It consists of six basic steps:

1. *Recognize and understand the problem.* Perhaps the most difficult part of problem solving is developing the ability to recognize and define the problem precisely. This is true at the beginning of the design process and when applying the engineering method to a subpart of the overall problem. Many academic problems that you will be asked to solve have this step completed by the instructor. For example, if your instructor asks you to solve a quadratic–algebraic equation and provides all the coefficients, the problem has been completely defined before it is given to you, and little doubt remains about what the problem is.

   If the problem is not well defined, considerable effort must be expended at the beginning in studying the problem, eliminating the things that are unimportant, and focusing on the root problem. Effort at this step pays great dividends by eliminating or reducing false trials, thereby shortening the time taken to complete later steps.

2. *Accumulate data and verify accuracy.* All pertinent physical facts, such as sizes, temperatures, voltages, currents, costs, concentrations, weights, times, and so on, must be ascertained. Some problems require that steps 1 and 2 be done simultaneously. In others, step 1 might automatically produce some of the physical facts. Do not mix or confuse these details with data that are suspect or only assumed to be accurate. Deal only with items that can be verified. Sometimes it will pay to verify data that you believe are factual but actually may be in error.

3. *Select the appropriate theory or principle.* Select appropriate theories or scientific principles that apply to the solution of the problem; understand and identify limitations or constraints that apply to the selected theory.

4. *Make necessary assumptions.* Perfect solutions to real problems do not exist. Simplifications need to be made if real problems are to be solved. Certain assumptions can be made that do not significantly affect the accuracy of the solution, yet other assumptions may result in a large reduction in accuracy.

   Although the selection of a theory or principle is stated in the engineering method as preceding the introduction of simplifying assumptions, there are cases when the order of these two steps should be reversed. For example, if you are solving a material balance problem, you often need to assume that the process is steady, uniform, and without chemical reactions so that the applicable theory can be simplified. Note that many of the engineering equations used in practice only apply when specific assumptions are made.

5. *Solve the problem.* If steps 3 and 4 have resulted in a mathematical equation (model), it is normally solved by an application of mathematical theory, although a trial-and-error solution that employs the use of a computer or perhaps some form of graphical solution also may be applicable. The results normally will be in numerical form with appropriate units. Make sure to show the resulting answer with appropriate significant digits.

6. **Verify and check results.** In engineering practice, the work is not finished merely because a solution has been obtained. It must be checked to ensure that it is mathematically correct and that units have been properly specified. Correctness can be verified by reworking the problem by using a different technique or by performing the calculations in a different order to be certain that the numbers agree in both trials. The units need to be examined to ensure that all equations are dimensionally correct. And finally, the answer must be examined to see if it makes sense. An experienced engineer will generally have a good idea of the order of magnitude to expect.

If the answer doesn't seem reasonable, there is probably an error in the mathematics, in the assumptions, or perhaps in the theory used. Judgment is critical. For example, suppose that you are asked to compute the monthly payment required to repay a car loan of $5000 over a 3-year period at an annual interest rate of 12%. Upon solving this problem, you arrived at an answer of $11 000 per month. Even if you are inexperienced in engineering economy, you know that this answer is not reasonable, so you should reexamine your theory and computations. Examination and evaluation of an answer's reasonableness are habits you should strive to acquire. Your instructor and employer alike will not accept results that you have indicated are correct if the results are obviously incorrect by a significant percentage.

4.4 Problem Presentation

The engineering method of problem solving as presented in the previous section is an adaptation of the well-known *scientific problem-solving method*. It is a time-tested approach to problem solving that should become an everyday part of the engineer's thought process. Engineers should follow this logical approach to the solution of any problem while at the same time learn to translate the information accumulated into a well-documented problem solution.

The following steps parallel the engineering method and provide reasonable documentation of the solution. If these steps are properly executed during the solution of problems in this text and all other courses, it is our belief that you will gradually develop an ability to solve and properly document a wide range of complex problems.

1. **Problem statement.** State, as concisely as possible, the problem to be solved. The statement should be a summary of the given information, but it must contain all essential material. Clearly state what is to be determined. For example, find the temperature (K) and pressure (Pa) at the nozzle exit.
2. **Diagram.** Prepare a diagram (sketch or computer output) with all pertinent dimensions, flow rates, currents, voltages, weights, and so on. A diagram is a very efficient method of showing given and needed information. It also is an appropriate way of illustrating the physical setup, which may be difficult to describe adequately in words. Most often a two-dimensional representation is adequate (e.g., a free-body diagram of a beam with associated loads and moments). Data that cannot be placed in a diagram should be listed separately.

3. ***Theory.*** The theory used should be presented. In some cases, a properly referenced equation with completely defined variables is sufficient. At other times, an extensive theoretical derivation may be necessary because the appropriate theory has to be derived, developed, or modified.

4. ***Assumptions.*** Explicitly list, in complete detail, any and all pertinent assumptions that have been made to realize your solution to the problem. This step is vitally important for the reader's understanding of the solution and its limitations. Steps 3 and 4 might be reversed or integrated in some problems.

5. ***Solution steps.*** Show completely all steps taken in obtaining the solution. This is particularly important in an academic situation because your reader, the instructor, must have the means of judging your understanding of the solution technique. Steps completed, but not shown, make it difficult for evaluation of your work and therefore difficult to provide constructive guidance.

6. ***Identify results and verify accuracy.*** Clearly identify (double underline) the final answer. *Assign proper units.* An answer without units (when it should have units) is meaningless. Remember, this final step of the engineering method requires an examination of the answer to determine if it is realistic, so check solution accuracy and, if possible, verify the results.

7. ***Discussion/Conclusion*** It is important to write a concise summary of your results. What do the results mean? Do you have any observations? This step should include whether the results are reasonable and what would happen if one or more of the dependent variable were changed (e.g., what if the temperature increased by five degrees?).

4.5 Standards of Problem Presentation

Once the problem has been solved and checked, it is necessary to present the solution according to some standard. The standard will vary from school to school and industry to industry.

On most occasions, your solution will be presented to other individuals who are technically trained, but you should remember that many times these individuals do not have an intimate knowledge of the problem. However, on other occasions, you will be presenting technical information to persons with nontechnical backgrounds. This may require methods that are different from those used to communicate with other engineers; thus, it is always important to understand who will be reviewing the material so that the information can be clearly presented.

One characteristic of engineers is their ability to present information with great clarity in a neat, careful manner. In short, the information must be communicated accurately to the reader. (Discussion of drawings or simple sketches will not be included in this chapter, although they are important in many presentations.)

Employers insist on carefully prepared presentations that completely document all work involved in solving the problems. Thorough documentation may be important in the event of legal considerations for which the details of the work might be introduced into court proceedings as evidence. Lack of such documentation may result in the loss of a case that might otherwise have been won. Moreover, internal company use of the work is easier and more efficient if all aspects of the work have been carefully documented and substantiated by data and theory.

Each industrial company, consulting firm, government agency, and university has established standards for presenting technical information. These standards vary slightly, but all fall into a basic pattern, which we will discuss. Each organization expects its employees to follow its standards. Details can be easily modified in a particular situation once you are familiar with the general pattern that exists in all of these standards.

It is not possible to specify a single problem layout or format that will accommodate all types of engineering solutions. Such a wide variety of solutions exists that the technique used must be adapted to fit the information to be communicated. In all cases, however, one must lay out a given problem in such a fashion that it can be easily grasped by the reader. No matter which technique is used, it must be logical and understandable.

We have listed guidelines for problem presentation. Acceptable layouts for problems in engineering also are illustrated. The guidelines are not intended as a precise format that must be followed but, rather, as a suggestion that should be considered and incorporated whenever applicable.

Two methods of problem presentation are typical in academic and industrial environments. Presentation formats can be either freehand or computer generated. As hardware technology and software developments continue to provide better tools, the use of the computer as a method of problem presentation will continue to increase. If you were working on a team, you may need to utilize shared document software in order to collaborate more effectively on a problem presentation (e.g., Google Docs).

If a formal report, proposal, or presentation is the choice of communication, a computer-generated presentation is the correct approach. The example solutions that are illustrated in Figures 4.1 through 4.4 include both freehand work and computer output. Check with your instructor to determine which method is appropriate for your assignments. Figure 4.1 illustrates the placement of information.

The following nine general guidelines should be helpful as you develop the freehand skills needed to provide clear and complete problem documentation. The first two examples, Figsures 4.1 and 4.2, are freehand illustrations. The third example, Figure 4.3, is computer generated with a word processor, and Figure 4.4 uses a spreadsheet for the computations and graphing.

These guidelines are most applicable to freehand solutions, but many of the ideas and principles apply to computer generation as well.

1. One common type of paper frequently used is called engineering problems paper. It is ruled horizontally and vertically on the *reverse* side, with only heading and margin rulings on the front. The rulings on the reverse side, which are faintly visible through the paper, help one maintain horizontal lines of lettering and provide guides for sketching and simple graph construction. Moreover, the lines on the back of the paper will not be lost as a result of erasures.

Figure 4.1

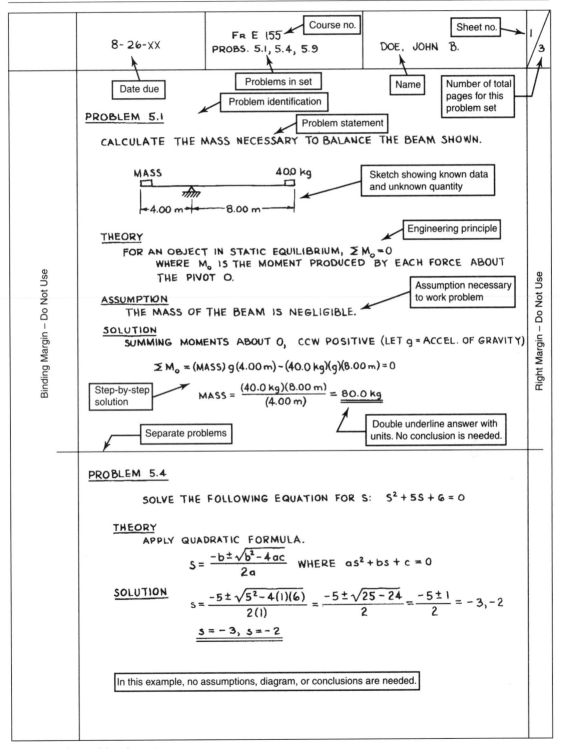

Elements of a problem layout.

Figure 4.2a

	8-22-XX	FR E 160 PROBLEM 13.1	DOE, JANE A.	1 / 2

PROBLEM 13.1   SOLVE FOR THE VALUE OF RESISTANCE R
                IN THE CIRCUIT SHOWN BELOW.

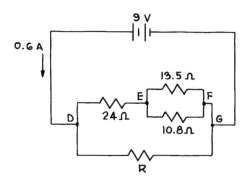

THEORY
- FOR RESISTANCES IN PARALLEL: $\frac{1}{R_{TOTAL}} = \frac{1}{R_1} + \frac{1}{R_2} + \frac{1}{R_3} + \cdots$

  THUS FOR 2 RESISTANCES IN PARALLEL

  $$R_{TOTAL} = \frac{R_1 R_2}{R_1 + R_2}$$

- FOR RESISTANCES IN SERIES: $R_{TOTAL} = R_1 + R_2 + R_3 + \cdots$

- OHM'S LAW: $E = RI$ WHERE E = ELECT. POTENTIAL IN VOLTS
                              I = CURRENT IN AMPERES
                              R = RESISTANCE IN OHMS

SOLUTION

- CALCULATE EQUIVALENT RESISTANCE BETWEEN POINTS E AND F.
  RESISTORS ARE IN PARALLEL.

  $$\therefore R_{EF} = \frac{R_1 R_2}{R_1 + R_2} = \frac{(13.5)(10.8)}{13.5 + 10.8} = \frac{145.8}{24.3} = 6.00 \ \Omega$$

- CALCULATE EQUIVALENT RESISTANCE OF UPPER LEG
  BETWEEN D AND G.

  SERIES CIRCUIT

  $$\therefore R'_{DG} = R_{24} + R_6 = 24 + 6 = 30 \ \Omega$$

> In this example, no assumptions were necessary.

Sample problem presentation done freehand.

Figure 4.2b

	8-22-XX	FR E 160 PROBLEM 13.1	DOE, JANE A.	2/2

- CALCULATE EQUIVALENT RESISTANCE BETWEEN D AND G.

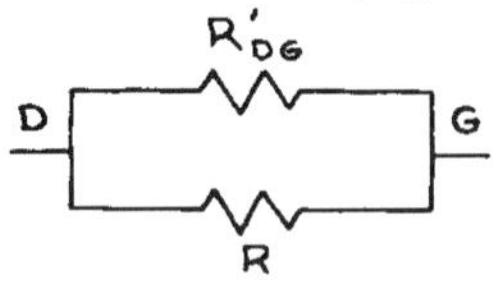

PARALLEL RESISTORS, SO

$$R_{DG} = \frac{(R'_{DG})(R)}{R'_{DG} + R}$$

- CALCULATE TOTAL RESISTANCE OF CIRCUIT USING OHM'S LAW.

$$R_{DG} = \frac{E}{I} = \frac{9\ V}{0.6\ A} = 15\ \Omega$$

- CALCULATE VALUE OF R FROM PREVIOUS EQUATIONS.

$$R_{DG} = 15\ \Omega = \frac{(R'_{DG})(R)}{R'_{DG} + R} = \frac{(30)(R)}{30 + R}$$

SOLVING FOR R:

$$(30 + R)(15) = 30R$$

$$30 + R = 2R$$

$$\underline{\underline{R = 30\ \Omega}}$$

Figure 4.3a

Problem

    A tank is to be constructed
that will hold 5.00×10^5 L when
filled. The shape is to be cylindrical,
with a hemispherical top. Costs to
construct the cylindrical portion will
be \$300/m², while costs for the
hemispherical portion are slightly
higher at \$400/m².

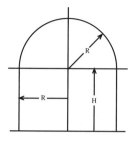

Find

    Calculate the tank dimensions that will result in the lowest dollar cost.

Theory

    Volume of cylinder is... $V_c = \pi R^2 H$

    Volume of hemisphere is... $V_H = \dfrac{2\pi R^3}{3}$

    Surface area of cylinder is... $SA_c = 2\pi RH$

    Surface area of hemisphere is... $SA_H = 2\pi R^2$

Assumptions

    Tank contains no dead air space
    Construction costs are independent of size
    Concrete slab with hermetic seal is provided for the base
    Cost of the base does not change appreciably with tank dimensions

Solution

    1. Express total volume in meters as a function of height and radius.

$$V_{Tank} = f(H, R)$$
$$= V_C + V_H$$
$$500 = \pi R^2 H + \frac{2\pi R^3}{3}$$

    Note: $1\text{m}^3 = 1000$ L

Sample problem presentation done with a word processor.

Figure 4.3b

2. Express cost in dollars as a function of height and radius

$$C = C(H, R)$$

$$= 300\,(SA_C) + 400\,(SA_H)$$

$$= 300\,(2\pi RH) + 400\,(2\pi R^2)$$

Note: Cost figures are exact numbers

3. From part 1 solve for $H = H(R)$

$$H = \frac{500}{\pi R^2} - \frac{2R}{3}$$

4. Solve cost equation, substituting $H = H(R)$

$$C = 300\left[2\pi R\left(\frac{500}{\pi R^2} - \frac{2R}{3}\right)\right] + 400\,(2\pi R^2)$$

$$C = \frac{300000}{R} + 400\pi R^2$$

Cost versus Radius

Radius, R, m	Cost, C, $
1.0	301 257
2.0	155 027
3.0	111 310
4.0	95 106
5.0	91 416
6.0	95 239
7.0	104 432
8.0	117 925
9.0	135 121
10.0	155 664

5. Develop a table of cost versus radius and plot graph.

6. From graph select minimum cost.

$$R = \underline{5.00\ \text{m}}$$
$$C = \underline{\$91\ 000}$$

7. Calculate H from part 3 above

$$H = \underline{3.033\ \text{m}}$$

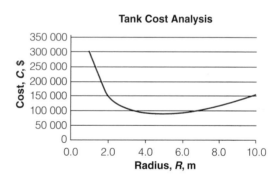

Tank Cost Analysis

8. Verification/check of results from the calculus:

$$\frac{dC}{dR} = \frac{d}{dR}\left[\frac{300\,000}{R} + 400\pi R^2\right]$$

$$= \frac{-300\,000}{R^2} + 800\pi R = 0$$

$$R^3 = \frac{300\,000}{800\pi}$$

$$R = \underline{4.92\text{m}}$$

9. Discussion/Conclusion: The minimum cost for the tank is found when the radius is 5.0 m and the height is 3.0 m. The height cost is found between a radius of 1.0 m and 2.0 m.

Figure 4.4

Problem 3-5

Analyze the buckling load for steel columns ranging from 50 to 100 ft long in increments of 5 ft.
The cross-sectional area is 7.33 in², the least radius of gyration is 3.19 in and modulus of elasticity is 30 × 10⁶ lb/in².
Plot the buckling load as a function of column length for hinged ends and fixed ends.

Theory

Euler's equation gives the buckling load for a slender column.

$$F_B = \frac{n\pi^2\,EA}{(L/r)^2}$$

where

F_B	= buckling load, lb	
E	= modulus of elasticity, lb/in²	3.00E+07
A	= cross-sectional area, in²	7.33
L	= length of column, in	
r	= least radius of gyration, in	3.19

The factor n depends on the end conditions: If both ends are hinged, $n = 1$;
if both ends are fixed, $n = 4$; if one end is fixed and the other is hinged, $n = 2$

Assumption: The columns being analyzed meet the slenderness criterion for Euler's equation

Solution

Length, ft	Buckling load (fixed), lb	Buckling load (hinged), lb
50	245394	61348
55	202805	50701
60	170412	42603
65	145204	36301
70	125201	31300
75	109064	27266
80	95857	23964
85	84911	21228
90	75739	18935
95	67976	16994
100	61348	15337

Discussion: The buckling load decreases with length for end conditions. The buckling load for the fixed ends condition is always higher, but becomes closer to the hinged condition with increased length.

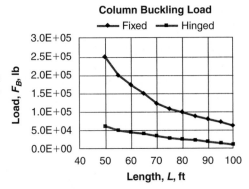

Sample problem presentation done with a spreadsheet.

2. The completed top heading of the problems paper should include such information as name, date, course number, and sheet number. The upper right-hand block should normally contain a notation such as a/b, where a is the page number of the sheet and b is the total number of sheets in the set.

3. Work should ordinarily be done in pencil using an appropriate lead hardness (HB, F, or H) so that the line work is crisp and not smudged. Erasures should always be complete, with all eraser particles removed. Letters and numbers must be dark enough to ensure legibility when photocopies are needed.

4. Either vertical or slant letters may be selected as long as they are not mixed. Care should be taken to produce good, legible lettering but without such care that little work is accomplished.

5. Spelling should be checked for correctness. There is no reasonable excuse for incorrect spelling in a properly done problem solution.

6. Work must be easy to follow and not crowded. This practice contributes greatly to readability and ease of interpretation.

7. If several problems are included in a set, they must be distinctly separated, usually by a horizontal line drawn completely across the page between problems. Never begin a second problem on the same page if it cannot be completed there. Beginning each problem on a fresh sheet is usually better, except in cases when two or more problems can be completed on one sheet. It is not necessary to use a horizontal separation line if the next problem in a series begins at the top of a new page.

8. Diagrams that are an essential part of a problem presentation should be clear and understandable. You should strive for neatness, which is a mark of a professional. Often a good sketch is adequate, but using a straight edge can greatly improve the appearance and accuracy of a diagram. A little effort in preparing a sketch to approximate scale can pay great dividends when it is necessary to judge the reasonableness of an answer, particularly if the answer is a physical dimension that can be seen on the sketch.

9. The proper use of symbols is always important, particularly when the International System (SI) of Units is used. It involves a strict set of rules that must be followed so that absolutely no confusion of meaning can result. There also are symbols in common and accepted use for engineering quantities that can be found in most engineering handbooks. These symbols should be used whenever possible. It is important that symbols be consistent throughout a solution and that they are all defined for the benefit of the reader and for your own reference.

The physical layout of a problem solution logically follows steps that are similar to those of the engineering method. You should attempt to present the process by which the problem was solved, in addition to the solution, so that any reader can readily understand all the aspects of the solution. Figure 4.1 illustrates the placement of the information.

Figures 4.2, 4.3, and 4.4 are examples of typical engineering problem solutions. You may find these examples to be helpful guides as you prepare your problem presentations.

Problems

4.1 The Cartesian components of a vector $\overline{B}$ are shown in Fig. 4.5. If B_x = 8.6 m and Δ = 35°, find α B_y, and $\overline{B}$.

4.2 Refer to Fig. 4.5. If α = 58° and B_y = 3.4 km, what are the values of Δ, B_x, and $\overline{B}$?

4.3 In Fig. 4.6, side YZ is 2.4 × 10⁴ m. Determine the length of side XZ.

4.4 Calculate the length of side AB in Fig. 4.7 if side AC = 2.75 × 10² m.

Figure 4.5

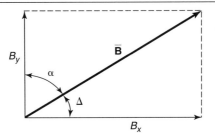

Figure 4.6

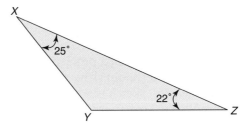

Figure 4.7

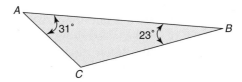

4.5 The vector $\overline{C}$ in Fig. 4.8 is the sum of vectors $\overline{A}$ and $\overline{B}$. Assume that vector $\overline{B}$ is horizontal. Given that α = 29°, β = 25°, and the magnitude of B = 29 m, find the magnitudes and directions of vectors $\overline{A}$ and $\overline{C}$.

4.6 Vector $\overline{R}$ in Fig. 4.9 is the difference between vectors $\overline{T}$ and $\overline{S}$. If $\overline{S}$ is inclined at 28° from the vertical and the angle between $\overline{S}$ and $\overline{T}$ is 32°, calculate the magnitude and direction of vector $\overline{R}$. The magnitudes of $\overline{S}$ and $\overline{T}$ are 19 cm and 36 cm, respectively.

4.7 An aircraft has a glide ratio of 12 to 1. (Glide ratio means that the plane drops 1 m in each 12 m it travels horizontally.) A building 45 m high lies directly in the glide path to the runway. If the aircraft clears the building by 12 m, how far from the building does the aircraft touch down on the runway?

Figure 4.8

87

Problems

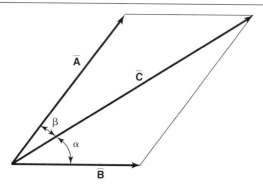

Figure 4.9

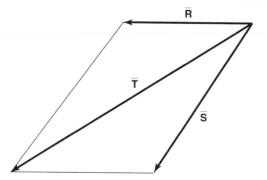

4.8 A pilot of an aircraft knows that the vehicle in landing configuration will glide 2.0×10^1 km from a height of 2.00×10^3 m. A TV transmitting tower is located in a direct line with the local runway. If the pilot glides over the tower with 3.0×10^1 m to spare and touches down on the runway at a point 6.5 km from the base of the tower, how high is the tower?

4.9 A simple roof truss design is shown in Fig. 4.10. The lower section, $VWXY$, is made from three equal length segments. UW and XZ are perpendicular to VT and TY, respectively. If $VWXY$ is 2.0×10^1 m and the height of the truss is 2.5 m, determine the lengths of XT and XZ.

4.10 An engineer is required to survey a nonrectangular plot of land but is unable to measure side UT directly due to a water obstruction (see Fig. 4.11). The following data are taken: $RU = 121.0$ m, $RS = 116.0$ m, $ST = 83.5$ m, angle $RST = 113°$, and angle $RUT = 82°$. Calculate the length of side UT and the area of the plot.

4.11 A park is being considered in a space between a small river and a highway as a rest stop for travelers (see Fig. 4.12). Boundary BC is perpendicular to the highway and boundary AD makes an angle of 75° with the highway. BC is measured to be 160.0 m, AD is 270.0 m, and the boundary along the highway is 190.0 m long. What are the length of side AB and the magnitude of angle ABC?

Figure 4.10

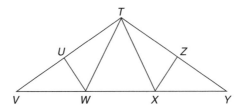

Figure 4.11

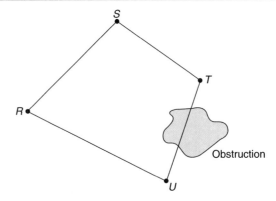

Figure 4.12

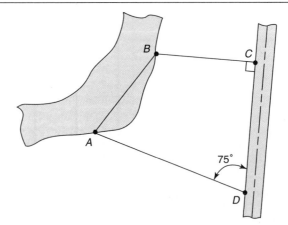

4.12 D, E, F, and G in Fig. 4.13 are surveyed points in a land development on level terrain so that each point is visible from each other. Leg DG is physically measured as 500.0 m. The angles at three of the points are found to be: angle $GDE = 55°$, angle $DEF = 92°$, angle $FGD = 134°$. Also, angle DGE is measured at $87°$. Compute the lengths of DE, EF, FG, and EG.

4.13 The height of an inaccessible mountain peak, C, in Fig. 4.14 must be estimated. Fortunately, two smaller mountains, A and B, which can be easily scaled, are located near the higher peak. To make matters even simpler, the three peaks lie on a single straight line. From the top of mountain A, altitude 2.000×10^3 m, the elevation angle to C is 12.32°. The elevation of C from mountain B is 22.73°. Mountain B is 1.00×10^2 m higher than A. The straight line (slant) distance between peaks A and B is 3.000×10^3 m. Determine the unknown height of mountain C.

4.14 A narrow belt is used to drive a 20.00-cm-diameter pulley from a 35.00-cm-diameter pulley. The centers of the two pulleys are 2.000 m apart. How long must the belt be if the pulleys rotate in the same direction? In opposite directions?

Figure 4.13

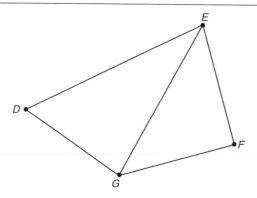

Figure 4.14

4.15 A motorcycle sprocket on the rear wheel has a diameter of 24 cm, and the driver sprocket has a diameter of 4.0 cm. The driver sprocket shaft and rear axle are 60 cm apart. What is the minimum chain length for this application?

4.16 A block of metal has a 90° notch cut from its lower surface. The notched part rests on a circular cylinder of diameter 2.0 cm, as shown in Fig. 4.15. If the lower surface of the part is 1.3 cm above the base plane, how deep is the notch?

4.17 A 1.00-cm-diameter circular gauge block is used to measure the depth of a 60° notch in a piece of tool steel. The gauge block extends a distance of 4.7 mm above the surface. How deep is the notch? See Fig. 4.16.

4.18 An aircraft moves through the air with a relative velocity of 3.00×10^2 km/h at a heading of N30°E. In a 35 km/h wind from the west,

    (*a*) Calculate the *true* ground speed and heading of the aircraft.

    (*b*) What heading should the pilot fly so that the *true* heading is N30°E?

4.19 To cross a river that is 1 km wide, with a current of 6 km/h, a novice boat skipper holds the bow of the boat perpendicular to the far riverbank, intending to cross to a point directly across the river from the launch point. At what position will the boat actually contact the far bank? What direction should the boat have been headed to actually reach a point directly across from the launch dock? The boat is capable of making 10 km/h.

4.20 What heading must a pilot fly to compensate for a 125 km/h west wind to have a ground track that is due south? The aircraft cruise speed is 6.00×10^2 km/h. What is the actual ground speed?

Figure 4.15

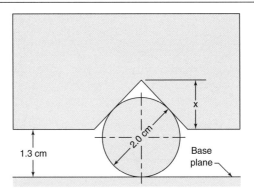

Figure 4.16

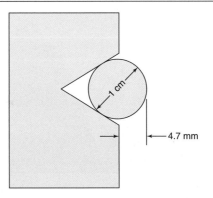

4.21 A tooling designer is designing a jig that will insert pins into the flip-up handles of a coffeemaker. The bin to hold the pins should hold enough pins for 3 h of work before reloading. The engineering estimate for the rate of inserting the pins is 300/h. The pin diameter is 0.125 in. Because of space constraints, the bin must be designed as shown in Fig. 4.17. The bin can be only the width of one pin. What is the minimum height that the bin should be if there is 0.250 in. at the top when it has been filled with the required number of pins?

4.22 Two friends are planning to go on RAGBRAI (Register's Annual Great Bike Ride Across Iowa) next year. One is planning to ride a mountain bicycle with 26-in. tires, and the other has a touring bicycle with 27-in. tires. A typical RAGBRAI is about 480 mi long.

    (*a*) How many more revolutions will the mountain bike tires make in that distance than the touring bike?

Figure 4.17

91

Problems

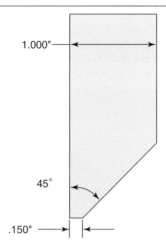

1.000"

45°

.150"

Figure 4.18

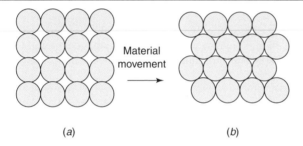

Material
movement

(a)

(b)

(b) Typical gearing for most bicycles ranges from 30 to 50 teeth on the chain wheel (front gears) and 12 to 30 teeth on the rear cog. Find how many more revolutions of the pedals the mountain biker will make during the trip, using an average pedaling time of 85% of the trip and a 42-tooth chain wheel and a 21-tooth rear cog for your calculations.

(c) If both cyclists have 170-mm cranks on their bikes, what will be the mechanical advantage (considering only the movement of the feet with a constant force for walking and riding), in percent, that each rider will have achieved over walking the same distance? (This can be found by dividing the distance walked by the distance the feet move in riding.)

4.23 An engineer has been given the assignment of finding how much money can be saved over a year's time by redesigning the press plates from the pattern shown in Fig. 4.18a to the pattern shown in Fig. 4.18b for stamping out 2.400-in.-diameter disks. The stamping material is 14-gauge sheet metal and can be purchased in 100-ft rolls in varying widths in 0.5-in. increments. One square foot of metal weighs 3.20 lb. The metal is sold for $0.20/lb. Do not consider the ends of the rolls. The company expects to produce 38 000 parts this year. How much can be saved?

4.24 Sally is making a sine bar, which is used to machine angles on parts (see Fig. 4.19). She has a 1.250-in. thick bar that needs 90° grooves machined into it for precision ground 1.0000-in. diameter cylinders. A sine bar is used by placing different thicknesses under one of the cylinders so that the proper angle is attained. Sally wants the distance between the centers of the cylinders to be 5.000 in.

(a) How deep should she mill the 90° grooves so that the top of the sine block is 2 in. tall?

(b) Once her sine block is finished, she wants to mill a 22.5° angle on a brass block. What thickness of gage blocks will produce this angle for this sine plate?

Figure 4.19

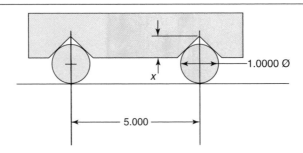

4.25 Standing at the edge of the roof of a tall building, you throw a ball upward with a velocity of 15 m/s (meters per second). The ball goes straight up and begins its downward descent just missing the edge of the building. The building is 40 m tall.

(a) What is the velocity of the ball at its uppermost position?

(b) How high above the building does the ball go before beginning its descent?

(c) What is the velocity of the ball as it passes the roof of the building?

(d) What is the speed of the ball just before it hits the ground?

(e) How long does it take for the ball to hit the ground after leaving your hand?

4.26 A stuntwoman is going to attempt a jump across a canyon 74 m wide. The ramp on the far side of the canyon is 25 m lower than the ramp from which she will leave. The takeoff ramp is built with a 15° angle from horizontal.

(a) If the stuntwoman leaves the ramp with a velocity of 28 m/s, will she make the jump?

(b) How many seconds will she be in the air?

4.27 An engineering student has been given the assignment of designing a hydraulic holding system for a hay-baling system. The system has four cylinders with 120-mm diameter pistons with a stroke of 0.320 m. The lines connecting the system are 1 cm id (inside diameter). There are 15.5 m of lines in the system. For proper design, the reserve tank should hold a minimum of 50% more than the amount of hydraulic fluid in the system. If the diameter of the reserve tank is 30.48 cm, what is the shortest height it should be?

4.28 The plant engineer for a large foundry has been asked to calculate the thermal efficiency of the generating plant used by the company to produce electricity for the aluminum melting furnaces. The plant generates 545.6 GJ of electrical energy daily. The plant burns 50 t (tons) of coal a day. The heat of combustion of coal is about 6.2×10^6 J/kg (joules/kilogram). What was the answer? (Efficiency $= W/J_{heat}$)

4.29 Using these three formulas

$$V = IR \qquad R = (\rho L)/A \qquad A = \pi(0.5d)^2$$

find the difference in current (I) that a copper wire ($\rho = 1.72 \times 10^{-8}\ \Omega \cdot$ m) can carry over an aluminum wire ($\rho = 2.75 \times 10^{-8}\ \Omega \cdot$ m) with equal diameters (d) of 0.5 cm and a length (L) of 10 000 m carrying 110 V (volts).

4.30 The light striking a pane of glass is refracted as shown in Fig. 4.20. The law of refraction states that $n_a \sin \theta_a = n_b \sin \theta_b$, where n_a and n_b are the refractive indexes of the materials through which the light is passing and the angles are from a line that is normal to the surface. The refractive index of air is 1.00. What is the refractive index of the glass?

Figure 4.20

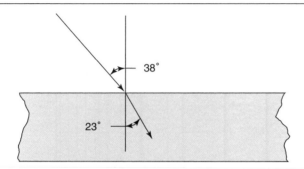

CHAPTER 5

Representation of Technical Information*

Chapter Objectives

When you complete your study of this chapter, you will be able to:

- Recognize the importance of collecting, recording, plotting, and interpreting technical data for engineering analysis and design
- Put into practice methods for graphical presentation of scientific data and graphical analysis of plotted data
- Develop the ability to graph data using uniform and nonuniform scales
- Apply methods of selected points and least squares for determining the equation that gives the best-fit line to the given data
- Determine the most appropriate family of curves (linear, power, or exponential) that gives the best fit to the given data

5.1 Introduction

This chapter begins with an example of an actual freshman engineering student team project. This team consisted of aerospace, electrical, and mechanical engineering students who were assigned to find how temperature and pressure varied with altitude in the atmosphere. The team was not given specific instructions as to how this might be accomplished, but once the information had been collected, they were expected to record, plot, and analyze the data.

After a bit of research, the team decided to request a university-owned plane that was equipped with the latest Rockwell Collins avionics gear, and since it was a class assignment, they asked if the university's Air Flight Service would consider helping them conduct this experiment free of charge. Because the pilots do periodic maintenance flights, they agreed to allow the students to ride along.

Using the plane's sophisticated data-acquisition equipment, the students were able to collect and record the data needed for the assignment (see Table 5.1) along the flight path as the plane ascended to 12 000 feet. They decided that one would record, one would make temperature readings, and the third would make pressure readings.

---

*Users will find Appendices A, F, and the inside covers useful reference material for this chapter.

Table 5.1

Height, H, ft	Temperature, T, °F	Pressure, P, lbf/in²
0	59	14.7
1000	55	14.2
2000	52	13.7
3000	48	13.2
4000	44	12.7
5000	41	12.2
6000	37	11.8
7000	34	11 3
8000	30	10.9
9000	27	10.5
10000	23	10.0
11000	19	9.7
12000	16	9.3

When the students returned to the hangar, they made a freehand plot (with straightedge) of the collected information (see Fig. 5.1).

Since the plot demonstrated linear results, the students returned to campus to prepare the required report, including an analysis of the data collected.

They decided the written report should include a computer-generated table, a computer-generated plot of the data, and a computer-generated least-squares curve fit to determine the mathematical relationship between the variables.

Table 5.2 shows the computer-generated table prepared after the students returned to campus. This could have been done with word processing, spreadsheet, or other commercially available software packages.

Figures 5.2 and 5.3 show how spreadsheet applications can be powerful and convenient for plotting once the fundamentals of good graph construction are understood. Figure 5.2 is an example of a Microsoft Excel spreadsheet using a scatter plot with each data point connected with a straight line. Figure 5.3 is an example of a scatter plot with only the data points plotted. A *trendline*

Table 5.2

Height, H, ft	Temperature, T, °F	Pressure, P, lbf/in²
0	59	14.7
1 000	55	14.2
2 000	52	13.7
3 000	48	13.2
4 000	44	12.7
5 000	41	12.2
6 000	37	11.8
7 000	34	11.3
8 000	30	10.9
9 000	27	10.5
10 000	23	10.0
11 000	19	9.7
12 000	16	9.3

Figure 5.1

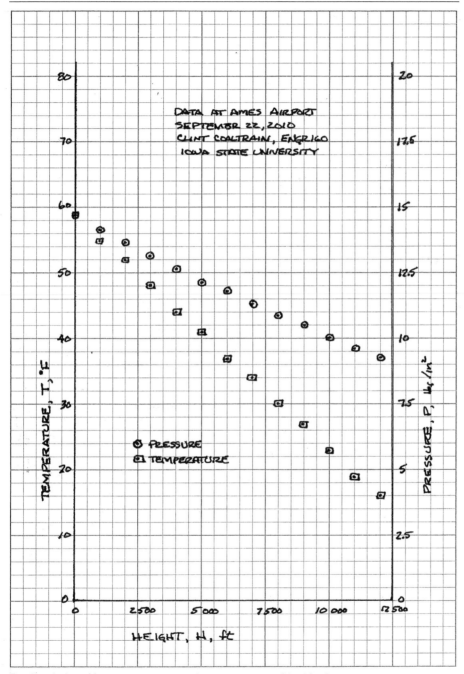

Freehand plot of how temperature and pressure vary with altitude.

Figure 5.2

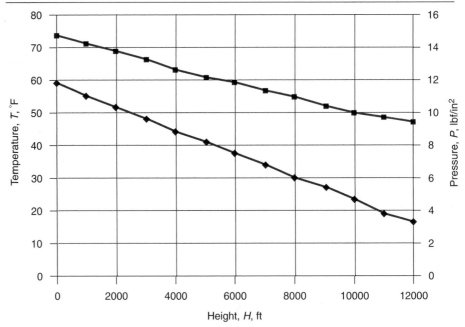

Excel spreadsheet hard copy of data displayed in Fig. 5.1.

Figure 5.3

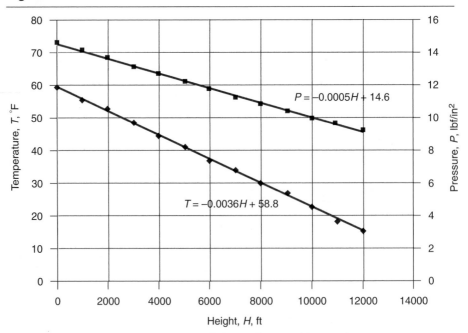

Excel spreadsheet hard copy of data displayed in Fig. 5.1 with trendline.

is then applied using the method of least squares with the equation of the line included. Further discussion of the use of the trendline/method of least squares will occur later in this chapter, with more details given in the chapter on statistics.

This chapter contains examples and guidelines as well as helpful information that will be needed when collecting, recording, plotting, and interpreting technical data. Two processes are: (1) graphical presentation of scientific data and (2) graphical analysis of plotted data.

Humans find it difficult to observe relationships among data shown in a table. We are much better at understanding relationships when the data are put into a graphical or pictorial format. We can immediately recognize the form of the relationship between two data sets (for example) when they are plotted on a graph. We can see whether the relationship is linear (straight line) or curved in some fashion. Perhaps we can even determine that the relationship appears to be in the form of a sine wave or a parabolic curve. Sometimes, just the impression of how the data relate is enough. More often, though, we need to obtain an equation that relates one variable to the other.

We have at our disposal computers and software that can make the production of a graph easy to accomplish and can help us to determine the relationship between variables in equation form. Much of this chapter is devoted to learning about engineering standards for graph production and methods for producing equations between variables that have been presented on a graph.

5.1.1 Software for Recording and Plotting Data

Data are recorded in the field as shown in Table 5.1. A quick freehand plot of the data is produced to provide a visual impression of the results while still in the field (see Fig. 5.1). This allows you to obtain corrected or additional data if the quick plot suggests it. Alternatively, the data could be entered into a laptop computer and processed initially in the field.

Upon returning to the laboratory, however, spreadsheet software, such as Excel, provides enormous recording and plotting capability. The data are entered into the computer, and by manipulation of software options, both the data and a graph of the data can be configured, stored, and printed (Table 5.2 and Figs. 5.2 and 5.3).

Programs such as Mathematica, MATLAB, and Mathcad provide a range of powerful tools designed to help analyze numerical and symbolic operations as well as to present a visual image of the results.

Software is also widely available to provide methods of curve fitting once the data have been collected and recorded.

Even though it is important for the engineer to interpret, analyze, and communicate different types of data, it is not practical to include in this chapter all forms of graphs and charts that may be encountered. For that reason, popular-appeal or advertising charts such as bar charts, pie diagrams, and distribution charts, although useful to the engineer, are not discussed here.

Even though commercial software is very helpful during the presentation and analysis process, the results are only as good as the original software design and its use by the operator. Some software provides a wide range of tools but only allows limited data applications and minimal flexibility to modify default outputs. Other software provides a high degree of in-depth analysis for a particular subject area with considerable latitude to adjust and modify parameters.

Inevitably, the computer together with its array of software will continue to provide an invaluable analysis tool. However, it is absolutely essential that you be knowledgeable of the software and demonstrate considerable care when manipulating the data. You need to understand the software's limitations and accuracies, but above all you must know what plotted results are needed and what the engineering standards are for producing them.

For this reason, the sections that follow are a combination of manual collection, recording, plotting, and analysis and computer-assisted collection, recording, plotting, and analysis.

5.2 Collecting and Recording Data

5.2.1 Manual Entry

Modern science was founded on scientific measurement. Meticulously designed experiments, carefully analyzed, have produced volumes of scientific data that have been collected, recorded, and documented. For such data to be meaningful, however, certain procedures must be followed. Field books, such as those shown in Figure 5.4, or data sheets should be used to record all observations. Information about equipment, such as the instruments and experimental apparatus used, should be recorded. Sketches illustrating the physical arrangement of equipment can be very helpful. Under no circumstances should observations be recorded elsewhere or data points erased. The data sheet or field book is the "notebook of original entry." If there is reason for doubting the value of any entry, it may be canceled (i.e., not considered) by drawing a line through it. The cancellation should be done in such a manner that the original entry is not obscured in case you want to consider it later.

As a general rule it is advantageous to make all measurements as carefully as time and the economics of the situation allow. Errors do enter into all experimental work regardless of the amount of care exercised.

It can be seen from what we have just discussed that the analysis of experimental data involves not only measurements and collection of data but also careful documentation and interpretation of results.

Experimental data once collected are normally organized into some tabular form, which is the next step in the process of analysis. Data, such as that shown in Table 5.1, should be carefully labeled and neatly lettered so that the data are not misunderstood. This particular collection of data represents atmospheric pressure and temperature measurements recorded at various altitudes by students during a flight in a light aircraft.

Although the manual tabulation of data is frequently a necessary step, you will sometimes find it difficult to visualize a relationship between variables when simply viewing a column of numbers. Therefore, a most important step in the sequence from collection to analysis is the construction of appropriate graphs or charts.

Figure 5.4

101
*Collecting and
Recording Data*

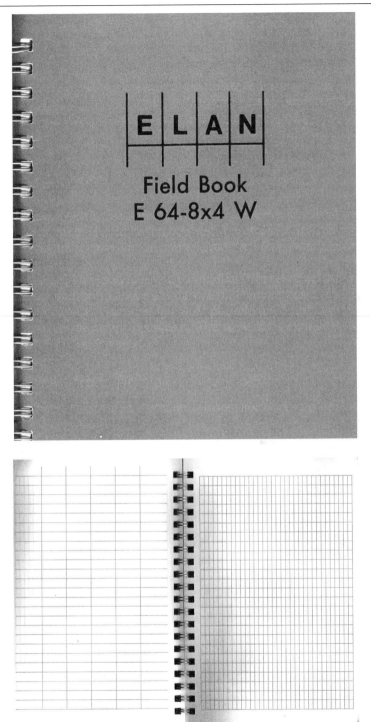

Field book typically used by civil engineers.

5.2.2 Computer-Assisted Techniques

A variety of equipment has been developed that will automatically sample experimental data for analysis. We expect to see expansion of these techniques along with visual displays that will allow us to interactively control the experiments. As an example, the flight data collected onboard the aircraft could be entered directly into a laptop computer through digital interfaces with the flight instruments and then printed as in Table 5.2.

5.3 General Graphing Procedures

Many examples appear throughout this chapter to illustrate methods of graphical presentation because their effectiveness depends to a large extent on the details of construction.

The proper manual construction of a graph from tabulated data can be described by a series of steps. Each of these steps will be discussed and illustrated in detail in the following sections. Once you understand the manual process of graph construction, the step to computer-generated graphs will be simple.

1. Select the type of graph paper (rectangular [also known as rectilinear], semilog, log-log) and grid spacing for best representation of the given data.
2. Choose the proper location of the horizontal and vertical axes.
3. Determine the scale units (range) for each axis to display the data appropriately.
4. Graduate and calibrate the axes using the 1, 2, 5 rule.
5. Identify each axis completely.
6. Plot points and use permissible symbols.
7. Check any point that deviates from the slope or curvature of the line.
8. Draw the curve or curves.
9. Identify each curve, add title, and include other necessary notes.
10. Darken lines for good reproduction.

5.3.1 Graph Paper

Printed coordinate graph paper is commercially available in various sizes with a variety of grid spacing. Rectangular ruling can be purchased in a range of lines per inch or lines per centimeter, with an overall paper size of 8.5 x 11 in. most typical.

Closely spaced coordinate ruling is generally avoided for results that are to be printed or photoreduced. However, for accurate engineering analyses requiring some amount of interpolation, data are normally plotted on closely spaced, printed coordinate paper. Graph paper is available in a variety of colors, weights, and grades. Translucent paper can be used when the reproduction system requires a material that is not opaque.

If the data require the use of log-log or semilog paper, such paper can also be purchased in different formats, styles, weights, and grades. Both log-log and semilog grids are available from 1 to 5 cycles per axis. (A later section will

Figure 5.5

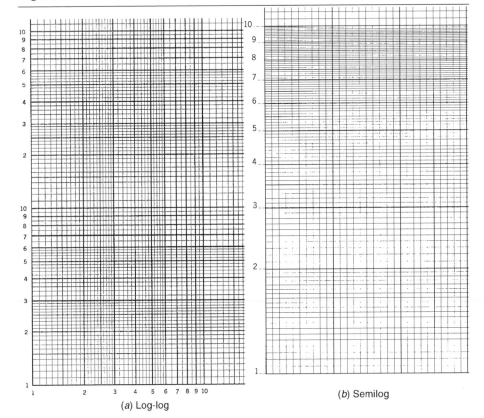

(a) Log-log

(b) Semilog

Commercial graph paper.

discuss different applications of log-log and semilog paper.). Examples of commercially available log and semilog paper are given in Figures 5.5a and 5.5b.

5.3.2 Axes Location and Breaks

The axes of a graph consist of two intersecting straight lines. The horizontal axis, normally called the *x-axis*, is the *abscissa*. The vertical axis, denoted by the *y-axis*, is the *ordinate*. Common practice is to place the independent variable values along the abscissa and the dependent variable values along the ordinate, as illustrated in Figure 5.6.

It is not always clear which variable is the independent variable and which is the dependent variable. You can think in terms of an experiment where one variable is set (independent variable) and another is determined (dependent variable). For example, in a test of an electrical circuit, if the voltage is set it is the independent variable and if the current is read from an instrument as a result of this voltage setting it is the dependent variable. You can also think in terms of reading from a graph. Normally you would find the value of the variable you set along the horizontal axis (independent variable) and read the value of the other (dependent variable) from the curve using the vertical axis.

Figure 5.6

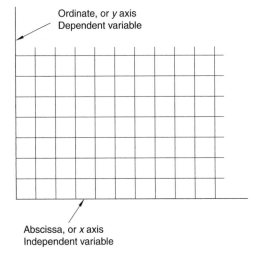

Ordinate, or *y* axis
Dependent variable

Abscissa, or *x* axis
Independent variable

Abscissa (*X*) and ordinate (*Y*) axes.

Sometimes mathematical graphs contain both positive and negative values of the variables. This necessitates the division of the coordinate field into four quadrants, as shown in Figure 5.7. Positive values increase toward the right and upward from the origin.

On any graph, a full range of values is desirable, normally beginning at zero and extending slightly beyond the largest value. To avoid crowding, one should use the entire coordinate area as completely as possible. However, certain circumstances require special consideration to avoid wasted space. For

Figure 5.7

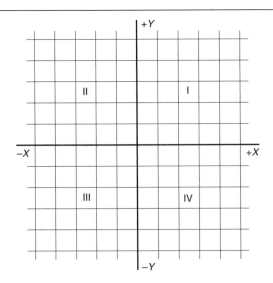

Coordinate axes.

Figure 5.8

105
General Graphing Procedures

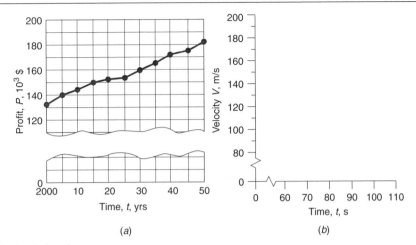

Typical axis breaks.

example, if values plotted along the axis do not range near zero, a "break" in the grid or the axis may be used, as shown in Figures 5.8a and 5.8b.

When judgments concerning relative amounts of change in a variable are required, the axis or grid should not be broken or the zero line omitted, with the exception of time in years, such as 2009, 2010, and so on, because that designation normally has little relation to zero.

Since most commercially prepared grids do not include sufficient border space for proper labeling, the axes should preferably be placed 20 to 25 mm (approximately 1 in.) inside the edge of the printed grid to allow ample room for graduations, calibrations, axes labels, reproduction, and binding. The edge of the grid may need to be used on log scales because it is not always feasible to move the axis inside the grid. However, with careful planning, the vertical and horizontal axes can usually be repositioned.

5.3.3 Scale Graduations, Calibrations, and Designations

The scale is a series of marks, called *graduations,* laid down at predetermined distances along the axis. Numerical values assigned to significant graduations are called *calibrations.*

A scale can be *uniform* or *linear,* with equal spacing along the axis, as found on the metric or engineer's scales. If the scale represents a variable whose exponent is not equal to 1 or a variable that contains trigonometric or logarithmic functions, the scale is called a *nonuniform,* or *functional, scale.* Examples of both these scales together with graduations and calibrations are shown in Figure 5.9. When you plot data, one of the most important considerations is the proper selection of scale graduations. A basic guide to follow is the *1, 2, 5 rule,* which only applies to uniform axes and can be stated as follows:

> Scale graduations are selected so that the smallest division of the axis is a positive or negative integer power of 10 times 1, 2, or 5.

Figure 5.9

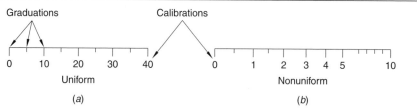

Scale graduations and calibrations.

Figure 5.10

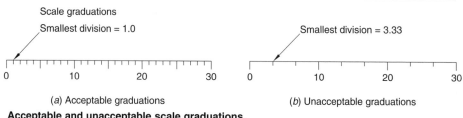

Acceptable and unacceptable scale graduations.

The justification and logic for this rule are clear. Graduation of an axis by this procedure allows better (more accurate) interpolation of data between graduations when plotting or reading a graph. Figure 5.10 illustrates both acceptable and unacceptable examples of scale graduations.

Violations of the 1, 2, 5 rule that are acceptable even for uniform axes involve certain units of time as a variable. Days, months, and years can be graduated and calibrated as illustrated in Figure 5.11.

Figure 5.11

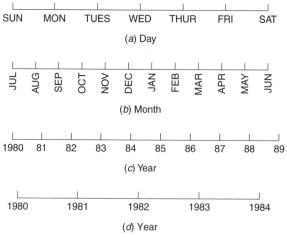

Time as a variable.

Figure 5.12

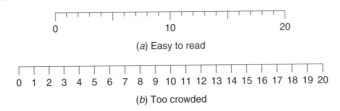

(a) Easy to read

(b) Too crowded

Acceptable and unacceptable scale calibrations.

Scale graduations follow a definite rule, but the number of calibrations included is a matter of good judgment. Each application requires consideration based on the scale length and range as well as the eventual use. Figure 5.12 demonstrates how calibrations can differ on a scale with the same range. Both examples obey the 1, 2, 5 rule, but as you can see, too many closely spaced calibrations make the axis difficult to read.

The selection of a scale deserves attention from another point of view. If the rate of change is to be depicted accurately, the slope of the curve should represent a true picture of the data. By compressing or expanding one of the axes, you could communicate an incorrect impression of the data. Such a procedure should be avoided. Figure 5.13 demonstrates how the equation $Y = X$ can be misleading if not properly plotted. Occasionally, distortion is desirable, but it should always be carefully labeled and explained to avoid misleading conclusions.

Figure 5.13

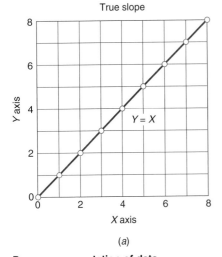

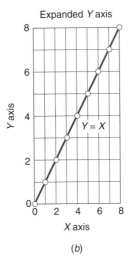

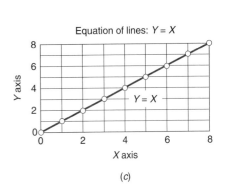

(a)

(b)

(c)

Proper representation of data.

Figure 5.14

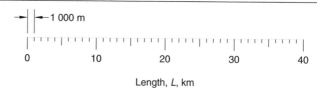

Length, *L*, km

Reading the scale.

If plotted data consist of very large or small numbers, the SI prefix names (milli-, kilo-, mega-, etc.) may be used to simplify calibrations. As a guide, if the numbers to be plotted and calibrated consist of more than three digits, it is customary to use the appropriate prefix; an example appears in Figure 5.14.

The length scale calibrations in Figure 5.14 contain only two digits, but the scale can be read by understanding that the distance between the first and second graduation (0 to 1) is a kilometer; therefore, the calibration at 10 represents 10 km.

Certain quantities, such as temperature in degrees Celsius and altitude in meters, have traditionally been tabulated without the use of prefix multipliers. Figure 5.15 depicts a procedure by which these quantities can be conveniently calibrated. Note in particular that the distance between 0 and 1 on the scale represents 1000°C.

The calibration of logarithmic scales is illustrated in Figure 5.16. Since log-cycle designations start and end with powers of 10 (i.e., 10^{-1}, 10^0, 10^1, 10^2, etc.) and since commercially purchased paper is normally available with each cycle printed 1 through 10, do not use the printed values as your calibrations. Instead, provide your own calibrations and use the printed numbers as a reference to be sure you understand what each line of the grid represents. Since the axes are nonuniform, it is sometimes difficult to determine what each grid line represents without those printed numbers. Figures 5.16a and 5.16b demonstrate two preferred methods of calibration.

Figure 5.15

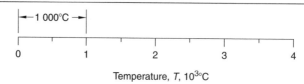

Temperature, *T*, 10^{3}°C

Reading the scale.

Figure 5.16

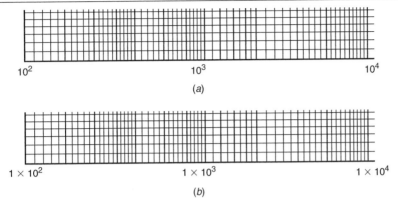

Calibration of log scales.

Figure 5.17

Axis identification.

5.3.4 Axis Labeling

Each axis should be clearly identified. At a minimum, the axis label should contain the name of the variable, its symbol, and its units. Since time is frequently the independent variable and is plotted on the x-axis, it has been selected as an illustration in Figure 5.17. Scale designations should preferably be placed outside the axes, where they can be shown clearly. Labels should be lettered parallel to the axis and positioned so that they can be read from the bottom or right side of the page, as illustrated in Figure 5.22.

5.3.5 Point-Plotting Procedure

Data can be described in one of three ways: as observed, empirical, or theoretical. Observed and empirical data points are usually located by various symbols, such as a small circle or square around each data point, whereas graphs of theoretical relations (equations) are normally constructed smooth, without the use of symbol designation. Figure 5.18 illustrates each type.

5.3.6 Curves and Symbols

On graphs prepared from observed data resulting from laboratory experiments, points are usually designated by various symbols (see Fig. 5.19). If more than one curve is plotted on the same grid, several of these symbols may be used (one type for each curve). To avoid confusion, however, it is good practice to label each curve. When several curves are plotted on the same grid, another way they can be distinguished from each other is by using different types of lines, as illustrated in Figure 5.20. Solid lines are normally reserved for single curves, and dashed lines are commonly used for extensions; however, a different line type can be used for each separate curve. The line weight of plotted curves should be heavier than the grid ruling.

A key, or legend, should be placed in an available portion of the grid, preferably enclosed in a border, to define point symbols or line types that are used for curves. Remember that the lines representing each curve *should never be drawn through the symbols,* so that the precise point is always identifiable. Figure 5.21 demonstrates the use of a key and the practice of breaking the line at each symbol.

Figure 5.18

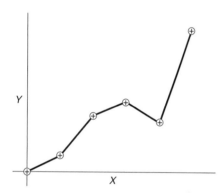

(*a*) Observed: Usually plotted with observed data points connected by straight, irregular line segments. Line does not penetrate the circles.

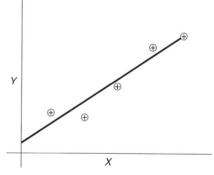

(*b*) Empirical: Reflects the author's interpretation of what occurs between known data points. Normally represented as a smooth curve or straight line fitted to data. Data points may or may not fall on curve.

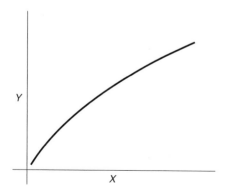

(*c*) Theoretical: Graph of an equation. Curves or lines are smooth and without symbols. Every point on the curve is a data point.

Plotting data points.

Figure 5.19

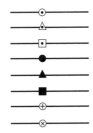

Symbols.

Figure 5.20

Line types.

Figure 5.21

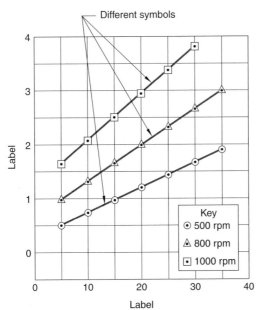

An example of a key.

5.3.7 Titles

Each graph must be identified with a complete title. The title should include a clear, concise statement of the data represented, along with items such as the name of the author, the date of the experiment, and any and all information concerning the plot, including the name of the institution or company. Titles may be enclosed in a border.

All lettering, the axes, and the curves should be sufficiently bold to stand out on the graph paper. Letters should be neat and of standard size. Figure 5.22 is an illustration of plotted experimental data incorporating many of the items discussed in the chapter.

5.3.8 Computer-Assisted Plotting

Several types of software are available to produce graphs (e.g., Mathcad, MATLAB, Minitab, Mathematica, Excel). The quality and accuracy of these computer-generated graphs vary depending on the sophistication of the

Figure 5.22

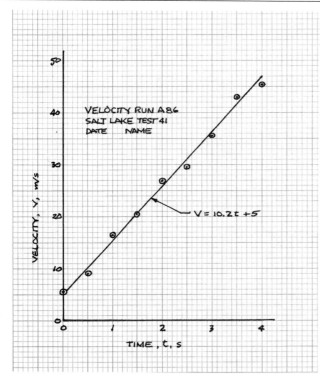

Necessary steps to follow when manually plotting a graph:

1. Select the type of graph paper (rectilinear, semilog, log-log) and grid spacing for best representation of the given data.

2. Choose the proper location of the horizontal and vertical axes.

3. Determine the scale units (range) for each axis to display the data appropriately.

4. Graduate and calibrate the axes using the 1, 2, 5 rule.

5. Identify each axis completely.

6. Plot points using permissible symbols.

7. Check any point that deviates from the slope or curvature of the line.

8. Draw the curve or curves.

9. Identify each curve, add title and necessary notes.

10. Darken lines for good reproduction.

Sample plot.

software as well as on the plotter or printer employed. Typically, the software will produce an axis scale graduated and calibrated to accommodate the range of data values that will fit the paper. This may or may not produce a readable or interpretable scale. Therefore, it is necessary to apply judgment depending on the results needed. For example, if the default plot does not meet needed scale readability, it may be necessary to specify the scale range to achieve an appropriate scale graduation, since this option allows greater control of the scale drawn.

Computer-produced graphs with uniform scales may not follow the 1, 2, 5 rule, particularly because the software plots the independent variable based on the data collected. If the software has the option of separately specifying the range—that is, plotting the data as an *X-Y* scatter plot—you will be able to achieve scale graduations and calibrations that do follow the 1, 2, 5 rule, making it easier to read values from the graph. The hand-plotted graph that was illustrated in Figure 5.22 is plotted using Excel with the results shown in Figure 5.23 using an *X-Y* scatter plot with a linear curve-fit and the equation of the line using the method of least squares (see section 5.6).

Figure 5.23

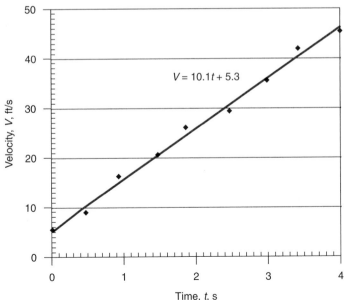

$V = 10.1t + 5.3$

Necessary steps to follow when using a computer-assisted alternative:
1. Input via keyboard or import data into spreadsheet.
2. Select independent (*x*-axis) and dependent variable(s).
3. Select appropriate graph (style or type) from menu.
4. Produce trial plot with default parameters.
5. Examine (modify as necessary) origin, range, graduation, and calibrations: Note, use the 1, 2, 5 rule.
6. Label each axis completely.
7. Select appropriate plotting-point symbols and legend.
8. Create complete title.
9. Examine plot and store the data.
10. Plot or print the data.

Excel spreadsheet hard copy of data displayed in Fig. 5.22.

5.4 Empirical Functions

Empirical functions are generally described as those based on values obtained by experimentation. Since they are arrived at experimentally, equations normally available from theoretical derivations are not always available. However, mathematical expressions can be modeled to fit experimental data, and it is possible to classify many empirical results into one of four general categories: (1) linear, (2) exponential, (3) power, or (4) periodic.

A linear function, as the name suggests, will plot as a straight line on uniform rectangular coordinate paper. Likewise, when a curve representing experimental data is a straight line or a close approximation to a straight line, the relationship of the variables can be expressed by a linear equation, such as $y = mx + b$.

Correspondingly, exponential functions, when plotted on semilog paper, will be linear. Why? Because the basic form of the equation is $y = be^{mx}$, all we do is take the log of both sides. If it is written in log (base 10) form, it becomes $\log y = mx \log e + \log b$. Alternatively, using natural logarithms, the equation becomes $\ln y = mx + \ln b$ because $\ln e = 1$. The independent variable x is plotted on the abscissa, and the dependent variable y is plotted on the functional ln (natural log) scale as $\ln y$.

The power equation has the form of $y = bx^m$. Written in log form, it becomes $\log y = m \log x + \log b$. This equation will plot as a straight line on log-log paper because the log of the independent variable x is plotted against the log of the dependent variable y.

The periodic type, often seen in alternating current, for example, is not covered in this text.

When the data represent experimental results and a series of points are plotted to represent the relationship between the variables, it is unlikely that a straight line can be constructed through every point because some error (instruments, readings, recordings) is inevitable. If all points do not lie on a straight line, an approximation technique or averaging method may be used to arrive at the best possible fit. This method of straight-line approximation is called curve fitting.

5.5 Curve Fitting

Different methods or techniques are available to arrive at the best "straight-line" fit. Two methods commonly employed for finding the best fit are

1. Method of selected points
2. Method of least squares

The most accurate method, least squares, is discussed in more detail in the chapter on statistics. However, several examples are presented in this chapter to demonstrate correct methods for plotting technical data using both the method of selected points and the method of least squares.

The method of selected points is a valid method of determining the equation that best fits data that exhibit a linear relationship. Once the data have been plotted and you have decided that a linear equation would be a good fit, a line is positioned that appears to best fit the data. This is most often accomplished by visually selecting a line that goes through as many data points as possible and has approximately the same number of data points on either side of the line.

Once the line has been drawn, two points, such as A and B, are selected *on the line* and at a reasonable distance apart (the further apart the better). The coordinates of both points $A(X_1, Y_1)$ and $B(X_2, Y_2)$ must satisfy the equation of the line because both are points on the line.

The method of least squares is a more accurate approach that will be illustrated as computer-assisted examples in most problems that follow. The method of least squares is a most appropriate technique for determination of the best-fit line. You should understand that the method presented represents a technique called *linear regression* and is valid only for *linear* relationships. The technique of least squares can, however, be applied to power ($y = bx^m$) and exponential ($y = be^{mx}$) relationships as well as $y = mx + b$, if done correctly. The power function can be handled by noting that there is a linear relationship between log y and log x (log y = m log x + log b, which plots as a straight line on log-log paper). Thus, we can apply the method of least squares to the variables log y and log x to obtain parameters m and log b.

The exponential function written in natural logarithm form is ln y = mx + ln b. Therefore, a linear relationship exists between ln y and x (this plots as a straight line on semilog paper). The next examples will demonstrate the use of the selected point method for power and experimental curves.

5.7 Empirical Equations: Linear

When experimental data plot as a straight line on rectangular grid paper, the equation of the line belongs to a family of curves whose basic equation is given by

$$y = mx + b \tag{5.1}$$

where m is the slope of the line, a constant, and b is a constant referred to as the *y intercept* (the value of y when $x = 0$).

To demonstrate how the method of selected points works, consider the following example.

Example problem 5.1 The velocity V of an experimental automobile is measured at specified time t intervals. Determine the equation of a straight line constructed through the points recorded in Table 5.3. Once an equation has been determined, velocities at intermediate values can be computed.

Procedure

1. Plot the data on rectangular paper. If the results form a straight line (see Fig. 5.24), the function is linear and the general equation is of the form

    $$V = mt + b$$

    where m and b are constants.

2. Select two points on the line, $A(t_1, V_1)$ and $B(t_2, V_2)$, and record the value of these points. Points A and B should be widely separated to reduce the effect on m and b of errors in reading values from the graph. Points A and B are identified in Figure 5.24 for instructional reasons. They should not be shown on a completed graph that is to be displayed.

    $$A(10, 60)$$
    $$B(35, 165)$$

Table 5.3

Time t, s	0	5	10	15	20	25	30	35	40
Velocity V, m/s	24	33	62	77	105	123	151	170	188

3. Substitute the points A and B into $V = mt + b$.

    Eq(1) $\qquad\qquad 60 = m(10) + b$
    Eq(2) $\qquad\qquad 165 = m(35) + b$

Figure 5.24

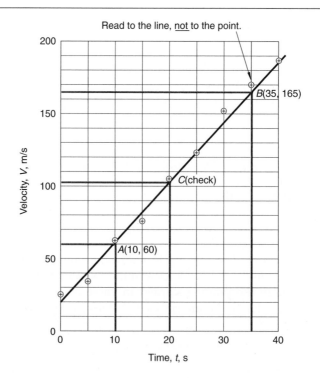

Data plot.

4. The equations are solved simultaneously for the two unknowns.

$$m = 4.2$$
$$b = 18$$

5. The equation of the line for this specific problem can be written as

$$V = 4.2t + 18$$

6. Using another point $C(t_3, V_3)$, check for verification:

$$C(20, 102)$$
$$102 = 4.2(20) + 18$$
$$102 = 84 + 18 = 102$$

7. A computer-assisted alternate:

The data set can be entered into a spreadsheet, and the software will provide a more precise solution. Figure 5.23 illustrates a software solution for data provided in the previous sample plot (Fig. 5.22). The computer solution provides a plot of data on rectangular paper with the equation of the line determined by the method of least squares. Once you understand the fundamentals, it can be very time efficient to use computer technology and commercial software.

5.8 Empirical Equations: Power Curves

When experimentally collected data are plotted on rectangular coordinate graph paper and the points do not form a straight line, you must determine which family of curves the line most closely approximates. If you have no idea as to the nature of the data, plot the experimentally collected points on log-log paper and/or semilog paper to determine if the data approximate a straight line. Consider the following familiar example. Suppose a solid object is dropped from a tall building. To anyone who has studied fundamental physics, it is apparent that distance and time should correspond to the general equation for a free-falling body (neglecting air friction): $s = 1/2\ gt^2$.

However, let's assume for a moment that we do not know this relationship and that all we have is a table of values experimentally measured on a free-falling body.

Example problem 5.2   A solid object is dropped from a tall building, and the values time versus distance are as recorded in Table 5.4.

Procedure

1. Make a freehand plot to observe the data visually (see Fig. 5.25). From this quick plot, the data points are more easily recognized as belonging to a family of curves whose general equation can be written as

$$y = bx^m \tag{5.2}$$

Figure 5.25

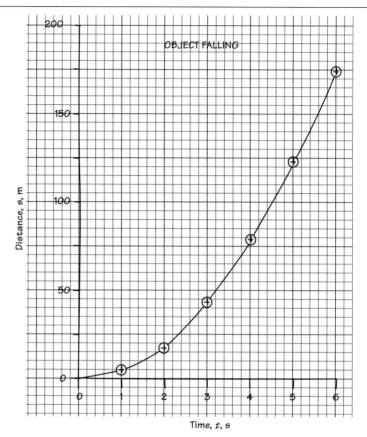

Rectangular graph paper (freehand).

Table 5.4

Time t, s	Distance s, m
0	0
1	4.9
2	19.6
3	44.1
4	78.4
5	122.5
6	176.4

Remember that before the method of selected points can be applied to determine the equation of the line, the plotted line must be straight because two points on a curved line do not uniquely identify the line. Mathematically, this general equation can be modified by taking the logarithm of both sides,

$$\log y = m \log x + \log b, \text{ or } \ln y = m \ln x + \ln b$$

This equation suggests that if the logs of all table values of y and x were computed and the results plotted on rectangular paper, the line would likely be straight.

Table 5.5

Time t, s	Distance s, m	Log t	Log s
0	0		
1	4.9	0.0000	0.6902
2	19.6	0.3010	1.2923
3	44.1	0.4771	1.6444
4	78.4	0.6021	1.8943
5	122.5	0.6990	2.0881
6	176.4	0.7782	2.2465

Realizing that the log of zero is undefined and plotting the remaining points that are recorded in Table 5.5 for log s versus log t, the results are shown in Figure 5.26.

Since the graph of log s versus log t does plot as a straight line, it is now possible to use the general form of the equation

$$\log s = m \log t + \log b$$

and apply the method of selected points.

When reading values for points A and B from the graph, we must remember that the logarithm of each variable has already been determined and the values plotted.

$$A(0.2, 1.09)$$
$$B(0.6, 1.89)$$

Points A and B can now be substituted into the general equation log $s = m$ log t + log b and solved simultaneously.

Figure 5.26

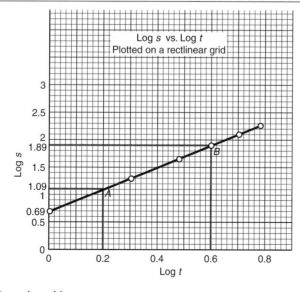

Log-log on rectangular grid paper.

$$1.89 = m(0.6) + \log b$$
$$1.09 = m(0.2) + \log b$$
$$m = 2.0$$
$$\log b = 0.69, \text{ or}$$
$$b = 4.9$$

As examination of Figure 5.26 shows, the value of log b (0.69) can be read from the graph where log $t = 0$. This, of course, is where $t = 1$ and is the y intercept for log-log plots.

The general equation can then be written as

$$s = 4.9t^{2.0}$$

or

$$s = 1/2gt^2,$$

where $g = 9.8 \ m/s^2$

Note: One obvious inconvenience is the necessity of finding logarithms of each variable and then plotting the logs of these variables. This step is not necessary since functional paper is commercially available with log x and log y scales already constructed. Log-log paper allows the variables themselves to be plotted directly without the need of computing the log of each value.

2.   An alternate method for the solution of this problem is as follows:

In the preceding part, once the general form of the equation is determined [Eq. (5.2)], the data can be plotted directly on log-log paper. Since the resulting curve is a straight line, the method of selected points can be used directly (see Fig. 5.27).

Figure 5.27

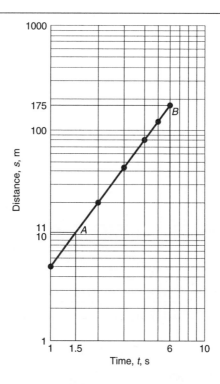

Log-log paper.

The log form of the equation is again used:

$$\log s = m \log t + \log b$$

Select points A and B on the line:

$A(1.5,11)$
$B(6,175)$

Substitute the values into the general equation $\log s = m \log t + \log b$, taking careful note that the numbers are the variables and *not* the logs of the variables.

$$\log 175 = m \log 6 + \log b$$
$$\log 11 = m \log 1.5 + \log b$$

Again, solving these two equations simultaneously results in the following approximate values for the constants b and m:

$$b = 4.8978 \cong 4.9$$
$$m = 1.9957 \cong 2.0$$

Identical conclusions can be reached:

$$s = 1/2gt^2$$

This time, however, one can use functional scales rather than calculate the log of each number.

3.   A computer-assisted alternate:

The data set can be entered into a spreadsheet, and the software will provide an identical solution. Figure 5.28 illustrates software that provides a plot

Figure 5.28

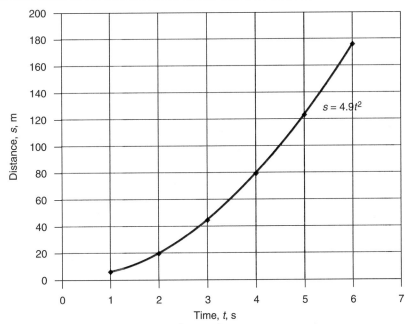

Graph of a free-falling object on rectangular graph paper.

Figure 5.29

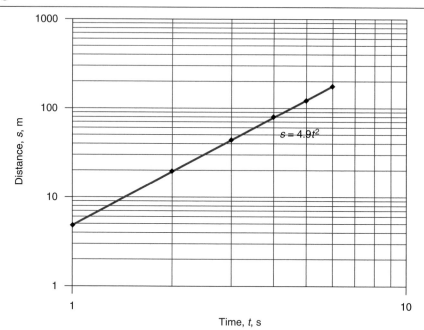

Graph of a free-falling object on log-log paper.

of data on rectangular paper, including the equation of the line. Figure 5.29 is an example of the data plotted on log-log paper with the software providing the equation.

5.9 Empirical Equations: Exponential Curves

Suppose your data do not plot as a straight line on rectangular coordinate paper or the line is not approximately straight on log-log paper. Without experience in analyzing experimental data, you may feel lost about how to proceed. Normally, when experiments are conducted, you have an idea as to how the parameters are related and you are merely trying to quantify that relationship. If you plot your data on semilog graph paper and it produces a reasonably straight line, then it has the general form

$$y = be^{mx} \tag{5.3}$$

Example problem 5.3   Vehicle fuel consumption is recorded as shown in Table 5.6. Determine the best-fit equation for the data by the method of selected points and by the method of least squares.

Procedure

1. The data (Table 5.6) when plotted on semilog paper produce the graph shown as Fig. 5.30. To determine the constants in the equation $y = be^{mx,}$ write it in linear form, either as

$$\log y = mx \log e + \log b$$

Table 5.6

Velocity V, m/s	Fuel Consumption (FC), mm³/s
10	25.2
20	44.6
30	71.7
40	115
50	202
60	367
70	608

or

$$\ln y = mx + \ln b$$

The method of selected points can now be employed for $\ln FC = mV + \ln b$ (choosing the natural log form). Points $A(15,33)$ and $B(65,470)$ are carefully selected on the line, so they must satisfy the equation. Substituting the values of V and FC at points A and B, we get

$$\ln 470 = 65\,m + \ln b$$

and

$$\ln 33 = 15\,m + \ln b$$

Solving simultaneously for m and b, we have

$$m = 0.0529$$

and

$$b = 14.8$$

Figure 5.30

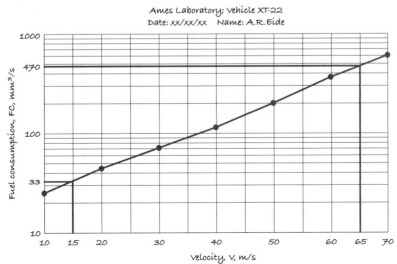

Ames Laboratory; Vehicle XT-22
Date: xx/xx/xx   Name: A.R. Eide

Semilog paper.

Figure 5.31

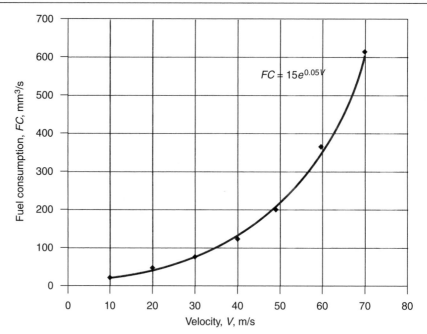

Graph of the fuel consumption of a rocket engine on rectangular graph paper.

The desired equation then is determined to be $FC = 15e^{(0.05V)}$. This determination can be checked by choosing a third point, substituting the value for V, and solving for FC.

2. A computer-assisted alternate:

The data set can be entered into a spreadsheet, and the software will provide an identical solution. Figure 5.31 illustrates software that provides a plot of data on rectangular paper, including the equation of the line. Figure 5.32 is an example of the data plotted on semilog paper with the software providing the equation.

Problems

5.1   The table shows data from a trial run on the Utah salt flats made by an experimental turbine-powered vehicle.

Time, t, s	Velocity, V, m/s
10.0	15.1
20.0	32.2
30.0	63.4
40.0	84.5
50.0	118.0
60.0	139.0

Figure 5.32

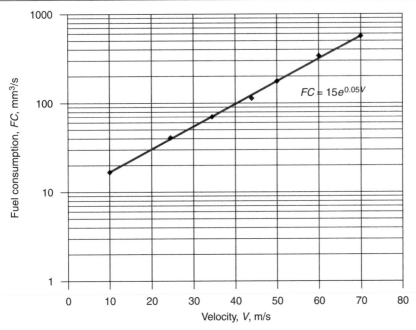

Fuel consumption, *FC*, mm³/s

Velocity, *V*, m/s

$$FC = 15e^{0.05V}$$

Graph of the fuel consumption of a rocket engine on semilog paper.

(*a*) Plot the data on rectangular graph paper.
(*b*) Determine the equation of the line using the method of selected points.
(*c*) Determine the equation of the line using computer-assisted methods.
(*d*) Interpret the slope of the line.

5.2 The table lists the values of velocity recorded on a ski jump in Colorado this past winter.

Time, *t*, s	Velocity, *V*, m/s
1.0	5.3
4.0	18.1
7.0	26.9
10.0	37.0
14.0	55.2

(*a*) Plot the data on rectangular graph paper.
(*b*) Determine the equation of the line using the method of selected points.
(*c*) Determine the equation of the line using computer-assisted methods.
(*d*) Give the average acceleration.

5.3 Below is a collection of data for an iron-constantan thermocouple. Temperature is in degrees Celsius, and the electromotive force (*emf*) is in millivolts.

Temperature, *T*, °C	Voltage, *emf*, mV
50.0	2.6
100.0	6.7
150.0	8.8
200.0	11.2
300.0	17.0
400.0	22.5
500.0	26.0
600.0	32.5
700.0	37.7
800.0	41.0
900.0	48.0
1 000.0	55.2

(a) Plot the graph using rectangular graph paper with voltage as the independent variable.

(b) Using the method of selected points, find the equation of the line.

(c) Using computer-assisted methods, find the equation of the line.

5.4 There are design specifications for the minimum sight distance (distance to see an approaching vehicle measured along the roadway from the intersection of the two roadways) that a driver stopped at a stop sign must have to safely enter a roadway where vehicles do not stop. Values in the table below are for safe entry to *cross* the roadway (not to turn onto the other roadway but to cross) where vehicles do not stop.

Major roadway design speed, *DS*, mph	Sight distance, *SD*, ft
25	240
30	290
35	335
40	385
45	430
50	480
60	575
65	625
70	670
75	720

(a) Plot a graph with design speed as the independent variable.

(b) Determine the equation of the relationship using the method of selected points.

(c) Determine the equation of the relationship using computer-assisted methods.

(d) Predict the slight distance required at 55 mph.

5.5 A spring was tested in Chicago last Thursday. The test of the spring, X-19, produced the following data.

Deflection, *D*, mm	Load, *L*, kN
2.25	35.0
12.0	80.0
20.0	120.0
28.0	160.0
35.0	200.0
45.0	250.0
55.0	300.0

(a) Plot the data on rectangular graph paper and determine the equation that expresses the deflection to be expected under a given load. Use both the method of selected points and computer-assisted methods.

(b) Predict the load required to produce a deflection of 75 mm.

(c) What load would be expected to produce a deflection of 120 mm?

5.6 An Acme furnace was tested 45 days ago in your hometown to determine the heat generated, expressed in thousands of British thermal units per cubic foot of furnace volume at varying temperatures. The results are shown in the table.

Heat released, H, 10^3 Btu/ft³	Temperature, T, °F
0.200	172
0.600	241
2.00	392
4.00	483
8.00	608
20.00	812
40.00	959
80.00	1 305

(a) Plot the data on log-log paper with temperature as the independent variable.

(b) Using the method of selected points, determine the equation that best fits the data.

(c) Using computer-assisted methods, plot the graph and determine the equation of the line.

5.7 The capacity of a 20-cm screw conveyor that is moving dry corn is expressed in liters per second and the conveyor speed in revolutions per minute. A test was conducted in Rock Island, IL, on conveyor model JD172 last week. The results of the test are given below.

Capacity, C, L/s	Angular velocity, V, r/min
3.01	10.0
6.07	21.0
15.0	58.2
30.0	140.6
50.0	245.0
80.0	410.0
110.0	521.0

(a) Plot the data on log-log paper with angular velocity as the independent variable.

(b) Determine the equation that expresses capacity as a function of angular velocity using the method of selected points.

(c) Repeat (a) and (b) using computer-assisted methods.

5.8 Electrical resistance for a given material can be a function of both area/unit thickness and material temperature. Holding temperature constant, a range of areas are tested to determine resistance. The measured resistance recorded in the table is expressed in milliohms per meter of conductor length.

Area, *A*, mm²	Resistance, *R*, mV/m
0.05	500
0.1	260
0.2	165
0.5	80
1.0	58
3.0	22
5.0	15
10	9.0

    (*a*) Plot the data on rectangular graph paper with area as the independent variable.

    (*b*) Plot the data on log-log graph paper with area as the independent variable.

    (*c*) Plot the data on semilog graph paper with area as the independent variable.

    (*d*) Compute equations from these three plots using the method of selected points.

    (*e*) Plot and find equations for parts (*a*), (*b*), and (*c*) using computer-assisted methods.

    (*f*) What would be the best curve fit for this application?

5.9 The area of a circle can be expressed by the formula $A = \pi R^2$. If the radius varies from 0.5 to 5 cm, perform the following:

    (*a*) Construct a table of radius versus area mathematically. Use radius increments of 0.5 cm.

    (*b*) Construct a second table of log *R* versus log *A*.

    (*c*) Plot the values from (*a*) on log-log paper and determine the equation of the line.

    (*d*) Plot the values from (*b*) on rectangular paper and determine the equation of the line.

    (*e*) Repeat parts (*c*) and (*d*) using computer-assisted methods.

5.10 The volume of a sphere is $V = 4/3\pi r^3$.

    (*a*) Prepare a table of volume versus radius allowing the radius to vary from 2.0 to 10.0 cm in 1-cm increments.

    (*b*) Plot a graph on log-log paper showing the relation of volume to radius using the values from the table in part (*a*) with radius as the independent variable.

    (*c*) Verify the equation given above by the method of selected points.

    (*d*) Repeat parts (*b*) and (*c*) using computer-assisted methods.

5.11 A 90° triangular weir is commonly used to measure flow rate in a stream. Data on the discharge through the weir were collected and recorded as shown below.

Height, *h*, m	Discharge, *Q*, m³/s
1	1.5
2	8
3	22
4	45
5	78
6	124
7	182
8	254

(*a*) Plot the data on log-log paper with height as the independent variable.

(*b*) Determine the equation of the line using the method of selected points.

(*c*) Plot and determine the equation using computer-assisted methods.

5.12 According to government statistics, the average selling price of a home in the United States varied as follows.

Year, Y	Price, P, $
1965	$21 500
1970	$26 600
1975	$42 600
1980	$76 400
1985	$100 800
1990	$149 800
1995	$158 700
2000	$207 000
2005	$292 200
2010	$265 500

(*a*) Plot the data on rectangular graph paper and find the best equation using the method of selected points.

(*b*) Plot the data on semilog paper and find the best equation using the method of selected points.

(*c*) Prepare linear and semilog plots using a computer-assisted method and determine the best equations for each.

(*d*) Which curve fit, linear or exponential, best fits these data?

5.13 The density of air is known to change with the temperature of the air. In a lab test, the following data were measured and recorded.

Temperature, T, K	Density, D, kg/m³
100	3.5
200	1.7
400	0.85
600	0.6
800	0.45
1 000	0.35
1 200	0.3
1 400	0.25
1 600	0.2

(*a*) Plot these data on linear, semilog, and log-log paper with temperature as the independent variable.

(*b*) Determine the equation that best fits these data using the method of selected points.

(*c*) Repeat parts (*a*) and (*b*) using a computer-assisted method.

5.14 Voltage across a capacitor during discharge was recorded as a function of time as shown below.

Time, t, s	Voltage, V, V
6	98
10	62
17	23
25	9.5
32	3.5
38	1.9
42	1.3

(a) Plot the data on semilog paper with time as the independent variable.

(b) Determine the equation of the line best representing the points using the method of selected points.

(c) Repeat parts (a) and (b) using a computer-assisted method.

5.15 When a capacitor is being discharged, the current flows until the voltage across the capacitor is zero. This current flow, when measured as a function of time, resulted in the data given in the following table.

Time, t, s	Current, I, A
0.1	1.81
0.2	1.64
0.3	1.48
0.4	1.34
0.5	1.21
1.0	0.73

(a) Plot the data on semilog paper with time as the independent variable.

(b) Determine the equation of the line best representing the points using the method of selected points.

(c) Repeat parts (a) and (b) using a computer-assisted method.

5.16 The density of water vapor in air changes rapidly with the change in air temperature. Data were recorded from an experimental test and are shown in the following table.

Air Temperature, T, K	Water Vapor Density, D, kg/m^3
400	0.55
450	0.49
500	0.44
550	0.39
600	0.36
650	0.34
700	0.33
750	0.29
800	0.27

(a) Plot the data on linear, semilog, and log-log paper with the air temperature as the independent variable.

(b) By the method of selected points, find the best equation relating water vapor density to air temperature.

(c) Repeat parts (a) and (b) using a computer-assisted method.

5.17 All materials are elastic to some extent. It is desirable that a part compresses when a load is applied to assist in making an airtight seal (e.g., a jar lid). The results in the following table are from a test conducted at the Smith Test Labs in Seattle on a material known as Zecon 5.

Pressure, P, MPa	Relative compression, R, %
1.12	27.3
3.08	37.6
5.25	46.0
8.75	50.6
12.3	56.1
16.1	59.2
30.2	65.0

(*a*) Plot the data on semilog and log-log paper with pressure as the independent variable.

(*b*) Using the method of selected points, determine the best equation to fit these data.

(*c*) Using a computer-assisted method, repeat the steps above.

(*d*) What pressure would cause compression of 10%?

5.18 The rate of absorption of radiation by metal plates varies with the plate thickness and the nature of the source of radiation. A test was conducted at Ames Labs on October 11, 2005, using a Geiger counter and a constant source of radiation; the results are shown in the following table.

Plate thickness, *W*, mm	Counter, *C*, counts per second
0.20	5 500
5.00	3 720
10.0	2 550
20.0	1 320
27.5	720
32.5	480

(*a*) Plot the data on semilog graph paper with plate thickness as the independent variable.

(*b*) Find the equation of the relationship between the parameters using the method of selected points.

(*c*) Repeat parts (*a*) and (*b*) using a computer-assisted method.

(*d*) What would you expect the counts per second to be for a 2-in.-thick plate of the metal used in the test?

5.19 It is expected that power functions represent the surface area and volume of a certain geometric shape. The values of the surface area and volume are given in the table below:

Radius, *R*, ft	Surface Area, *SA*, ft^2	Volume, *V*, ft^3
1	12.6	4.19
2	50.3	33.5
3	113	113
4	200	268
5	314	524
6	450	905
7	616	1 437
8	800	2 145
9	1 018	3 054
10	1 257	4 189

(*a*) Using a computer-assisted method, determine the equations of SA and V as functions of R.

(*b*) What is the geometric shape?

5.20 According to the United States Department of Labor, the Consumer Price Index for several household expense items are shown in the table below. The time period 1982–1984 is established as the basis with an index of 100.

Year	Food	Apparel	Housing	Transportation	Medical Care	Total
1995	148.4	132.0	148.5	139.1	220.5	152.4
1996	153.3	131.7	152.8	143.0	228.2	156.9
1997	157.3	132.9	156.8	144.3	234.6	160.5
1998	160.7	133.0	160.4	141.6	242.1	163.0
1999	164.1	131.3	165.9	144.4	250.6	166.6
2000	167.8	129.6	169.6	153.3	260.8	172.2
2001	173.1	127.3	176.4	154.3	272.8	177.1
2002	176.2	124.0	180.3	152.9	285.6	179.9
2003	180.0	120.9	184.8	157.6	297.1	184.0
2004	186.2	120.4	189.5	163.1	310.1	188.9

(*a*) Using a computer-assisted method, plot all of these data on the same graph. Be sure to format the graph according to chapter guidelines.

(*b*) Describe in a couple of paragraphs what the plotted information suggests to you.

5.21 Data from the Federal Reserve, the average prime interest rate, the average home mortgage rate, and the average 6-month CD rate for the years 1972 to 2009 are shown in the table below:

Year	Ave. prime rate, %	Ave. mortgage rate, %	Ave. 6-month CD rate, %
1972	5.25	7.38	5.01
1973	8.03	8.04	9.05
1974	10.81	9.19	10.02
1975	7.86	9.04	6.9
1976	6.84	8.86	5.63
1977	6.83	8.84	5.91
1978	9.06	9.63	8.6
1979	12.67	11.19	11.42
1980	15.26	13.77	12.94
1981	18.87	16.63	15.79
1982	14.85	16.08	12.57
1983	10.79	13.23	9.28
1984	12.04	13.87	10.71
1985	9.93	12.42	8.24
1986	8.33	10.18	6.5
1987	8.21	10.2	7.01
1988	9.32	10.34	7.91
1989	10.87	10.32	9.08
1990	10.01	10.13	8.17
1991	8.46	9.25	5.91
1992	6.25	8.4	3.76
1993	6	7.33	3.28
1994	7.15	8.35	4.96
1995	8.83	7.95	5.98
1996	8.27	7.8	5.47
1997	8.44	7.6	5.73
1998	8.35	6.94	5.44
1999	8	7.43	5.46
2000	9.23	8.06	6.59
2001	6.91	6.97	3.66
2002	4.67	6.54	1.81
2003	4.12	5.82	1.17
2004	4.34	5.84	1.74
2005	6.19	5.86	3.73
2006	7.96	6.41	5.24
2007	8.05	6.34	5.23
2008	5.09	6.04	3.14
2009	3.25	5.04	0.87

(a) Using a computer-assisted method, plot all of these data on the same graph. Be sure to format the graph according to chapter guidelines.

(b) In a short narrative, describe in general terms the relationship among the three interest rates as you see it from your graph.

Source: http://www.federalreserve.gov/releases/h15/data.htm

5.22 Performance data for an experimental truck engine were recorded in a recent laboratory test as shown in the table below.

Engine speed, R, rpm	Rated power output, P, bhp	Full load torque, T, ft-lb	Fuel consumption, F, gal/hr
800	148	959	3.2
900	203	1 155	5.3
1 000	249	1 341	6.3
1 100	330	1 601	8.4
1 200	414	1 762	10.3
1 300	499	2 012	13.5
1 400	545	2 072	16.6
1 500	589	2 049	19.2
1 600	605	2 028	22.9
1 700	644	1 981	27.3
1 800	670	1 919	34.1

As a team

(a) Plot each data set on linear, semilog, and log-log graph paper with engine speed as the independent variable.

(b) Using the method of selected points, compute the best equation fit to these data.

(c) Repeat parts (a) and (b) using computer-assisted methods.

(d) Discuss your results in a form that could be understood by other engineering students who have not solved this problem.

Hint: For data that plot as a convex upward curve, try reversing the independent and dependent variables in a plot before applying a curve fitting method. Once the equation has been found, simply solve for the desired dependent variable as a function of the independent variable.

5.23 As a team, conduct an Internet search for data in a technical field of interest to your team that can be graphed by a computer-assisted method. Find a data set (or sets) of a size that would be difficult to enter by hand, download the data, and directly enter them into your graphing software.

For each set, prepare a graph that follows engineering graphing standards as closely as your software allows. Within the graph, demonstrate the following:

■ Multiple curves on a graph complete with a legend

■ Two separate y-axes

Provide documentation to include:

■ Explanation and interpretation of the data you selected

■ Source of data

■ Hard copy from your software with data and graph

CHAPTER 6

Engineering Measurements and Estimations

Chapter Objectives

When you complete your study of this chapter, you will be able to:

- ■ Determine the number of significant digits in a measurement
- ■ Perform numerical calculations with measured quantities and express the answer with the appropriate number of significant digits
- ■ Define accuracy and precision in measurements
- ■ Define systematic and random errors and explain how they occur in measurements
- ■ Solve problems involving estimations of the required data and assumptions to enable a solution
- ■ Develop and present problem solutions, involving finding or estimating the necessary data, that enable others to understand your method of solution and to determine the validity of the numerical work

6.1 Introduction

The nineteenth-century physicist Lord Kelvin stated that knowledge and understanding are not of high quality unless the information can be expressed in numbers. Numbers are the operating medium for most engineering functions. In order to perform analysis and design, engineers must be able to measure physical quantities and express these measurements in numerical form. Furthermore, engineers must have confidence that the measurements and subsequent calculations and decisions made based on the measurements are reasonable.

In this chapter, we will describe how to properly use measurements (numbers) in engineering calculations. In your specific engineering discipline you will gain experience in selecting the correct measuring device for a particular situation.

6.2 Measurements: Accuracy and Precision

In measurements, "accuracy" and "precision" have different meanings and cannot be used interchangeably. *Accuracy* is a measure of the nearness of a value to

135

Figure 6.1

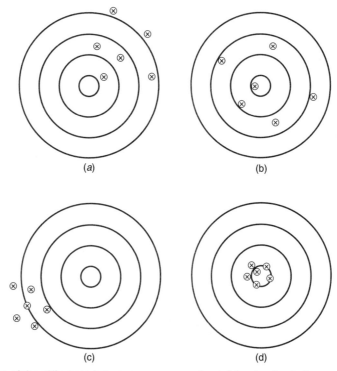

(a)   (b)

(c)   (d)

Illustration of the difference between accuracy and precision in physical measurements.

the correct or true value. *Precision* refers to the repeatability of a measurement, that is, how close successive measurements are to each other. Figure 6.1 illustrates accuracy and precision of the results of four dart throwers. Thrower (a) is both inaccurate and imprecise because the results are away from the bull's-eye (accuracy) and widely scattered (precision). Thrower (b) is accurate because the throws are evenly distributed about the desired result but imprecise because of the wide scatter. Thrower (c) is precise with the tight cluster of throws but inaccurate because the results are away from the desired bull's-eye. Finally, thrower (d) demonstrates accuracy and precision with tight cluster of throws around the center of the target. Throwers (a), (b), and (c) can improve their performance by analyzing the causes for the errors. Body position, arm motion, and release point could cause deviation from the desired result.

As engineers perform computations in analysis and design, the accuracy and precision of data gathered for the computations must be ascertained. For a quantity that is measured by a physical instrument, the exact numerical value of the quantity is likely to remain unknown. Therefore the measurement must be recorded with the known limitations in accuracy and precision taken into account. To do this engineers apply accepted practices and rules and carefully note the conditions under which the data was obtained. The practices and rules for performing numerical computations are discussed in Section 6.3. A brief discussion of the identification of errors in measurements is found in Section 6.4.

6.3 Measurements: Significant Digits

Numbers used for calculation purposes in engineering design and analysis may be integers (exact) or real (exact or approximate). For example, six one-dozen cartons of eggs include a countable number of eggs, exactly 72. There are 2.54 centimeters in one inch, an example of an exact real number. Thus if the conversion of inches to centimeters is required in a calculation, 2.54 is not a contributing factor to the precision of the result of the calculation. The ratio of the circumference of a circle to its diameter, ϖ, is an approximate real number that may be written as 3.14, 3.142, 3.14159, . . . , depending on the precision required in a numerical calculation.

Any physical measurement that is not a countable number will be approximate. Errors are likely to be present regardless of the precautions used when making the measurement. Let us look at measuring the length of the metal bar in Figure 6.2 with a scale graduated in tenths of inches. At first glance it is obvious that the bar is between 2 and 3 inches in length. We could write down that the bar is 2.5 ± 0.5 inches long. Upon closer inspection we note the bar is between 2.6 and 2.7 inches in length, or 2.65 ±0.05 inches. What value would we use in a computation? 2.64? 2.65? 2.66? We might select 2.64 as the "best" measurement, realizing that the third digit in our answer is doubtful and therefore our measurement must be considered approximate.

It is clear that a method of expressing results and measurements is needed that will convey how "good" these numbers are. The use of significant digits gives us this capability without resorting to the more rigorous approach of computing an estimated percentage error to be specified with each numerical result or measurement. Before we introduce significant digits, it is necessary to discuss the presentation of numerical values in formats that leave no doubts in interpretation.

The following are accepted conventions for the presentation of numbers in engineering work:

1. For numbers less than one, a zero is written in front of the decimal point to omit any possible errors due to copy processes or careless reading. Therefore, we write 0.345 and not .345.
2. A space, not a comma, is used to divide numbers of three orders of magnitude or more. We write 4 567.8 instead of 4,567.8 and 0.678 91 instead of 0.678,91.

Figure 6.2

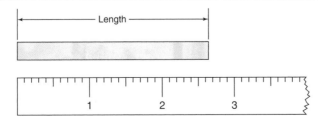

Making an approximate measurement.

3. For very large or very small numbers we use scientific notation to reduce the unwieldy nature of these numbers. For example, supercomputer calculating rates are compared by using the Linpack benchmark performance criteria. One of the criteria used by Linpack is the solving of a very large set of simultaneous linear equations. In the 2009 rankings of supercomputers, an IBM/DOE supercomputer named BlueGene/L, installed at the Lawrence Livermore National Laboratory in California, was benchmarked at 478 200 000 000 000 flops/s, seventh best sustainable performance in the world. The unit flops/s stands for *floating-point operations* per *second*, which is computer terminology for calculations using real numbers. In scientific notation this number would be 478.2×10^{12}, an obviously more compact representation. We will see that scientific notation is of great assistance in the determination and representation of significant digits.

Another convenient method of representing measurements is with prefix names that denote multipliers by factors of 10. Table 6.1 illustrates decimal multipliers and their corresponding prefixes and symbols. Thus the BlueGene/L computer performance could also be represented as 478.2 teraflops/s, or 478.2 Tflops/s. The highest computational performance in the world in 2009 was established by the Jaguar, a Cray XT5-HE Opteron system, measured at 1 759 Tflops/s or 1.759 Pflops/s.

The use of prefixes enables us to express any measurement as a number between 0.1 and 1 000 with a corresponding prefix applied to the unit. For example, it is clearer to the reader if the distance between two cities is expressed as 35 kilometers (35 km) rather than 35 000 meters (35 000 m). More on prefixes may be found in the Chapter 7 discussion of the International System of Units (SI).

A *significant digit,* or *significant figure,* is defined as any digit used in writing a number, *except* those zeros that are used only for location of the decimal

Table 6.1 Decimal Multiples

Multiplier	Prefix name	Symbol
10^{18}	exa	E
10^{15}	peta	P
10^{12}	tera	T
10^{9}	giga	G
10^{6}	*mega	M
10^{3}	*kilo	k
10^{2}	hecto	h
10^{1}	deka	da
10^{-1}	deci	d
10^{-2}	centi	c
10^{-3}	*milli	m
10^{-6}	*micro	—
10^{-9}	nano	n
10^{-12}	pico	p
10^{-15}	femto	f
10^{-18}	atto	a

*Most often used.

Figure 6.3

139

Measurements:
Significant Digits

Quantity	Number of Significant Figures
4784	4
36	2
60	1 or 2
600	1, 2, or 3
6.00×10^2	3
31.72	4
30.02	4
46.0	3
0.02	1
0.020	2
600.00	5

point or those zeros that do not have any nonzero digit on their left. When you read the number 0.001 5, only the digits 1 and 5 are significant, since the three zeros have no nonzero digit to their left. We would say this number has two significant figures. If the number is written 0.001 50, it contains three significant figures; the rightmost zero is significant.

Numbers 10 or larger that are not written in scientific notation and that are not counts (exact values) can cause difficulties in interpretation when zeros are present. For example, 2000 could contain one, two, three, or four significant digits; it is not clear which. If you write the number in scientific notation as 2.000×10^3, then clearly four significant digits are intended. If you want to show only two significant digits, you would write 2.0×10^3. It is our recommendation that if uncertainty results from using standard decimal notation, you switch to scientific notation so your reader can clearly understand your intent. Figure 6.3 shows the number of significant figures for several quantities.

You may find yourself as the user of measurements where the writer was not careful to properly show significant figures. Assuming that the number is not a count or a known exact value, you should establish a reasonable number of significant figures based on the context of the measurement and on your experience. Once you have decided on a reasonable number of significant digits in each measured value, you can then use that number in any calculations that are required making sure you clearly explain your actions.

When an instrument, such as an engineer's scale, analog thermometer, or fuel gauge, is read, the last digit will normally be an estimate. That is, the instrument is read by estimating between the smallest graduations on the scale to get the final digit. In Figure 6.4a, the reading is between 1.2 and 1.3, or 1.25 ± 0.05. For calculation purposes we might select 1.27 as a best value, with the 7 being a doubtful digit of the three significant figures. **It is standard practice to count one doubtful digit as significant, thus the 1.27 reading has three significant figures.** Similarly, the thermometer reading in Figure 6.4b is noted as 52.5° ± 0.5, and we estimate a best value of 52.8° with the 8 being doubtful.

In Figure 6.4c, the graduations create a more difficult task for reading a fuel level. Each graduation is one-sixth of a full tank. The reading is between

Figure 6.4

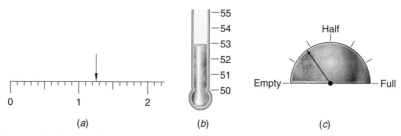

(a)   *(b)*   *(c)*

Reading graduations on instruments will include a doubtful or estimated value.

1/6 and 2/6 full, or 3/12 ± 1/12. How many significant figures are there? If we convert the reading to 0.25 ± 0.0833, a "best" estimate might be 0.30. In any case you cannot justify more than one significant figure and the answer would be expressed as 0.3. The difficulty in this example is not the significant figures but the scale of the fuel gauge. It is meant to convey a general impression of the fuel level and not a numerically significant value. Therefore, the selection of the instrument is an important factor in physical measurements.

Calculators and computers maintain countable numbers (integers) in exact form up to the capacity of the machine. Real numbers are kept at the precision level of the particular device. This is true no matter how many significant digits an input value or calculated value should have. Therefore, you will need to exercise care in reporting values from a calculator display or from a computer output (a spreadsheet for example). Calculators and most high-level computer languages allow you to control the number of digits that are to be displayed or printed. If a computer output is to be a part of your final solution presentation, you will need to carefully control the output form. If the output is only an intermediate step, you can round the results to a reasonable number of significant figures in your presentation.

As you perform arithmetic operations, it is important that you not lose the significance of your measurements or, conversely, imply precision that does not exist. Rules for determining the number of significant figures that should be reported following computations have been developed by engineering associations. The following rules customarily apply.

1. **Rounding.** General Rule: Round a value to the proper number of significant figures, *increase the last digit retained by 1 if the first figure dropped is 5 or greater.* This is the rule normally built into a calculator display control or a control language.

   Examples
   a. 23.650 rounds to 23.7 for three significant figures.
   b. 0.014 3 rounds to 0.014 for two significant figures.
   c. 827.48 rounds to 827.5 or 827 for four and three significant digits, respectively.
   (*Note:* You must decide the number of significant figures before you round. For example, rounding 827.48 to three significant figures yields

827. However, if you first round to four figures, obtaining 827.5, and then round that number to three figures, the result would be 828—which is not correct.)

2. **Multiplication and Division.** General Rule: The product or quotient should contain the same number of significant digits as are contained in the number with the fewest significant digits.

Examples

a. $(2.43)(17.675) = 42.950\,25$

   If both the multiplier and multiplicand are exact, the answer should be reported as 42.950 25. If one or both of the numbers are not exact, as is normally the case, the rule is applied using the inexact number with the fewest significant figures. In this example, assuming both numbers are inexact, 2.43 has three significant figures and 17.675 has five. Applying the rule, the answer should contain three significant figures and be reported as 43.0 or 4.30×10^1.

b. $(2.479\text{ h})(60\text{ min/h}) = 148.74\text{ min}$

   In this case, the conversion factor is exact (a definition) and could be thought of as having an infinite number of significant figures (60.00000. . ..). Thus, 2.479, which has four significant figures, controls the precision, and the answer is 148.7 min, or 1.487×10^2 min.

c. $(4.00 \times 10^2\text{ kg})(2.204\,6\text{ lbm/kg}) = 881.84\text{ lbm}$

   Here, the conversion factor is not an exact number, but you should not let the conversion factor dictate the precision of the answer if it can be avoided. You should attempt to maintain the precision of the value being converted (4.00×10^2); you cannot improve its precision. If you need to maintain precision greater than what is available in the conversion factor, we recommend that you locate or calculate the conversion factor to one or two more significant figures than the value you are converting. For example, the conversion factors given in the appendix of this text contain five significant figures. If you require more for a calculation, you would need to locate a more precise conversion factor from another source or derive the conversion factor and calculate it to the desired number of significant figures. For example (c), three significant figures should be reported, yielding 882 lbm.

d. $589.62/1.246 = 473.210\,27$

   The answer, to four significant figures, is 473.2.

3. **Addition and Subtraction.** General Rule: The answer should show significant digits only as far to the right as is seen in the least precise number in the calculation. Remember that the last number recorded is doubtful, that is, an estimate.

Example

a.   1 725.463
       189.2
        16.73
    ————————
    1 931.393

The least precise number in this group is 189.2 because the (2) is an estimate, so, according to the rule, the answer should be reported as 1 931.4. Using alternative reasoning, suppose these numbers are instrument readings, which means the last reported digit in each is a doubtful digit. A column addition that contains a doubtful digit will result in a doubtful digit in the sum. As a result, all three digits to the right of the decimal in the answer are doubtful. We keep just one doubtful digit in the answer; thus the answer is 1 931.4 after rounding.

   b. 897.0

      $\underline{- 0.092\ 2}$

      896.907 8

   Application of the rule results in an answer of 896.9.

4. **Combined Operations.** General rule: If products or quotients are to be added or subtracted, perform the multiplication or division first, establish the correct number of significant figures in the intermediate answer, then perform the addition or subtraction, and round to proper significant figures. Note, however, that in calculator or computer applications it is not practical to perform intermediate rounding. **It is normal practice to perform the entire calculation and then report a reasonable number of significant figures.**

    If results from additions or subtractions are to be multiplied or divided, an intermediate determination of significant figures can be made when the calculations are performed manually. For calculator or computer answers, use the suggestion already mentioned.

    Subtractions that occur in the denominator of a quotient can be a particular problem when the numbers to be subtracted are very nearly the same. For example, $39.7/(772.3 - 772.26)$ gives 992.5 if intermediate rounding is not done. If, however, the subtraction in the denominator is reported with one digit to the right of the decimal, the denominator becomes zero and the result becomes undefined. Commonsense application of the rules is necessary to avoid problems.

6.4 Errors

It is important to realize that all measurements of physical quantities have a certain amount of uncertainty (error) associated with them. This error must be made as small as possible. Therefore we must determine, as best we can, what errors are present and account for them in the measurement. Consider the thermometer in Figure 6.4b. After we make the reading of 52.8°, can we say for sure that this is the true temperature with the 8 being a doubtful digit? The answer is no. For example, what if the thermometer has not been calibrated properly and reads high by a degree or two? Maybe the thermometer has been smudged with oil, dirt, or grease and a mistake is made in reading the calibrations. Perhaps the sensor connecting the environment being measured to the thermometer is not connected properly. These and other known error possibilities must be accounted for to obtain an acceptable accuracy in the reading.

However, even if you carefully account for possible errors, the measurement will still have some error present. Depending upon the level of accuracy and preciseness required in the measurement, further effort may be needed in error determination. To clarify the presence of error, measurements can be expressed in two parts, (1) a number representing a mean value of the physical quantity measured, and (2) an amount of doubt (error) in this mean value. The amount of doubt (error) provides the accuracy of the measurement. For example, our thermometer reading in Figure 6.4b could be expressed as 52.5 ± 0.5, which brackets the upper and lower limits of the measurement based on the calibration of the instrument. To perform computations with measurements we carefully estimate a "best" value of 52.8 with the 8 being in doubt. You should note that error (deviation from the true temperature) is still present and that the true temperature is not known exactly.

A common application of measurement error is in pressure gauges. Suppose a gauge on an air tank is labeled ±2% at 150 lb/in². The range of the gauge error would be $(0.02)(150) = 3.0$ lb/in². Therefore a reading of 190 would indicate that the true pressure falls between 187 and 193 lb/in².

Errors can be classified into two broad categories for analysis: systematic and random.

6.4.1 Systematic Errors

A systematic error tends to shift a measurement consistently in the same direction from the true value. Examples include errors in the calibration of the measuring instrument, failure to account for some external effect on the instrument and improper use of the instrument. Large systematic errors must be corrected for in any situation. Consider a 25 m steel tape that is compared with the standard in the U.S. Bureau of Standards in Washington, D.C. If the tape is not exactly 25.000 m long then each time the tape is used, the known error can be incorporated in the measurement. Similarly large temperature changes in a steel tape can be corrected by a known coefficient of thermal expansion. However, if all known systematic errors are accounted for, there always remain small errors. For example, no instrument can be calibrated perfectly.

6.4.2 Random Errors

Random errors are those that fluctuate from one measurement to another for the same instrument. One cause is the sensitivity of the instrument. A small change in the quantity measured may not be picked up by the instrument. Such errors are usually distributed equally around the true value. Another random error may occur when an instrument is read by more than one person. Consider a water barometer that measures atmospheric pressure. Close inspection of the water level shows a meniscus which can easily result in different readings from several observers. These readings will very likely divide above and below a mean, or true, value. Statistical analysis is used to provide some insight into random errors. Chapters 10 and 11 provide an introduction to the statistical concepts of central tendency which are a part of the analysis of random errors.

6.5 Estimations

Engineers strive for a high level of precision in their work. However, it is also important to be aware of an acceptable precision and the time and cost of attaining it. There are many instances where an engineer is expected to estimate the result to a problem with reasonable accuracy but under tight time and cost constraints. To do this engineers rely on their basic understanding of the problem under discussion coupled with their previous experience. This knowledge and experience is what distinguishes an "estimation" from a "guess." If greater accuracy is needed, the initial estimation can be refined when time, funds, and the necessary additional data for refining the result are available.

Initial estimates may be in error by perhaps 10 to 20% or even more. The accuracy of these estimates depends strongly on what reference materials we have available, how much time is allotted for the estimate, and, of course, how experienced we are with similar problems. The first example we present will attempt to illustrate what an engineer might be called upon to do in a few minutes with little in the way of references.

Example Problem 6.1   An aerospace engineer is asked to sit in on a meeting of executives of an airline considering the purchase of the new Boeing 787 Dreamliner. The engineer is asked if she could give the group a quick estimate of the average cruise fuel consumption of the 787 on a per-mile basis. The executives can use the result to compare the 787 with competitive aircraft as they proceed toward a purchase decision.

Discussion   The engineer has reviewed preliminary specifications published by Boeing to prospective buyers of the 787. The base model of the aircraft (787-8) has an estimated range of 7 650 to 8 200 nautical miles (nmi), or 8 803 to 9 436 miles, and a cruise Mach number of $M = 0.85$. Mach number is the ratio of the speed of the aircraft to the speed of sound at the flight altitude. The fuel capacity is about 33 000 gallons or about 220 000 lb. Assuming that 10% of the fuel is used during taxi, takeoff, climb to cruise altitude, descent, and final taxi and a 10% reserve is required upon arrival, and using the high end of the mileage range, the engineer quickly estimates cruise fuel consumption per mile as 26 400/9 436 = 2.8 gal/mi or 176 000/9 436 = 18.7 lb/mi. From this quick estimate, the executives can estimate a fuel cost per mile at current fuel prices and the overall fuel cost for any flight leg. This problem does not require specialized knowledge in any particular engineering discipline, but does require an ability to understand the problem and put together the available data to obtain a solution. Experience in the aerospace discipline speeds the understanding of the problem and enables the quick estimate to be done. Some additional data and estimation problems involving the 787 are presented in the chapter problems.

Example Problem 6.2 is an illustration of a problem you might be assigned. You have the necessary experience to perform the estimation without special knowledge. Not counting the final written presentation, you should be able to do a similar problem in one-half to 1 hour.

The Boeing 787 Dreamliner

The new Boeing 787 Dreamliner commercial aircraft. *(Courtesy of The Boeing Company.)*

The latest commercial aircraft being designed, developed, and manufactured by The Boeing Company is the 787 Dreamliner. This aircraft is in response to preferences expressed by airlines around the world for a super-efficient airplane. The 787 is the ninth airplane in the Boeing 7x7 commercial fleet, which began in the 1950s with the 707 and continued with the 727, 737, 747, 757, 767, 717, 777, and now the 787, due to go into service in late 2010.

The 787 family will initially consist of two models. The 787-8 will carry 210–250 passengers on routes of 7650 to 8200 nautical miles (nmi), or 14 200 to 15 200 kilometers (km). The 787-8 was test-flown for the first time on December 15, 2009. Delivery to customers is expected to begin in late 2010. The second family member is the 787-9. This Dreamliner will carry 250–290 passengers on routes of 8 000 to 8 500 nmi (14 800 to 15 750 km). Each of the family members is specifically optimized for its design range, passenger capacity, and cargo capacity.

Advancing technologies, design procedures, and manufacturing improvements have led to an aircraft of unprecedented efficiency. The primary structure of the aircraft will be 50% composites and 20% aluminum, compared with 12% composites and 50% aluminum on the 777, which went into service in 1995. The great increase in composites will eliminate 1 500 aluminum sheets and 40 000 to 50 000 fasteners per fuselage section, thus saving assembly time, material handling, and weight. The empty weight of the 787 will run 40 000 to 50 000 lb less than other aircraft in its class, giving airlines added cargo revenue. Maximum takeoff weight consists of empty aircraft plus fuel weight plus cargo (passengers and freight). If the empty weight can be reduced, cargo can increase correspondingly for the same takeoff weight, generating increased revenue.

The Rolls-Royce Trent 1000 and the General Electric GEnx engines, each rated at 64 000 lb of thrust have been selected for installation on the 787-8. Advances in engine technology will create an 8% increase in efficiency and the engines will use up to 20% less fuel for comparable routes than any other aircraft of similar size. Engine emissions are therefore reduced by an equivalent amount. The engines will propel the aircraft at speeds equivalent to current large jets, Mach 0.85.

Another major area of improvement is the replacement of pneumatic systems with electrical systems to run auxiliary devices. In previous aircraft, bleed air was taken from the engine compressor to, among other uses, power the pneumatic systems for cabin pressure. Now the engines will drive electric generators that will in turn provide the power for the auxiliary devices. This results in a 35% drop in the power extracted from the engines.

Passengers will experience improved conditions for flights. The Dreamliner's windows are 65% larger than those in competitive aircraft. Climate control in the cabin will include higher humidity and improved air handling. Seats are ergonomically designed to provide improved comfort and convenience.

The 787 Dreamliner is truly an international effort. The Boeing Commercial Airplanes in the Seattle, Washington, area will handle 787 development integration, final assembly, and program leadership. Major components of the aircraft will be manufactured in 26 foreign countries, including Japan (wing box and main landing gear well), France (landing gear structure and electric brakes), the United Kingdom (landing gear actuation and control system and Rolls-Royce engines), Korea (raked wing tips), Sweden (cargo and access doors), and Germany (main cabin lighting and metal tubing and ducting). Other contributing companies in the United States include Boeing subsidiaries in Washington and Kansas, Rockwell Collins in Iowa, Honeywell in Arizona, Goodrich in North Carolina, General Electric in Ohio (engine), Moog Inc. in New York, and Hamilton Sundstrand in Connecticut.

Example Problem 6.2   Suppose your instructor assigns the following problem: Estimate the height of two different flagpoles on your campus. This will be done on a cloudy day so no shadow is present. The poles are in the ground (not on top of a building) and the bases of the poles are accessible. One of the poles is on level ground and you have available a carpenter's level, straight edge, protractor, and masking tape. The other pole is on ground that slopes away from the base and you have available a carpenter's level, straight edge, protractor, and a 12-ft. tape. See Figure 6.5 for the response of one student (whom we will call Dave).

Discussion   To estimate the height of the flagpole on level ground, Dave recognizes that he does not have a normal distance measuring device but that he must know some distance in order to use trigonometry to solve the problem. He knows that he is 6'1" tall, and he can mark that height on the pole with masking tape, which he can see from a location several feet from the pole. He can use the level, protractor, and straight edge to estimate angles from the horizontal. The distance away from the pole for measuring the angles is arbitrary, but he chooses a distance that will provide angles that are neither too large nor too small, both of which would be difficult to estimate. Then from this point on the ground he estimates the angles to his 6'1" masking tape mark and to the top of the pole. Note that Dave has kept all of the significant figures through the calculations and rounded only at the end, reasoning that he does not want intermediate rounding to affect his answer. Based on the method used for estimation, he believes his answer is not closer than the nearest foot so he rounds to that value.

When estimating the height of the pole on sloping ground, Dave had a tape available so it was not necessary to mark his height on the pole. Again, Dave kept all significant figures through the calculation procedure and rounded only the final result.

Example Problem 6.3   A homeowner has asked you to estimate the number of gallons of paint required to prime and finish-coat her new garage. You are told that paint is applied about 0.004 in. thick on smooth surfaces. The siding is to be gray and the roof overhang and trim are to be white. See Figure 6.6 for one approach done by Laura.

Discussion   From experience, Laura knew that one coat of primer would be needed and that two coats of finish paint would be required for a lasting outcome. She also noted that the garage doors were painted by the manufacturer before installation and therefore would not need further paint. She decided to neglect the effect of windows in the garage because of their small size. She observed that the siding was a rough vertical wood type and that the soffit (underside of the roof overhang) was smooth plywood. Since she had limited experience with rough siding, she contacted a local paint retailer and learned that rough siding would take approximately three times as much primer as smooth siding and that because of the siding roughness the finish paint would cover only about 3/4 of the normal area. She carefully documented this fact in her presentation.

Figure 6.5a

	9 – 24 – XX	ENGR 160 FLAGPOLE PROBLEM	DAVE DOE	1⁄2

PROBLEM 5.2

ESTIMATE THE HEIGHT OF 2 FLAGPOLES ON YOUR CAMPUS. ASSUME THE SUN IS NOT
SHINING AND THAT THE BASES OF THE POLES ARE AT GROUND LEVEL (NOT ON TOP OF A
BUILDING) AND THAT THE BASES ARE ACCESSIBLE.

A. DO THE ESTIMATE FOR POLE 1 WHERE YOU HAVE AVAILABLE A CARPENTER'S
   LEVEL, STRAIGHT EDGE, PROTRACTOR AND MASKING TAPE. POLE 1 SITS ON LEVEL
   GROUND.

B. FOR POLE 2, YOU HAVE A CARPENTER'S LEVEL, STRAIGHT EDGE, PROTRACTOR AND
   A 12' TAPE. THE GROUND AROUND POLE 2 SLOPES AWAY FROM THE BASE.

PART A

ASSUMPTIONS
   1. GROUND APPROXIMATELY LEVEL AROUND FLAGPOLE BASE

PROCEDURE
   • PLACE A PIECE OF MASKING TAPE AT MY HEIGHT ON THE FLAGPOLE.
   • CHOOSE POSITION ABOUT FLAGPOLE HEIGHT AWAY FROM THE BASE AND
     MEASURE ELEVATION ANGLES TO TOP OF POLE (β) AND TO TAPE (α). SEE FIG. 1.

DATA

   1. MY HEIGHT KNOWN TO BE 6' 1",
   2. LEVEL, STRAIGHT EDGE, AND PROTRACTOR
       CAN SERVE AS A SYSTEM FOR MEASURING
       ELEVATION ANGLES (FIG. 2)
   3. $\alpha = 7.5°$, $\beta = 48°$

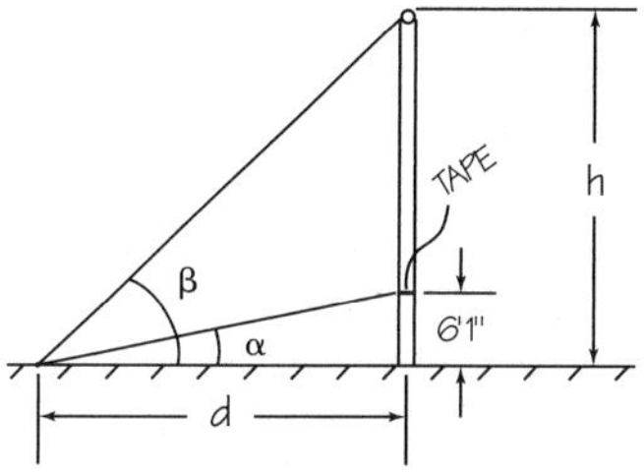

FIGURE 1

SOLUTION

$$\text{TAN } \alpha = \frac{6' 1''}{d} = \text{TAN } 7.5°$$

$$\text{TAN } \beta = \frac{h}{d} = \text{TAN } 48°$$

$$h = d (\text{TAN } 48°) = \left(\frac{6' 1''}{\text{TAN } 7.5°} \right) \left(\text{TAN } 48° \right) = 51.319 \text{ FT}$$

APPROXIMATE $h = 51'$

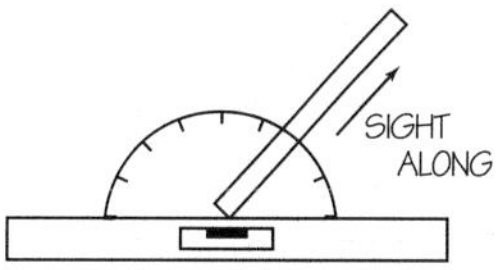

FIGURE 2

Student presentation for Example Problem 6.2.

Figure 6.5b

	9 – 24 – XX	ENGR 160 FLAGPOLE PROBLEM	DAVE DOE	2/2

PART B

ASSUMPTIONS

1. GROUND HAS CONSTANT SLOPE FROM BASE TO MEASURING POINT.

PROCEDURES

- CHOOSE POINT ABOUT FLAGPOLE HEIGHT AWAY FROM BASE AND MEASURE ELEVATION ANGLES TO TOP(γ) AND TO BASE(Δ)
- MEASURE DISTANCE (d) FROM CHOSEN POINT TO BASE. FIG. 3.

DATA

1. $\Delta = 3.5°$
2. $\gamma = 44°$
3. $d = 93'4''$

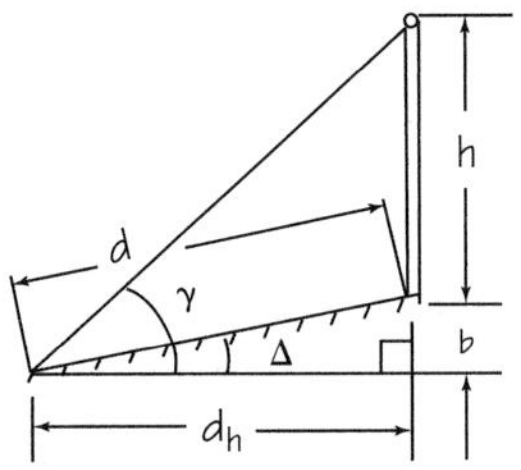

FIGURE 3

SOLUTION

$$\text{SIN } \Delta = \frac{b}{d}$$

$$b = d \text{ SIN } \Delta = (93'4'') \text{ SIN } 3.5° = 5.6979'$$

$$\text{TAN } \Delta = \frac{b}{d_h}$$

$$d_h = \frac{b}{\text{TAN } \Delta} = \frac{5.6979}{\text{TAN } 3.5°} = 93.1598'$$

$$\text{TAN } \gamma = \frac{h+b}{d_h}$$

$$h = d_h \text{ TAN } \gamma - b = (93.1598) \text{ TAN } 44° - 5.6979$$

$$= 84.265'$$

APPROXIMATE $h = 84'$

DISCUSSION/CONCLUSIONS

1. THE ACCURACY OF THE ESTIMATED HEIGHTS COULD BE VERIFIED IF AT LEAST TWO OTHER STUDENTS MADE INDEPENDENT MEASUREMENTS WITH THE SAME INSTRUMENTS
2. TOTAL TIME REQUIRED: 70 MINUTES TO SET UP THE PROCEDURE, TAKE MEASUREMENTS AND DO THE CALCULATIONS; 35 MINUTES FOR WRITEUP.

Figure 6.6a

	3 – 16 – XX	ENGR 161 PROBLEM 5.3	LAURA LYNN	1/2

PROBLEM 5.3

A HOMEOWNER HAS ASKED FOR AN ESTIMATE OF THE NUMBER OF GALLONS OF PAINT REQUIRED TO PRIME AND FINISH COAT HER NEW GARAGE. PAINT SHOULD BE APPLIED ABOUT 0.004 IN. THICK ON SMOOTH SURFACES. THE SIDING IS TO BE GRAY AND THE TRIM WHITE.

ASSUMPTIONS

1. 1 COAT OF PRIMER, 2 FINISH COATS
2. GARAGE DOORS ARE NOT PAINTED.
3. NEGLECT AREA OF SMALL WINDOWS IN GARAGE.

PROCEDURE

MEASURE GARAGE SURFACES TO OBTAIN TOTAL AREA TO BE PAINTED. OBSERVE "SMOOTHNESS" OF SIDING TO ESTIMATE PAINT COVERAGE. COMPUTE AMOUNT OF EACH TYPE OF PAINT.

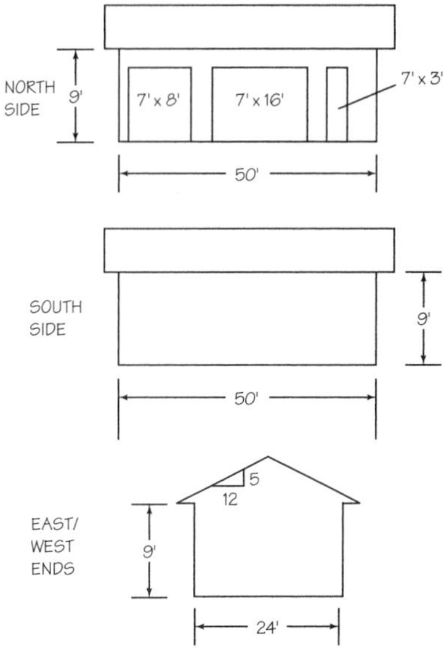

COLLECTED DATA

1. SINCE 1 GAL = 231 IN^3, PAINT THICKNESS OF 0.004 IN RESULTS IN 1 GAL COVERING ≅ 400 FT^2 OF SMOOTH SURFACE.
2. OVERHANGS ARE 18 IN.
3. SIDING IS OBSERVED TO BE ROUGH WOOD/VERTICAL. SOFFIT IS SMOOTH PLYWOOD.
4. LOCAL PAINT STORE REPRESENTATIVE SUGGESTED THAT PRIMER ON ROUGH WOOD SIDING COVERS ONLY 1/3 NORMAL AREA AND THAT TOP COAT COVERS ABOUT 3/4 NORMAL AREA.

SOLUTION

NORTH SIDE AREA $= (9)(50) - (7)(8) - (7)(16) - (7)(3) = 261 \text{ FT}^2$

SOUTH SIDE AREA $= (9)(50) = 450 \text{ FT}^2$

EAST/WEST END AREA $= 9\,(24) + \frac{1}{2}(24)(5) = 276 \text{ FT}^2\text{/END}$

$$\text{OVERHANG AREA} = (53)(1.5)(2) + 2\,(2)(1.5)\left[(13.5)^2 + \left\{\left(\tfrac{5}{12}\right)(13.5)\right\}^2\right]^{1/2}$$
$$= 159 + 87.75 \cong 247 \text{ FT}^2$$

Student presentation for Example Problem 6.3.

Figure 6.6b

	3 – 16 – XX	ENGR 161 PROBLEM 5.3	LAURA LYNN	2/2

TOTAL AREA OF SIDING $= 261 + 450 + 2(276) = 1263 \text{ FT}^2$

TOTAL OVERHANG AREA $= 247 \text{ FT}^2$

PRIMER NEEDED FOR SIDING $= \left(\frac{1263}{400}\right) 3 = 9.47$ GAL

PRIMER NEEDED FOR OVERHANG $= \frac{247}{400} = 0.62$ GAL

TOTAL PRIMER NEEDED $= 9.47 + 0.62 = 10.1$ GAL

GRAY FINISH COAT FOR SIDING $= (2)\left(\frac{1263}{400}\right)\left(4/3\right) = 8.42$ GAL

WHITE FINISH COAT FOR OVERHANG/TRIM $= (2)\left(\frac{247}{400}\right) = 1.24$ GAL

RECOMMENDED PURCHASE :

PRIMER: 10 GAL
GRAY TOP COAT: 9 GAL
WHITE TOP COAT: 2 GAL

DISCUSSION/CONCLUSIONS

1. HOMEOWNER SHOULD BE INFORMED THAT HER PAINTING EXPERIENCE MAY AFFECT THE AMOUNT OF PAINT REQUIRED. MAINTAINING A CONSISTENT THICKNESS IS DIFFICULT.
2. A FRIEND HELPED ME OBTAIN THE MEASUREMENTS USING A 20 FT STEEL TAPE.
3. TOTAL TIME REQUIRED; 25 MINUTES FOR MEASUREMENTS, A STEP-LADDER WAS NEEDED; 40 MINUTES FOR CALCULATIONS AND WRITEUP.

Laura took all necessary measurements and computed the areas that must be painted. She determined that the paint film thickness of 0.004 in. corresponds to approximately 400 ft²/gal coverage. Like Dave in the previous example, Laura retained extra significant figures until finally rounding at the end of the estimate. In this case, except for the primer, she correctly rounded up rather than to the nearest gallon, as the paint would be purchased in whole gallons.

Example Problem 6.4 Estimate the cost of the concrete that is in the roadbed of Interstate 17 running from Flagstaff, Arizona, to Interstate 10 near the Phoenix airport. Consider only the actual roadbed and not stopping lanes or interchanges. Keep track of the time to develop a solution and perform the write-up.

Discussion Figure 6.7 is a write-up on a word processor of the solution performed. The assumptions are listed and simplify the data collection process considerably. Two telephone calls were made to obtain information on interstate highway construction and the cost of concrete. Thus the resulting estimate is based on current design practice and costs. A similar estimation problem relaxing some of the assumptions is provided in Problem 6.29.

Problems

6.1 How many significant digits are contained in each of the following quantities?
- (a) 0.724 70
- (b) 7 247.0
- (c) 0.031
- (d) 24 000
- (e) 0.10 × 10⁴
- (f) 0.320 00
- (g) 200.07
- (h) 1.3200 × 10⁻³
- (i) 2 420 000.0
- (j) 3.026 7 × 10²

6.2 How many significant digits are contained in each of the following quantities?
- (a) 2.0345
- (b) 0.100 30
- (c) 0. 023
- (d) 2.2046 kg/lbm
- (e) 300 003.0
- (f) 2.54 cm/in.
- (g) 1.002
- (h) 1.000 × 10⁸
- (i) 60 s/min
- (j) 4.030 × 10¹²

6.3 Perform the computations below and report with the answer rounded to the proper number of significant digits.
- (a) 56.3 × 372.5
- (b) 37.35 − 1.4300
- (c) (6.231 827)(4.23 × 10⁷)
- (d) 2.5 / 0.50
- (e) 31.05 / 2.0
- (f) (1.456 × 10⁴)(1.03 × 10⁻³)
- (g) 1.845 7 × 0.700 25
- (h) 17.546 78 / 0.024 35
- (i) 4 300 200 / 784
- (j) (450.3 + 372 − 112.1) / 86.00

6.4 Using the conversion factors given in each problem, perform the calculations below using exact conversions or with enough significant digits that it does not affect the accuracy of the answer.
- (a) 76 200 feet to miles   1 mi = 5 280 ft
- (b) 3.4 radians to degrees   1 radian = 57.296 degrees
- (c) 650 kg to pound mass   1 kg = 2.204 6 lbm
- (d) 7.358 inches to centimeters   1 in. = 2.54 cm
- (e) 7.062 cubic meters to cubic feet 1 m = 3.280 8 ft
- (f) 12.5 days to seconds   1 day = 86 400 s

Figure 6.7

10-21-XX ENGR 160: Prob 4.4 R. George Page 1 of 1

Estimate the cost of concrete in the roadbed of the entire length of Interstate Highway 17 (I-17), from its beginning at Flagstaff, Arizona to its terminus at Interstate Highway 10 (I-10) in Phoenix near the Phoenix airport. In the presentation, specify the approximate amount of time spent on the solution.

ASSUMPTIONS

1. Four lanes (two lanes each direction) from Flagstaff to Loop 101 in Phoenix
2. Eight lanes in Phoenix
3. Assume entire roadbed is concrete
4. Neglect on and off ramps, bridge supports and railings, and emergency stop lanes

COLLECTED DATA

1. Concrete costs \$95 per cubic yard delivered to site (estimate from contractor)
2. Average depth of roadbed is 12 inches (Department of Transportation)
3. Lane width is 12 feet (Department of Transportation)
4. Unit relationships: 1 mi = 5280 ft
   1 yd = 3 ft
   1 ft = 12 in
5. Data from 2009 State Farm Road Atlas

	miles	lanes
a. Flagstaff to Phoenix Loop 101	124	4
b. Loop 101	21	8

CALCULATIONS

Volume (V) = Length (L) x Width (W) x Depth (D)
Cost (C) = Volume (V) x (cost/cubic yard)

$V_a = (124\text{ mi})(5280\text{ ft/mi})(4)(12\text{ ft})(12\text{ in})(1\text{ ft}/12\text{ in})(1\text{ yd}^3/27\text{ ft}^3) = 1.164(10)^6\text{ yd}^3$

$V_b = (21\text{ mi})(5280\text{ ft/mi})(8)(12\text{ ft})(12\text{ in})(1\text{ ft}/12\text{ in})(1\text{ yd}^3/27\text{ ft}^3) = 0.3942(10)^6\text{ yd}^3$

$V = \underline{1.558(10)^6\text{ yd}^3}$

$C = (1.558(10)^6\text{ yd}^3)(\$95/\text{yd}^3) = \underline{\$148{,}000{,}000}$

Discussion/Conclusions

1. Cost of concrete may vary considerably in a short time period. \$95 cost per yard used was an average of several reports found on Internet.

2. Time estimate: 40 min (obtaining data and calculations) + 30 min (write-up) = 70 min

Student presentation, produced on a word processor for Example Problem 6.4.

6.5 Solve the following problems and give the answers rounded to the proper number of significant digits.
- (*a*) v = 0.0214t² + 0.363 5t + 2.25 for t = 32
- (*b*) 24.56 ft × 12 in./ft = ? inches
- (*c*) $400 a plate × 24 guests = $?
- (*d*) V = [ϖ(4.62 cm)²(7.53 cm)]/3 = ? cm³ (volume of a cone)
- (*e*) 325.03 + 527.897 - 615.0 =
- (*f*) 32¢ per part × 45 250 parts = $?

6.6 A pressure gauge on an air tank reads 210 pounds per square inch (psi). The face of the gauge says ±3% at 180 psi.
- (*a*) What is the range of air pressure in the tank when the gauge reads 210 psi?
- (*b*) What is the range of air pressure in the tank when the gauge reads 87 psi?

6.7 A vacuum gauge reads 86 kPa. The face of the gauge says ± 3.3 kPa at 85 kPa.
- (*a*) What is the actual range of vacuum?
- (*b*) What is the range when the gauge reads 130 kPa?

6.8 What is the percent of error if you use a pair of calipers on a 6-in. precision gauge block and get a reading of 5.988?

For problems 6.9 to 6.32, develop and present a solution in a manner demonstrated in Example Problems 6.2, 6.3, or 6.4. Your solution discussion should indicate the amount of time required for developing and preparing the solution. The problems are grouped into four categories: individual in-class, individual homework, team in-class, and team homework. Groups of two or three students are best for the team problems.

Individual in-class problems

6.9 Estimate the amount of time during a typical class day that you spend walking. Also estimate the number of steps you take during this time.

6.10 Estimate the number of hours that you spend watching television in a typical week during the academic semester.

6.11 Estimate the number of tennis balls that will fit in a cubic box 3 feet on a side.

6.12 Estimate the number of quarters that will fit in a box 16 inches by 10 inches by 12 inches.

6.13 Estimate the number of basketballs that will fit into your classroom. Assume room is empty of students and furniture.

6.14 Estimate the number of hours you spent on a computer during a typical week this semester. Include "surfing the net" as well as research and class requirements. Carefully document each of the categories of use.

Team in-class problems

6.15 Estimate the amount of paint required to change the color of your classroom walls.

6.16 Estimate the volume required to store 15 000 basketballs.

6.17 Estimate the number of regular M&M's needed to fill a two-liter bottle.

6.18 Estimate the volume of water used to take showers by the members of this class in one academic semester.

The following are general specifications for the Boeing 787-8 Dreamliner. Use the data to perform the estimations required in problems 6.19–6.21.

Maximum takeoff weight	502 500 lb (227 930 kg)
Empty weight	242 000 lb (109 770 kg)
Maximum fuel capacity	33 528 gal (126 917 L)
Passengers	224*
Range	7650–8200 nmi
Cruise speed (Mach number)	M = 0.85

*Occupancy ranges from 210 to 250, depending upon the configuration for first, business, and tourist classes

6.19 Estimate the weight of the crew, passengers, cargo, and fuel for a sold-out flight.

6.20 Using the information in Example Problem 4.1, estimate the flight time and fuel consumed on a flight from Tokyo to San Francisco. Assume the speed of sound at cruising altitude is 975 feet per second (fps).

6.21 Estimate the fuel consumption on a per-hour basis for a maximum range flight at an altitude where the speed of sound is 975 fps.

Individual homework problems

6.22 Estimate the number of minutes students in your engineering college spend on their cell phones in a typical academic week. A survey of a representative segment of students is necessary. Compare your results with other members of this class.

6.23 Estimate the weight, in pounds, of cars in the parking lot closest to the building where this class is held. Assume all parking spots are occupied.

6.24 Estimate the weight of water in a swimming pool on or near your campus.

6.25 Estimate the area of the running surface of an outdoor track on or near your campus.

6.26 Estimate the volume of a conical pile of sand that you have approximated the base circumference to be 210 feet. Hint: consider the angle of repose.

6.27 Estimate the weight of concrete in a 4-lane highway segment that begins as a 3% grade climb from 1000 feet of altitude to 4000 feet and then a 6% downgrade to 2500 feet.

Team homework problems (these problems will involve significant research, a specific plan for activity is strongly recommended)

6.28 Estimate the volume and cost of water used by a family of five living in a detached home during a one-year period.

6.29 Estimate the cost of concrete for a segment of interstate highway designated by your instructor. Include the rural and city components (number of lanes), interchanges, and extended width of lanes for emergency stopping.

6.30 Estimate the number of soccer balls that can be transported in a railroad box car.

6.31 Estimate how much carpet would be needed to carpet the building in which this class is held.

6.32 Your trucking company has been asked to transport a huge pile of sand from a pit to a construction site 35 miles away. The pit and construction site are both within a mile of the same two-lane paved road. The base of the sand pile covers approximately half an acre. Local laws allow only single-axle dump trucks on the highway. Your company has 14 single-axle trucks and drivers available. Estimate the time, in work days, to move the sand.

CHAPTER 7

Dimensions, Units, and Conversions*

Chapter Objectives

When you complete your study of this chapter, you will be able to:

- Identify physical quantities in terms of dimensions and units
- Differentiate between fundamental and derived dimensions
- Understand the use of non-SI dimensional systems (gravitational and absolute)
- Recognize base, supplementary, and derived SI units
- Apply the appropriate SI symbols and prefixes
- Describe the relationship between U.S. Customary, Engineering System, and SI
- Systematically convert units from one system to another
- Use knowledge of dimensions and units, along with conversion rules, in the solution of engineering problems

7.1 Introduction

Years ago when countries were more isolated from one another, individual governments tended to develop and use their own set of measures. Today, primarily through the development of technology, global communication has brought countries closer together. As countries from one corner of the world to the other interact with each other, the need for a universal system of measurement has become abundantly clear. A standard set of dimensions, units, and measurements is vital if today's wealth of information and knowledge is to be shared and benefit all. The move toward a universal system first requires a thorough understanding of existing systems of measurement. This chapter begins with an explanation of the importance of physical quantities in engineering and explains the difference between dimensions and units. This is followed by the development of procedures for orderly conversion from one system of units to another, ultimately enabling measurements to be expressed in one system—that is, a metric international standard.

---

*Users will find Appendix A useful reference material for this chapter.

7.2 Progress in the United States toward Metrification

The United States Congress considered adoption of the metric system in the 1850s. In fact, the metric system was made legal in the United States in 1866, but its use was not made compulsory. In spite of many attempts in the intervening 150 years, full conversion to the metric system has not yet been realized.

In 1875, the United States together with 16 other nations signed an agreement called the Treaty of the Meter. These signatory nations established a governing body and gave that agency authority and overall responsibility for the metric system. The governing body is called the General Conference of Weights and Measures. That body approved an updated version of the metric system in 1960 called the international metric system, or Système International d'Unités, abbreviated SI. These units are a modification and refinement of earlier versions of the metric system (MKS) that designated the meter, kilogram, and second as fundamental units.

That 1960 standard is currently accepted in all industrial nations but still optional in the United States. In fact, the United States is the only industrialized nation on the globe that does not use the metric system as its predominant system of measurement. Current estimates suggest that the United States is at best 50% metric, so engineers must be well versed in the use of a variety of units as well as fluent in SI.

Some corporations within the U.S. economy have become almost 100% metric, others are somewhere in the middle, and many have far to go. That is not necessarily the fault of a given company or agency. For example, how many metric highway road signs have you seen lately driving across the United States? Figure 7.1 is an illustration of highway signs in Canada. Canada has a large population of French Canadians; as it happens, France was the first country, in 1840, to legislate adoption of the metric system and mandate its use. It takes a generation or two under a given system for a country's population to find it second nature. How many drivers in this country would actually know what 100 km/h really means?

To continue that train of thought, do you as a student give your weight in kilograms and buy gasoline by the liter? If someone tells you it is 30 degrees outside, do you immediately think in Celsius or Fahrenheit? How many of you knew that 30°C is 86°F? It is a difficult two-way street: Manufacturers, wholesalers, and distributors are not going to flood the market with products the consumer does not understand and therefore will not buy. As time marches on, a gradual increase in the use of SI will evolve, but in the meantime, be prepared to handle whatever system of units you encounter.

As a reminder of how important it is to understand the information presented in this chapter, let us review the NASA mission to send the Mars Climate Orbiter to the planet Mars. The $125 million spacecraft had almost completed a ten-month fight plan before it was lost on September 23, 1999, just as the spacecraft reached the planet's atmosphere. NASA convened three panels of experts to investigate what led to the loss of the orbiter. The reason for the loss was simple. One engineering team on a key navigational maneuver of the orbiter used metric units while a different team working on the same project used English units. The result was a major error in the entry trajectory resulting in the destruction of the vehicle.

Figure 7.1

157
Physical Quantities

Look for the **km/h** tab below the maximum speed limit sign, indicating that this is the new speed in metric.

100km/h This speed limit will likely be the most common on freeways. On most rural two-lane roadways, **80 km/h** will be typical.

50km/h A **50 km/h** speed limit will apply in most cities.

Actual speed limits will be established in accordance with local regulations.

Metric Commission Canada

Commission du système métrique Canada

Highway signs in Canada. (Metric Commission of Canada.)

7.3 Physical Quantities

Engineers are constantly concerned with the measurement of fundamental physical quantities such as length, time, temperature, force, and so on. In order to specify a physical quantity fully, it is not sufficient to indicate merely a numerical value. The magnitude of physical quantities can be understood only when they are compared with predetermined reference amounts, called *units*. Any measurement is, in effect, a comparison of how many (a number) units are contained within a physical quantity. Given length (L) as the physical quantity and 20.0 as the numerical value, with meters (m) as the designated unit, then a general relation can be represented by the expression

$$\text{Length } (L) = 20.0 \text{ m}$$

For this relationship to be valid, the exact reproduction of a unit must be theoretically possible at any time. Therefore standards must be established and maintained. These standards are a set of fundamental unit quantities kept under normalized conditions in order to preserve their values as accurately as possible. We shall speak more about standards and their importance later.

7.4 Dimensions

Dimensions are used to describe physical quantities; however, the most important concept to remember is that dimensions are independent of units. As mentioned in Section 7.3, the physical quantity "length" can be represented by the dimension L, for which there are a large number of possibilities available when selecting a unit. For example, in ancient Egypt, the cubit was a unit related to the length of the arm from the tip of the middle finger to the elbow. At that time in history measurements were a function of physical stature, with variation from one individual to another. Much later, in Britain, the inch was specified as the distance covered by three barley corns, round and dry, laid end to end.

Today we require considerable more precision. For example, the meter is defined in terms of the distance traveled by light in a vacuum during a specified amount of time. We can draw two important points from this discussion: (1) Physical quantities must be accurately measured and be reproducible, and (2) these units (cubit, inch, and meter), although distinctly different, have in common the quality of being a length and not an area or a volume.

A technique used to distinguish between units and dimensions is to call all physical quantities of length a specific dimension (for example, L). In this way, each new physical quantity gives rise to a new dimension, such as T for time, F for force, M for mass, and so on. (Note that there is a dimension for each kind of physical quantity.)

However, to simplify the process, dimensions are divided into two areas—fundamental and derived. A fundamental dimension is a dimension that can be conveniently and usefully manipulated when expressing all physical quantities of a particular field of science or engineering. Derived dimensions are a combination of two or more fundamental dimensions. Velocity, for example, could be defined as a fundamental dimension V, but it is more customary as well as more convenient to consider velocity as a combination of fundamental dimensions, so that it becomes a derived dimension, $V = (L)(T)^{-1}$. L and T are fundamental dimensions, and V is a derived dimension because it is made up of two fundamental dimensions (L,T).

For simplicity it is advantageous to use as few fundamental dimensions as possible, but the selection of what is to be fundamental and what is to be derived is not fixed. In actuality, any dimension can be selected as a fundamental dimension in a particular field of engineering or science; for reasons of convenience, it may be a derived dimension in another field.

Once a set of primary dimensions has been adopted, a base unit for each primary dimension must then be specified. So let's look at how this works.

A *dimensional system* can be defined as the smallest number of fundamental dimensions that will form a consistent and complete set for a field of science. For example, three fundamental dimensions are necessary to form a complete mechanical dimensional system. Depending on the discipline, these dimensions may be specified as either length (L), time (T), and mass (M) or length (L),

Table 7.1   Two Basic Dimensional Systems

Quantity	Absolute	Gravitational
Length	L	L
Time	T	T
Mass	M	$FL^{-1}T^2$
Force	MLT^{-2}	F
Velocity	LT^{-1}	LT^{-1}
Pressure	$ML^{-1}T^{-2}$	FL^{-2}
Momentum	MLT^{-1}	FT
Energy	ML^2T^{-2}	FL
Power	ML^2T^{-3}	FLT^{-1}
Torque	ML^2T^{-2}	FL

time (T), and force (F). If temperature is important to the application, a fourth fundamental dimension may be added.

The *absolute system* (so called because the dimensions used are not affected by gravity) has as its fundamental dimensions L, T, and M. An advantage of this system is that comparisons of masses at various locations can be made with an ordinary balance, because the local acceleration of gravity has no influence upon the results.

The *gravitational system* has as its fundamental dimensions L, T, and F. It is widely used in many engineering branches because it simplifies computations when weight is a fundamental quantity in the computations. Table 7.1 lists the dimensions used in the absolute and gravitational systems; a number of other dimensional systems are commonly used depending on the specific discipline.

7.5   Units

After a consistent dimensional system has been identified, the next step is to select a specific unit for each fundamental dimension. The problem one encounters when working with units is that there can be a large number of unit systems to choose from for any given dimensional system. It is obviously desirable to limit the number of systems and combinations of systems. The Système International d'Unités (SI) is intended to serve as an international standard that will provide worldwide consistency.

There are three fundamental systems of units commonly used today. The metric system, used in almost every industrial country of the world, is a decimal-absolute system based on the meter, kilogram, and second (MKS) as the units of length, mass, and time, respectively.

In the United States, however, there are two other system of units commonly used. The first, called the U.S. Customary System (formerly known as the British gravitational system), has the fundamental units of foot (ft) for length, pound (lb) for force, and second (s) for time. The second system of units, called the Engineering System, is based on the foot (ft) for length, pound-force (lbf) for force, and second (s) for time. More information regarding the Customary and Engineering Systems will be presented in Section 7.8.

Numerous international conferences on weights and measures over the past 40 years have gradually modified the MKS system to the point that all countries previously using various forms of the metric system are beginning to standardize. SI is now considered the international system of units. Although the United States has officially adopted this system, as indicated earlier, full implementation will be preceded by a long and expensive period of change. During this transition period, engineers will have to be familiar with not only SI but also other systems and the necessary conversion process between or among systems. This chapter will focus on the international standard (SI units and symbols); however, examples and explanations of the Engineering System and the U.S. Customary System will be included.

7.6 SI Units and Symbols

SI, developed and maintained by the General Conference on Weights and Measures (Conférence Générale des Poids et Mesures, CGPM), is intended as a basis for worldwide standardization of measurements. The name and abbreviation were set forth in 1960.

This new international system is divided into three classes of units:

1. Base units
2. Supplementary units
3. Derived units

There are seven base units in the SI. The units (except the kilogram) are defined in such a way that they can be reproduced anywhere in the world.

Table 7.2 lists each base unit along with its name and proper symbol.

In the following list, each of the base units were defined and adopted at various meetings of the General Conference on Weights and Measures held between 1889 and 1983:

1. *Length.* The meter (m) is a length equal to the distance traveled by light in a vacuum during 1/299 792 458 s.
2. *Time.* The second (s) is the duration of 9 192 631 770 periods of radiation corresponding to the transition between the two hyperfine levels of the ground state of the cesium-133 atom.
3. *Mass.* The standard for the unit of mass, the kilogram (kg), is a cylinder of platinum-iridium alloy kept by the International Bureau of Weights and Measures in France. A duplicate copy is maintained in the United States. It is the only base unit that is nonreproducible in a properly equipped lab.
4. *Electric current.* The ampere (A) is a constant current that, if maintained in two straight parallel conductors of infinite length and of negligible circular cross-sections and placed one meter apart in volume, would produce between these conductors a force equal to 2×10^{-7} newton per meter of length.
5. *Temperature.* The kelvin (K), a unit of thermodynamic temperature, is the fraction 1/273.16 of the thermodynamic temperature of the triple point of water.

Table 7.2   Basic Units

Quantity	Name	Symbol
Length	meter	m
Mass	kilogram	kg
Time	second	s
Electric current	ampere	A
Thermodynamic temp.	kelvin	K
Amount of substance	mole	mol
Luminous intensity	candela	cd

6. *Amount of substance.* The mole (mol) is the amount of substance of a system that contains as many elementary entities as there are atoms in 0.012 kg of carbon-12.
7. *Luminous intensity.* The base unit candela (cd) is the luminous intensity in a given direction of a source that emits monochromatic radiation of frequency 540×10^{12} hertz and has a radiant intensity in that direction of 1/683 watts per steradian.

The units listed in Table 7.3 are called *supplementary units* and may be regarded either as base units or as derived units.

The unit for a plane angle is the radian (rad), a unit that is used frequently in engineering. The steradian is not as commonly used. These units can be defined in the following way:

1. Plane angle: The radian is the plane angle between two radii of a circle that cut off on the circumference of an arc equal in length to the radius.
2. Solid angle: The steradian (sr) is the solid angle which, having its vertex in the center of a sphere, cuts off an area of the sphere equal to that of a square with sides of length equal to the radius of the sphere.

As indicated earlier, derived units are formed by combining base, supplementary, or other derived units. Symbols for them are carefully selected to avoid confusion. Those that have special names and symbols, as interpreted for the United States by the National Bureau of Standards, are listed in Table 7.4 together with their definitions in terms of base units.

Additional derived units, such as those listed in Table 7.5, have no special SI unit names or symbols but are nevertheless combinations of base units and units with special names.

Table 7.3   Supplementary Units

Quantity	Name	Symbol
Plane angle	radian	rad
Solid angle	steradian	sr

Table 7.4   Derived Units

Quantity	SI Unit Symbol	Name	Base Units
Frequency	Hz	hertz	s^{-1}
Force	N	newton	$kg \cdot m \cdot s^{-2}$
Pressure or stress	Pa	pascal	$kg \cdot m^{-1} \cdot s^{-2}$
Energy or work	J	joule	$kg \cdot m^2 \cdot s^{-2}$
Quantity of heat	J	joule	$kg \cdot m^2 \cdot s^{-2}$
Power radiant flux	W	watt	$kg \cdot m^2 \cdot s^{-3}$
Electric charge	C	coulomb	$A \cdot s$
Electric potential	V	volt	$kg \cdot m^2 \cdot s^{-3} \cdot A^{-1}$
Potential difference	V	volt	$kg \cdot m^2 \cdot s^{-3} \cdot A^{-1}$
Electromotive force	V	volt	$kg \cdot m^2 \cdot s^{-3} \cdot A^{-1}$
Capacitance	F	farad	$A^2 \cdot s^4 \cdot kg^{-1} \cdot m^{-2}$
Electric resistance	Ω	ohm	$kg \cdot m^2 \cdot s^{-3} \cdot A^{-2}$
Conductance	S	siemens	$kg^{-1} \cdot m^{-2} \cdot s^3 \cdot A^2$
Magnetic flux	Wb	weber	$m^2 \cdot kg \cdot s^{-2} \cdot A^{-1}$
Magnetic flux density	T	tesla	$kg \cdot s^{-2} \cdot A^{-1}$
Inductance	H	henry	$kg \cdot m^2 \cdot s^{-2} \cdot A^{-2}$
Luminous flux	lm	lumen	cd
Illuminance	lx	lux	$cd \cdot m^{-2}$
Celsius temperature*	°C	degree Celsius	K
Activity (radionuclides)	Bq	becqueret	s^{-1}
Absorbed dose	Gy	gray	$m^2 \cdot s^{-2}$
Dose equivalent	S_v	sievert	$m^2 \cdot s^{-2}$

*The thermodynamic temperature (T_K) expressed in kelvins is related to Celsius temperature (t_C) expressed in degrees Celsius by the equation $t_C = T_K - 273.15$.

Table 7.5   Additional Derived Units

Quantity	Units	Quantity	Units
Acceleration	$m \cdot s^{-2}$	Molar entropy	$J \cdot mol^{-1} \cdot K^{-1}$
Angular acceleration	$rad \cdot s^{-2}$	Molar heat capacity	$J \cdot mol^{-1} \cdot K^{-1}$
Angular velocity	$rad \cdot s^{-1}$	Moment of force	$N \cdot m$
Area	m^2	Permeability	$H \cdot m^{-1}$
Concentration	$mol \cdot m^{-3}$	Permittivity	$F \cdot m^{-1}$
Current density	$A \cdot m^{-2}$	Radiance	$W \cdot m^{-2} \cdot sr^{-1}$
Density, mass	$kg \cdot m^{-3}$	Radiant intensity	$W \cdot sr^{-1}$
Electric charge density	$C \cdot m^{-3}$	Specific heat capacity	$J \cdot kg^{-1} \cdot K^{-1}$
Electric field strength	$V \cdot m^{-1}$	Specific energy	$J \cdot kg^{-1}$
Electric flux density	$C \cdot m^{-2}$	Specific entropy	$J \cdot kg^{-1} \cdot K^{-1}$
Energy density	$J \cdot m^{-3}$	Specific volume	$m^3 \cdot kg^{-1}$
Entropy	$J \cdot K^{-1}$	Surface tension	$N \cdot m^{-1}$
Heat capacity	$J \cdot K^{-1}$	Thermal conductivity	$W \cdot m^{-1} \cdot K^{-1}$
Heat flux density	$W \cdot m^{-2}$	Velocity	$m \cdot s^{-1}$
Irradiance	$W \cdot m^{-2}$	Viscosity, dynamic	$Pa \cdot s$
Luminance	$cd \cdot m^{-2}$	Viscosity, kinematic	$m^2 \cdot s^{-1}$
Magnetic field strength	$A \cdot m^{-1}$	Volume	m^3
Molar energy	$J \cdot mol^{-1}$	Wavelength	m

Being a decimal system, the SI is convenient to use because by simply affixing a prefix to the base, a quantity can be increased or decreased by factors of 10 and the numerical quantity can be kept within manageable limits. The proper selection of prefixes will also help eliminate nonsignificant zeros and leading zeros in decimal fractions. One rule to follow is that the numerical value of any measurement should be recorded as a number between 0.1 and 1 000. This rule is suggested because it is easier to make realistic judgments when working with numbers between 0.1 and 1 000. For example, suppose that you are asked the distance to a nearby town. It would be more understandable to respond in kilometers than meters. That is, it is easier to visualize 10 km than 10 000 m.

The use of prefixes representing powers of 1 000, such as kilo, mega, milli, etc., are preferred over multipliers such as deci, deka, etc. However, the three exceptions listed below are still in common used because of convention.

1. When expressing area and volume, the prefixes hecto-, deka-, deci-, and centi- may be used; for example, cubic centimeter.
2. When discussing different values of the same quantity or expressing them in a table, calculations are simpler to perform when you use the same unit multiple throughout.
3. Sometimes a particular multiple is recommended as a consistent unit even though its use violates the 0.1 to 1 000 rule. For example, many companies use the millimeter for linear dimensions even when the values lie far outside this suggested range. The cubic decimeter (commonly called liter) is also used in this manner.

Recalling the importance of significant figures, we see that SI prefix notations can be used to a definite advantage. Consider the previous example of 10 km versus 10 000 m. In an estimate of distance to the nearest town, a round number certainly implies an approximation. Suppose that we were talking about a 10 000 m Olympic track and field event. The accuracy of such a distance must certainly be greater than something between 5 000 and 15 000 m, which would be the implied accuracy with one significant figure. If, however we use prefix multipliers, such as 10.000 km, then all five numbers are in fact significant, and the race length is accurate to within 1 m (9 999.5 to 10 000.5). If only four numbers are significant (10.00 km), then the race length is accurate to within 10 m (9 995 to 10 005).

There are two logical and acceptable methods available for eliminating confusion concerning zeros or the correct number of significant figures:

1. Use proper prefixes to denote intended significance.

Distance	Precision	Number of significant figures
10.000 km	9 999.5 to 10 000.5 m	5
10.00 km	9 995 to 10 005 m	4
10.0 km	9 950 to 10 050 m	3
10 km	5 000 to 15 000 m	1

How would you express a degree of significance between 10.0 km and 10 km? One viable solution is to use scientific notation.

2. Use scientific notation to indicate significance.

Distance	Precision	Number of significant figures
$1.000\ 0 \times 10^4$ m	9 999.5 to 10 000.5 m	5
$1.000\ 0 \times 10^4$ m	9 995 to 10 005 m	4
1.00×10^4 m	9 950 to 10 050 m	3
1.0×10^4 m	9 500 to 10 500 m	2
1×10^4 m	5 000 to 15 000 m	1

Selection of a proper prefix is customarily the logical way to handle problems of significant figures; however, there are conventions that do not lend themselves to the prefix notation. An example would be temperature in degrees Celsius; that is, $4.00(10^3)°C$ is the conventional way to handle it, not 4.00 k°C.

7.7 Rules for Using SI Units

Along with the adoption of SI comes the responsibility to thoroughly understand and properly apply the new system. Obsolete practices involving both English and metric units are widespread. This section provides rules that should be followed when working with SI units.

7.7.1 Unit Symbols and Names

1. Periods are never used after SI symbols unless the symbol is at the end of a sentence (that is, SI unit symbols are not abbreviations).

2. Unit symbols are written in lowercase letters unless the symbol derives from a proper name, for example, Ampere (A) or Kelvin (K), in which case the first letter is capitalized.

Lowercase	Uppercase
m, kg, s, mol, cd	A, K, Hz, Pa, C

3. Symbols rather than self-styled abbreviations should always be used to represent units.

Correct	Not correct
A	amp
s	sec

4. An s is never added to the symbol to denote plural.

5. A space is always left between the numerical value and the unit symbol.

Correct	Not correct
43.7 km	43.7km
0.25 Pa	0.25Pa

Exception: No space should be left between numerical values and the symbols for degree, minute, and second of angles and for degree Celsius.

6. There should be no space between the prefix and the unit symbols.

Correct	Not correct
mm, MΩ	k m, μ F

7. A unit name is written in lowercase (except at the beginning of a sentence), even if the unit is derived from a proper name.

8. Plurals are used as required when writing unit names. For example, henries is plural for henry. The following exceptions are noted:

Singular	Plural
lux	lux
hertz	hertz
siemens	siemens

With these exceptions, unit names form their plurals in the usual manner.

9. No hyphen or space should be left between a prefix and the unit name. In three cases the final vowel in the prefix is omitted: megohm, kilohm, and hectare.

10. The symbol should be used following a number in preference to the unit name because unit symbols are standardized. An exception to this is made when a number is written in words preceding the unit; for example, we would write *nine meters,* not *nine m.* The same is true the other way, for example, 9 m, not 9 meters.

7.7.2 Multiplication and Division

1. When writing unit names as a product, always use a space (preferred) or a hyphen.

Correct usage
newton meter or newton-meter

2. When expressing a quotient using unit names, always use the word *per* and not a solidus (/). The solidus, or slash mark, is reserved for use with symbols.

Correct usage	Not correct
meter per second	meter/second

3. When writing a unit name that requires a power, use a modifier, such as squared or cubed, after the unit name. For area or volume, the modifier can be placed before the unit name.

Correct usage
millimeter squared or square millimeter

4. When expressing products using unit symbols, the center dot is preferred.

Correct usage
N · m for newton meter

5. When denoting a quotient by unit symbols, any of the following are accepted form:

Correct usage

$$\text{m/s} \quad \text{or} \quad \text{m} \cdot \text{s}^{-1} \quad \text{or} \quad \frac{\text{m}}{\text{s}}$$

In more complicated cases, consider using negative powers or parentheses. For acceleration use m/s^2 or m $\cdot$ s^{-2} but not m/s/s. For electrical potential use kg $\cdot$ m^2/(s^3 $\cdot$ A) or kg $\cdot$ s^{-3} $\cdot$ A^{-1} but not kg $\cdot$ m^1/s^3/A.

7.7.3 Numbers

1. To denote a decimal point, use a period on the line. When expressing numbers less than 1, a zero should be written before the decimal marker.

   Example

   15.6
   0.93

2. Since a comma is used in many countries to denote a decimal point, its use is to be avoided in grouping data. To avoid confusion, separate the digits into groups of three, counting from the decimal to the left or right, and use a small space to separate the groups.

   Correct and recommended procedure

   6.513 824   76 851   7 434   0.187 62

7.7.4 Calculating with SI Units

Before we look at some suggested procedures that will simplify calculations in SI, let us review the following positive characteristics of the system.

Only one unit is used to represent each physical quantity, such as the meter for length, the second for time, and so on. The SI metric units are *coherent*; that is, each new derived unit is a product or quotient of the fundamental and supplementary units without any numerical factors. Since coherency is a strength of the SI system, it would be worthwhile to demonstrate this characteristic by the following two examples. The relationship among force, mass and time can be illustrated by Newton's second law, $F \propto ma$. To satisfy coherency the newton (N) becomes a derived unit. Its magnitude is defined as the force required to impart an acceleration of one meter per second squared to a mass of one kilogram. It was not arbitrary determined independent of mass and time. Thus,

$$1.0 \text{ N} = (1.0 \text{ kg})(1.0 \text{ m/s}^2)$$

Newton's second law can now be written in equation form as follows:

$$F = \frac{ma}{g_C}, \text{ where } g_C = \frac{ma}{F} \text{ or } g_C = \frac{1.0 \text{ kg} \cdot 1.0 \text{ m}}{\text{N} \cdot \text{s}^2}$$

This constant of proportionality serves as a reminder that the units are in fact coherent and that the conversion factor is 1.0.

Consider next the joule, the SI equivalent of the British thermal unit, the calorie, foot-pound-force, the electron volt, and the horsepower-hour, intended to represent most forms of energy. The joule is defined as the amount of work done when an applied force of one newton acts through a distance of one meter in the direction of the force. Thus,

$$1.0\,J = (1.0\,N)(1.0\,m)$$

To maintain coherency of units, however, time must be expressed in seconds rather than minutes or hours, since the second is the base unit. Once coherency is violated, then a conversion factor must be included and the advantage of the system is diminished.

But there are certain units *outside* SI that are accepted for use in the United States, even though they diminish the system's coherence. These exceptions are listed in Table 7.6.

Calculations using SI can be simplified if you;
1. Remember that fundamental relationships are simple and easier to use because of coherence.
2. Recognize how to manipulate units and gain a proficiency in doing so. Since watt = J/s = $N \cdot m/s$, you can algebraically rearrange the units to produce $N \cdot m/s = (N/m^2)(m^3/s) = $ (pressure)(volume flow rate).
3. Understand the advantage of occasionally adjusting all variables to base units; for example, replacing N with $kg \cdot m/s^2$ and Pa with $kg \cdot m^{-1}s^{-2}$.
4. Develop a proficiency with exponential notation of numbers to be used in conjunction with unit prefixes.

$$1\,mm^3 = (10^{-3}\,m)^3 = 10^{-9}\,m^3$$

$$1\,ns^{-1} = (10^{-9}\,s)^{-1} = 10^9\,s^{-1}$$

When calculating with SI the term "weight" can be confusing. Frequently we hear statements such as "The person weighs 100 kg." A correct statement

Table 7.6   Non-SI Units Accepted for Use in the United States

Quantity	Name	Symbol	SI equivalent
Time	minute	min	60 s
	hour	h	3 600 s
	day	d	86 400 s
Plane angle	degree	°	$\pi/180$ rad
	minute	′	$\pi/10\ 800$ rad
	second	″	$\pi/648\ 000$ rad
Volume	liter	L*	$10^{-3}\,m^3$
Mass	metric ton	t	10^3 kg
	unified atomic mass unit	u	$1.660\ 57 \times 10^{-27}$ kg (approx)
Land area	hectare	ha	$10^4\,m^2$
Energy	electronvolt	eV	1.602×10^{-19} J (approx)

*Both "L" and "l" are acceptable international symbols for liter. The uppercase letter is recommended for use in the United States because the lowercase "l" can be confused with the numeral 1.

would be "The person has a mass of 100 kg." To clear up any confusion, let's look at some basic definitions.

First, the term *mass* should be used to indicate only a quantity of matter. Mass is measured in kilograms (kg) or pound-mass (lbm) and is always measured against a standard.

Force, as defined by the International Standard of Units, is measured in newtons. By definition the newton was established as the force required to accelerate a mass of one kilogram to one meter per second squared.

The acceleration of gravity varies at different points on the surface of the Earth as well as distance from the Earth's surface. The accepted standard value of gravitational acceleration is 9.806 650 m/s² at sea level and 45 degrees latitude.

Gravity is instrumental in measuring mass with a beam balance or scale. If you use a beam balance to compare an unknown quantity against a standard mass, the effect of gravity on the two masses cancels out. If you use a spring scale, mass is measured indirectly, since the instrument responds to the local force of gravity. Such a scale can be calibrated in mass units and be reasonably accurate when used where the variation in the acceleration of gravity is not significant.

The following example problem clarifies the confusion that exists in the use of the term *weight* to mean either force or mass. In everyday use, the term *weight* nearly always means mass; thus, when a person's weight is discussed, the quantity referred to is mass.

Example Problem 7.1 A "weight" of 100.0 kg (the unit itself indicates mass) is suspended by a cable from an I-beam. Calculate the force or tension in the cable in newtons to hold the mass stationary when the local gravitational acceleration is (a) 9.807 m/s² and (b) 1.63 m/s² (approximate value for the surface of the Moon).

Theory Tension in the cable or force required to hold the object when the mass is at rest or moving at constant velocity is

$$F = \frac{mg_L}{g_c}$$

where g_L is the local acceleration of gravity and replaces acceleration in Newton's equation $F = ma$, g_c is the proportionality constant, and m is the mass of object. Remember that due to coherence

$$g_c = 1.0 \left[\frac{\text{kg} \cdot \text{m}}{\text{N} \cdot \text{s}^2} \right]$$

Assumption Neglect the mass of the cable.

Solution

(a) For $g_L = 9.807 \text{ m/s}^2$

$$F = \frac{mg_L}{g_c}$$

$$F = \frac{(100.0 \text{ kg})(9.807 \text{ m})}{s^2} \times \frac{N \cdot s^2}{1.0 \text{ kg} \cdot m} = 980.7 \text{ N}$$

$$= 0.980\ 7 \text{ kN}$$

(b) For $g_L = 1.63 \text{ m/s}^2$

$$F = \frac{(100.0 \text{ kg})(1.63 \text{ m})}{s^2} \times \frac{N \cdot s^2}{1.0 \text{ kg} \cdot m} = 0.163\ 0 \text{ kN}$$

7.8 U.S. Customary and Engineering Systems

Before you study the material in this section, ask yourself why it is necessary to consider any system of dimensions and units other than SI. Next, think about a few common products that you might purchase for a home remodeling project: threaded fasteners, lumber, nails, paint, and so on. How many of these are available in metric units?

Although SI is ultimately intended to be adopted worldwide, at the present time many segments of the U.S. industrial complex regularly use other systems. For many years to come, engineers in the United States will have to be comfortable and proficient with a variety of unit systems.

As noted earlier in this chapter, there are two systems of units other than SI commonly used in the United States.

7.8.1 U.S. Customary System

The first, the U.S. Customary System has the fundamental units of foot (ft) for length, pound (lb) for force, and second (s) for time (see Table 7.7). However, in this system mass (m) is not a fundamental unit; it is a new derived unit called the slug. Since it is a new derived unit, its magnitude can be established. A slug is defined as a specific amount of mass. In fact, it is the amount of mass that would be accelerated to one foot per second squared given a force of one pound. This system works perfectly well as long as mass is derived totally independent of force. In fact, since we define the derived unit mass as the slug and establish its mass as a quantity of matter that will be accelerated to 1.0 ft/s^2 when a force of 1.0 lb is applied, we have a coherent system of units. We can once again write Newton's second law as an equality,

$$F = \frac{ma}{g_c}, \text{ where } g_c = \frac{ma}{F} \text{ or } g_c = \frac{(1.0) \text{ slug} \cdot 1.0 \text{ ft}}{\text{lbf} \cdot s^2}$$

Note that the constant of proportionality, g_c, is included to clarify units, but the conversion factor is (1.0).

Table 7.7 The U.S. Customary System

Quality	Unit	Symbol
Mass	slug	slug
Length	foot	ft
Time	second	s
Force	pound	lb

7.8.2 The Engineering System

The Engineering System, uses length, time, mass, and force as the fundamental dimensions (see Table 7.8). In the Engineering System, the fundamental dimension force was established independently from the other primary dimensions. Recall that in the SI system of units force was determined with relation to Newton's second law. However, in this case, sometime during the fourteenth century, a quantity of matter was selected to be one pound-mass (lbm). At a later time it was decided that one pound-force (lbf) would be the force required to hold a one pound-mass in a gravitational field where the local acceleration of gravity was the standard value of 32.174 0 ft/ s². In fact one pound-force (lbf) is the amount of force that would be required to accelerated a mass of 32.174 0 (lbm) to one foot per second squared. Unfortunately, in the Engineering System of units the independent selection of four fundamental dimensions require that we insert a conversion factor in the constant of proportionality (g_c). It is for this reason that we included (g_c) in equations that relate pound-mass and pound-force. It provides us with a visual reminder that

$$g_C = 32.1740 \, \frac{\text{lbm} \cdot \text{ft}}{\text{lbf} \cdot \text{s}^2}$$

When calculating in the Engineering System, the constant of proportionality g_c and the local gravitational constant g_L can be particularly confusing when using the term "weight." If you were to hold a child in your arms, you might say that this child is heavy, and ask the question, how much does this child weigh? Does the question refer to the amount of force exerted to hold the child (lbf) or the child's mass (lbm)?

Normally the term "weight" refers to pound-mass. In other words the child's mass is 50.0 lbm. Holding the child requires a force of 50.0 lbf where the local acceleration of gravity is exactly 32.1740 ft/s².

$$F = \frac{mg_L}{g_C} = \frac{50.0 \text{ lbm}}{1.0} \times \frac{32.174 \text{ ft}}{\text{s}^2} \times \frac{\text{lbf} \cdot \text{s}^2}{32.1740 \text{ lbm} \cdot \text{ft}} = 50.0 \text{ lbf}$$

Table 7.8 The Engineering System

Quality	Unit	Symbol
Mass	pound-mass	lbm
Length	foot	ft
Time	second	s
Force	pound-force	lbf

If the local gravitational constant were any value other than 32.1740, then the force required to hold the child would either be greater than or less than the force required in the example. For instance, on planet x the local acceleration of gravity is 8.72 ft/s². In this case the force required to hold the 50.0 lbm child (mass never changes) would be determined as follows.

$$F = \frac{mg_L}{g_C} = \frac{50.0 \text{ lbm}}{1.0} \times \frac{8.72 \text{ ft}}{s^2} \times \frac{lbf \cdot s^2}{32.1740 \text{ lbm} \cdot \text{ft}} = 13.6 \text{ lbf}$$

So the next time someone asks how much you can bench press, say, "About 600 lbm,"—just don't mention on what planet.

Once again a word of caution when using the Engineering System in expressions such as Newton's second law (F α ma). This particular combination of units—lbf, lbm, ft, and s²—do not constitute a coherent set. Recall a coherent set of non-SI units involving lbf, slug, and ft/s² was the U.S Customary system. So you have a choice. You can either include the conversion factor or always convert mass quantities from lbm to slugs (1.0 slug = 32.174 0 lbm). For example, in the problem above it would be necessary to first convert 50.0 lbm to slugs and then simply use the U.S. Customary system.

$$50.0 \text{ lbm} = \frac{50.0 \text{ lbm}}{1.0} \times \frac{slugs}{32.174 \text{ lbm}} = 1.554 \text{ slugs}$$

$$F = \frac{mg_L}{g_C} = \frac{1.554 \text{ slugs}}{1.0} \times \frac{32.174 \text{ ft}}{s^2} \times \frac{lbf \cdot s^2}{1.0 \text{ slugs} \cdot \text{ft}} = 50.0 \text{ lbf}$$

7.9 Conversion of Units

The two dimensional systems listed in Table 7.1 can be further divided into the four systems of mechanical units presently encountered in the United States. See Table 7.9. The table does not provide a complete list of all possible quantities; this list is presented to demonstrate the different units that are associated with each unique system. As an example, the physical quantity L (length) can be expressed in a variety of units. Fortunately, it is a simple matter to convert the units from any system to the one in which you are working. To do this, the basic conversion for the units involved must be known and a logical series of steps must be followed. This procedure is often referred to as dimensional analysis or the unit-factor method.

Mistakes can be minimized if you remember that a conversion factor simply relates the same physical quantity in two different unit systems. For example, 1.0 in. and 25.4 mm each describe the same length quantity. So let's say you wish to convert 65.7 in to mm. The five steps outlined below summarizes a method to follow during any conversion of units.

1. Write the following identity starting with the quantity you wish to convert

$$65.7 \text{ in} = \frac{65.7 \text{ in}}{1.0}$$

2. Recall or lookup appropriate conversion factor

$$1.0 \text{ in} = 25.4 \text{ mm}$$

Table 7.9 Mechanical Units

Quality	Absolute System		Gravitational System	
	MKS	CGS	Type I	Type II
Length	m	cm	ft	ft
Mass	kg	g	slug	lbm
Time	s	s	s	s
Force	N	dyne	lbf	lbf
Velocity	$m \cdot s^{-1}$	$cm \cdot s^{-1}$	$ft \cdot s^{-1}$	$ft \cdot s^{-1}$
Acceleration	$m \cdot s^{-2}$	$cm \cdot s^{-2}$	$ft \cdot s^{-2}$	$ft \cdot s^{-2}$
Torque	$N \cdot m$	$dyne \cdot cm$	$lbf \cdot ft$	$lbf \cdot ft$
Moment of inertia	$kg \cdot m^2$	$g \cdot cm^2$	$slug \cdot ft^2$	$lbm \cdot ft^2$
Pressure	$N \cdot m^{-2}$	$dyne \cdot cm^{-2}$	$lbf \cdot ft^{-2}$	$lb \cdot ft^{-2}$
Energy	J	erg	$ft \cdot lbf$	$ft \cdot lbf$
Power	W	$erg \cdot s^{-1}$	$ft \cdot lbf \cdot s^{-1}$	$ft \cdot lbf \cdot s^{-1}$
Momentum	$kg \cdot m \cdot s^{-1}$	$g \cdot cm \cdot s^{-1}$	$slug \cdot ft \cdot s^{-1}$	$lbm \cdot ft \cdot s^{-1}$
Impulse	$N \cdot s$	$dyne \cdot s$	$lbf \cdot s$	$lbf \cdot s$

Type I—U.S. Customary
Type II—Engineering

Remember, you can divide both sides by 1.0 in or you can divide both sides by 25.4 mm.

3. Arrange this equality such that the desired unit conversion is in the numerator, i.e., mm.

$$\frac{1.0 \text{ in}}{1.0 \text{ in}} = \frac{25.4 \text{ mm}}{1.0 \text{ in}} = 1.0$$

4. Multiply the two identities to obtain desired answer

$$65.7 \text{ in} = \frac{65.7 \text{ in}}{1.0} \times \frac{25.4 \text{ mm}}{1.0 \text{ in}} = 1\,668.78 \text{ mm}$$

5. Consider significant figures

$$65.7 \text{ in} = 1\,670 \text{ mm}$$

Thus, when using the conversion factor 25.4 mm/in. to convert a quantity in inches to millimeters, you are multiplying a factor that is not numerically equal to 1 but is physically identical. This fact allows you to readily avoid the most common error, that of using the reciprocal of a conversion. Let's say you are to convert ft to m. From the conversion tables in the Appendix we obtain 1.0 ft = 0.3048 m or 0.3048 m/ft. Just imagine that the value in the numerator of the conversion must describe the same physical quantity as that in the denominator. When so doing, you will never use the incorrect factor 0.3048 ft/m, since 0.3048 ft is clearly not the same length as 1 m.

The following five example problems review the procedure outlined above and presents a systematic series of steps that can be used when performing a unit conversion. Often times the five steps outlined are condensed to some smaller number, i.e., individual steps are combined, nevertheless, the construction of a series of individual steps will aid the thought process and help ensure a correct unit analysis. Notice the units to be eliminated will cancel algebraically, leaving the desired results. The final answer should be checked to make sure it is reasonable.

Example Problem 7.2 Convert 375 lbm/s to slugs/hr

Step 1: Write the identity and list conversion factors

$$375 \text{ lbm/s} = \frac{375 \text{ lbm}}{1.0 \text{ s}}$$

$$1.0 \text{ slug} = 32.174 \text{ 0 lbm} \quad \text{and} \quad 1.0 \text{ hr} = 3\ 600 \text{ s}$$

Step 2: Multiply this identity by the appropriate conversion factors

$$375 \text{ lbm/s} = \frac{375 \text{ lbm}}{1.0 \text{ s}} \times \frac{\text{slugs}}{32.174 \text{ lbm}} \times \frac{3\ 600 \text{ s}}{\text{hr}} = 41\ 959 \text{ slugs/hr}$$

Notice that by the correct positioning of conversion factors the desired answer can be realized, i.e., lbm and seconds cancel leaving slugs per hr.

Step 3: Check for reasonable answer and significant figures

$$375 \text{ lbm/s} = 4.20 \times (10)^4 \text{ slugs/hr}$$

Example Problem 7.3 Convert 85.0 lbm/ft³ to kilograms per cubic meter.

Solution

Step 1: Write the identity and list conversion factors

$$85.0 \text{ lbm/ft}^3 = \frac{85.0 \text{ lbm}}{1.0 \text{ ft}^3}$$

$$1.0 \text{ ft} = 0.304 \text{ 8 m} \qquad \text{and} \qquad 1.0 \text{ lbm} = 0.453 \text{ 6 kg}$$

Step 2: Multiply this identity by the appropriate conversion factors

$$85.0 \text{ lbm/ft}^3 = \frac{85.0 \text{ lbm}}{1.0 \text{ ft}^3} \times \frac{0.4536 \text{ kg}}{1.0 \text{ lbm}} \times \frac{(1.0 \text{ ft})^3}{(0.3048 \text{ m})^3} = 1.36 \times 10^3 \text{ kg/m}^3$$

Step 3: Check for reasonable answer and significant figures

Example Problem 7.4 Determine the gravitation force (in newtons) on an automobile with a mass of 3 645 lbm. The acceleration of gravity is known to be 32.2 ft/s².

Solution A Force, mass, and acceleration of gravity are related by

$$F = \frac{mg_L}{g_C}$$

Convert lbm to kilograms

$$m = \frac{3\ 645 \text{ lbm}}{1.0} \times \frac{1 \text{ kg}}{2.204 \text{ 6 lbm}} = 1\ 653.36 \text{ kg}$$

$$g_L = \frac{32.2 \text{ ft}}{s^2} \times \frac{0.304 \text{ 8 m}}{1.0 \text{ ft}} = 9.814 \text{ 6 m/s}^2$$

$$F = \frac{mg_L}{g_C} = (1\ 653.36 \text{ kg})\left(\frac{9.814 \text{ 6 m}}{s^2}\right)\left(\frac{1.0 \text{ kg} \cdot \text{m}}{\text{N} \cdot s^2}\right)$$

$$= 16\ 227 \text{ N} = 16.2 \text{ kN}$$

Note: Intermediate values were not rounded to final precision, and we have used either exact or conversion factors with at least one more significant figure than contained in the final answer.

Solution B (combine steps in Solution A into one step)

$$F = \frac{mg_L}{g_c} = \frac{3\,645\,\text{lbm}}{1} \times \frac{32.2\,\text{ft}}{1\text{s}^2} \times \frac{1\,\text{kg}}{2.2046\,\text{lbm}} \times \frac{0.304\,8\,\text{m}}{1\,\text{ft}} \times \frac{\text{N}\cdot\text{s}^2}{1.0\,\text{kg}\cdot\text{m}}$$

$$= 16\,227\,\text{N} = 16.2\,\text{kN}$$

Note: It is often convenient to include conversions with the appropriate engineering relationship in a single calculation.

Example Problem 7.5   Convert a mass flow rate of 195 kg/s (typical of the airflow through a turbofan engine) to slugs per minute.

Solution: $195\,\text{kg/s} = \dfrac{195\,\text{kg}}{1\,\text{s}} \times \dfrac{1\,\text{slug}}{14.954\,\text{kg}} \times \dfrac{60\,\text{s}}{1\,\text{min}} = 782\,\text{slug/min}$

Example Problem 7.6   Compute the power output of a 225-hp engine in (a) British thermal units per minute and (b) kilowatts.

Solution

(a) $225\,\text{hp} = \dfrac{225\,\text{hp}}{1} \times \dfrac{2.5461 \times 10^3 \text{Btu}}{1\,\text{hp}\cdot\text{h}} \times \dfrac{1\,\text{hr}}{60\,\text{min}}$

$= 9.55 \times 10^3\,\text{Btu/min}$

(b) $225\,\text{hp} = \dfrac{225\,\text{hp}}{1} \dfrac{0.745\,70\,\text{kW}}{1\,\text{hp}} = 168\,\text{kW}$

7.10   Celsius, Fahrenheit, and Absolute Scales

Temperature scales also appear in two common forms, the Celsius scale (previously called centigrade) and the Fahrenheit scale. The Celsius scale has the same temperature increment as its absolute thermodynamic scale, the Kelvin scale. However, the zero point on the Celsius scale is 273.15 K above absolute zero.

$$t(°C) = T(K) - 273.15 \tag{7.1}$$

Scales more commonly used in the United States are the Fahrenheit scale and its corresponding absolute thermodynamic scale, the Rankine scale. A unit degree on the Fahrenheit scale is precisely the same as a unit degree on the Rankine scale. However, the zero point on the Fahrenheit scale is 459.67°R above absolute zero. (See Table 7.10.)

$$t(°F) = T(°R) - 459.67 \tag{7.2}$$

When these relationships are combined, a convenient equation can be developed for conversion between Celsius and Fahrenheit or vice versa.

$$t(°F) = 9/5t(°C) + 32°F \tag{7.3}$$

$$t(°C) = 5/9[t(°F) - 32°F] \tag{7.4}$$

Table 7.10 Temperature scales

	°R	K	°F	°C
Abs Zero	0	0	−459.67	−273.15
°F = °C	419.67	233.15	−40	−40
Zero on °F	459.67	255.37	0	−17.78
Zero on °C	491.67	273.15	32	0
Boiling	671.67	373.15	212	100

If you were to construct the above Rankine and Kelvin scales parallel to each other, they would begin at absolute zero and be graduated and calibrated to the boiling point of water. To simplify the comparison consider only the portion of the scales from freezing to boiling. On the Rankine scale there are 180 degrees (671.67 − 491.67) and on the Kelvin scale only 100 (373.15 − 273.15). When you divide these two quantities by 20 (180/20) and (100/20) it becomes apparent that 9 degrees on the Rankine scale is equivalent to 5 degrees on the Kelvin scale or 9°R = 5 K. So if you are converting from one scale to the other you must use the conversion factor 9°R = 5 K, the factor that makes the scales equivalent.

Example Problem 7.7   Convert 373.15 K to a Rankine temperature.

$$373.15 \text{ K} = \frac{373.15 \text{ K}}{1.0} \times \frac{9°\text{R}}{5 \text{ K}} = 671.67 \text{ °R}$$

Example Problem 7.8   The Universal Gas Constant (UGC) in the different unit systems has been determined as follow:

UGC = 1545 ft · lbf/lbmol · °R = 1.986 Btu/lbmol · °R = 8.314 kJ/kmol · K

Show how you would convert from 1545 · lbf/lbmol · °R to 8.314 kJ/kmol · K

$$1\,545 \frac{\text{ft} \cdot \text{lbf}}{\text{lbmol} \cdot °\text{R}} = \frac{1\,545 \text{ ft} \cdot \text{lbf}}{\text{lbmol} \cdot °\text{R}} \times \frac{4.448\,2 \text{ N}}{\text{lbf}} \times \frac{\text{lbmol}}{0.453\,59 \text{ kmol}} \times \frac{\text{m}}{3.280\,8 \text{ ft}}$$

$$\times \frac{9°\text{R}}{5 \text{ K}} \times \frac{\text{J}}{\text{N} \cdot \text{m}} \times \frac{\text{kJ}}{1\,000 \text{ J}} = 8.313 \text{ kJ/kmol} \cdot \text{K}$$

Example Problem 7.9   Convert 98.6°F to °C.

Solution A

$$t(°\text{C}) = [(t°\text{F}) − 32]5/9 = 333/9 = 37.0°\text{C}$$

Solution B

$$T(°\text{R}) = 98.6 + 459.67 = 558.27°\text{R}$$

$$558.27°\text{R} = \frac{558.27°\text{R}}{1.0} \times \frac{5 \text{ K}}{9°\text{R}} = 310.15 \text{ K}$$

$$t(°\text{C}) = 310.15 − 273.15 = 37.0°\text{C}$$

The problem of unit conversion becomes more complex if an equation has a constant with hidden dimensions. It is necessary to work through the equation converting the constant K_1 to a new constant K_2 consistent with the equation units.

Consider the following example problem.

Example Problem 7.10 The velocity of sound in air (c) can be expressed as a function of temperature (T):

$$c = 49.02\sqrt{T}$$

where c is in feet per second and T is in degrees Rankine.

Find an equivalent relationship when c is in meters per second and T is in kelvins.

Procedure

1. First, the given equation must have consistent units; that is, it must have the same units on both sides. Squaring both sides we see that

$$c^2\,(ft^2/s^2) = 49.02^2\ T°R$$

From this equation it is apparent that the constant $(49.02)^2$ must have units in order to maintain unit consistency. (The constant must have the same units as c^2/T.)

Solving for the constant,

$$(49.02)^2 = c^2\,\frac{ft^2}{s^2}\left[\frac{1}{T\cdot°R}\right] = \frac{c^2}{T}\left[\frac{ft^2}{s^2\,°R}\right]$$

2. The next step is to convert the constant $49.02^2\ ft^2/(s^2°R)$ to a new constant that will allow us to calculate c in meters per second given T in kelvins. We recognize that the new constant must have units of square meters per second squared per kelvin.

$$\frac{(49.02)^2ft^2}{s^2°R} = \frac{(49.02)^2ft^2}{1.0\ s^2°R} \times \frac{(0.3048\,m)^2}{(1\,ft)^2} \times \frac{9°R}{5\,K} = \frac{401.84\,m^2}{1\,s^2K}$$

3. Substitute this new constant 401.84 back into the original equation

$$c^2 = 401.84T$$
$$c = 20.05\sqrt{T}$$

where c is in meters per second and T is in kelvins. If you wish to verify this new equation, take a temperature in Fahrenheit (80°F) and convert to Rankine (540°R). Take the original equation and compute the value of c (1 139 ft/s).

Next convert 80°F to kelvins (26.67 + 273.15), calculate c (347.17 m/s), and convert back to ft/s.

$$1\,139\ ft/s = \frac{1\,139\ ft}{1.0\ s} \times \frac{0.3048\,m}{1.0\ ft} = 347.17\,m/s$$

You will have verified the new constant.

Problems

7.1 Using the correct number of significant figures, convert the following physical quantities into the proper SI units.
  (a) 215 hp
  (b) 960 acres
  (c) 3.7×10^4 ft^3
  (d) 212 ft/s
  (e) 65 mph
  (f) 72 slugs
  (g) 155 lbm
  (h) 2 140 ft
  (i) 4 325 gal
  (j) 1 535 miles

7.2 Convert the following to SI units, using the correct significant figures.
  (a) 615 slugs/min
  (b) 14.7 Btu/min
  (c) 115 °C
  (d) 61 oz
  (e) 7.91 atm
  (f) 217 bushels/acre
  (g) 185 hp·hr
  (h) 18.5 lbm/ft^3
  (i) 3.77×10^4 ft^3/hr
  (j) 4 695 lbf

7.3 Convert as indicated giving the answer using proper significant figures.
  (a) 9.72 ft to millimeters
  (b) 65.4 ft^2 to L
  (c) 255 K to degrees Rankin
  (d) 7 595 bushels to cubic meters
  (e) 105 250 Btu/h to kilowatts

7.4 Convert as indicated giving the answer using proper significant figures.
  (a) 7 980 ft·lbf to joules
  (b) 14.7 lbf/in^2 to pascals
  (c) 29 028 ft to m
  (d) 94.5 slugs/ft^3 to grams per cubic centimeter
  (e) 212°F to Kelvin

7.5 Using the rules for expressing SI units, correct each of the following if given incorrectly.
  (a) 12 amps
  (b) 12.5 cm's
  (c) 250 degrees Kelvin
  (d) 125.0 m m
  (e) 152 KW/hours
  (f) 6.7 m/s/s
  (g) 86.3 j
  (h) 3500 K
  (i) 375 n
  (j) 4 225 pa

7.6 Using the rules for expressing SI units, correct each of the following if given incorrectly.
  (a) 5..50 N
  (b) 63.5 C
  (c) 108 farads
  (d) 65 nM
  (e) 17 m per s
  (f) 725 N/m/m
  (g) 72.0 Kg
  (h) 750 J/sec
  (i) 95 A's
  (j) 1.5 m · m

7.7 If a force of 1.15×10^3 N is required to lift an object with a uniform velocity and an acceleration of gravity shown as follows, determine the mass, in kg, of the object:
  (a) 32.5 ft/s^2
  (b) 7.86 m/s^2

7.8 If you were on another planet, say, Mars, which of the following, g_C or g_L, would change and which would stay constant? Explain the difference.

7.9 Determine the acceleration of gravity required (meters per second squared) to lift a 1 500 kg object at a uniform velocity when the force exerted is
  (a) 3.580×10^3 lbf
  (b) 8.90×10^3 N

7.10 The average density of Styrofoam is 1.00 kg/m^3. If a Styrofoam cooler is made with outside dimensions of $50.0 \times 35.0 \times 30.0$ cm and the uniform thickness of the Styrofoam is 3.00 cm (including the lid), what is the volume of the Styrofoam used in cubic inches? What is the mass in lbm? How many gallons of liquid could be stored in the cooler?

7.11 A small town purchased a 25 ft diameter cylindrical tank for potable water in the event of an emergency. The town consists of 3 600 family units. If each family were to collect 25 gallons of water, what would be the minimum height of the tank?

7.12 The *Eurostar* provides international high speed train service between Paris, London, and Brussels through the English Channel Tunnel. *Eurostar* trainsets can operate at maximum speeds of 3.00×10^3 km/h. Assume that the resistance between the train and the track is 115×10^3 newtons. If air resistance adds an additional 25.5×10^3 newtons, determine the horsepower needed to power the engine at maximum speed. [Note: Power = (force) × (velocity)]

7.13 *Eurostar's* nose is computer-optimized for running in the Channel Tunnel where pressure waves can affect passenger comfort. The tunnel itself is passed at a reduced speed of 1.6×10^2 km/h. Determine the length of time it takes to complete the 23 mile underwater portion of the trip from London to Paris.

7.14 Assume that your hometown is growing so rapidly that an additional water tower will be necessary to meet the needs of the community. Engineers predict that the water tower will need to hold 1.25×10^6 kilograms of water, with a density of 999 kilograms per cubic meter.
(a) What will the volume of the tower have to be?
(b) Determine the vertical force, in kN, acting on each of the four evenly spaced legs from the weight of the water alone.
(c) If the tower is spherical, what would be the diameter of the tower?

7.15 The Hubert H. Humphrey Metrodome in Minneapolis, Minnesota opened in 1982 at a cost of $55 million. To prepare for the footings, etc. 300 000 cubic yards of dirt were removed. Its inflatable roof covers 10.0 acres and its interior volume is 60.0×10^6 cubic feet. During construction, 40.0×10^3 cubic yards of concrete, 11.9×10^3 tons of reinforcing steel and 5.00×10^2 tons of structural steel were utilized. The roof material includes an outer layer of Teflon-coated fiberglass and an inner layer of woven fiberglass. Given this information, answer the following:
(a) If removed evenly over the 10 acres, how deep of a hole would be formed as a result of the excavation?
(b) What is the total mass of concrete and steel in lbm?
(c) If the two-layer fiberglass roofing material has a mass of 0.6667 lbm/ft^2, what is the total mass of the roof?
(d) If 10.0 inches of wet snow collected on the roof (this happened in 1982), what is the added mass to the roof? (Assume 1 inch of water equals 10 inches of snow.)

7.16 The U.S. currently imports 15.0×10^6 barrels of oil each day. If a cylindrical storage tank were to be constructed with a base of 50.0 ft.
(a) What would be the height of the container to store this daily consumption?
(b) Recall that Mt. Everest is 29 028 ft. What would the diameter of the cylinder have to be to match the height of Mt. Everest?

7.17 Construction sand is piled in a right cone that has a height of 25.0 feet and a diameter of 110.0 feet. If this sand has a density of 97.0 lbm/ft^3,
(a) determine the volume of the cone

(b) determine the mass of the sand in the cone

(c) during the winter months the sand will be spread on county roads. Using a spreadsheet, determine the volume and mass of the sand remaining if the height decreases from 25.0 to 5.00 ft in increments of 0.500 ft. Assume that the base remains constant.

7.18 A cylindrical underground storage tank with a diameter of 22.0 ft and a height of 30.0 ft is filled with gasoline. Given a density of 675 kg/m³, determine the mass of the gas in the tank in both lbm and kg. If the average automobile gas tank holds 25.0 gal, calculate the number of autos that may be filled from this underground facility.

7.19 A cylindrical tank is 25.0 ft long and 10.0 ft in diameter is oriented such that is longitudinal axis is horizontal. Develop a table that will

(a) Show how many gallons of diesel fuel are in the tank if the fluid level is measured in 1.00 ft increments from the bottom of the tank

(b) Show the corresponding mass at each increment, in kg, if the specific gravity is 0.73.

7.20 A southwest rancher constructs a spherical water tank that is 10 ft in diameter. Develop a table that will

(a) Show how many gallons of water are contained in the tank if the volume is measured in 1.00 ft increments from the bottom to the top of the container.

(b) Determine the mass of water at each increment in lbm.

7.21 The ideal gas law shows the relationship among some common properties of ideal gases.

$$pV = nRT$$

where

p = pressure

V = volume

n = number of moles of the ideal gas

R = universal gas constant = 8.314kJ/(kmol K)

T = absolute temperature

If you have 5 moles of an ideal gas at twenty-two degrees Celsius and it is stored in a container that is 0.650 meters on each side, calculate the pressure in Pa.

7.22 Approximately 50 000 years ago a meteorite hit the earth near Winslow, Arizona. The impact crater is 1 200 m in diameter and 170 m deep. Determine the volume of earth in ft³ removed assuming the crater to be a spherical segment. Verify that the radius of the sphere is approximately 1 150 m.

7.23 Conservation of energy suggests that potential energy is converted to kinetic energy when an object falls in a vacuum. $KE = \dfrac{mV^2}{2g_C}$ and $PE = \dfrac{mg_Lh}{g_C}$ Velocity at impact can be determined as follows:

$$V = Constant \sqrt{h}$$

(a) Determine the constant so that the equation is valid for h in ft and V in feet per second (ft/s)

(b) Determine the constant so that the equation is valid for h in ft but V in mph.

(c) If you drop an object from 225 ft, what is the velocity on impact with the ground in mph and ft/s?

7.24 A small portable cylindrical pressure tank has inside dimensions of 14.0 inches (dia) and 36.0 inches end to end. The maximum recommended safety pressure at 70°F is 200 psi. The device has a safety release value set at 250 psi.

(a) Determine the inside volume of the pressure tank in ft³.

(b) Calculate the specific volume of the tank at 70°F and 200 psi. (See below)

(c) Find the mass of air in the tank from part b.

(d) If the homeowner inserts 3.5 lbm of air into the tank at 70°F, what would be the reading on the tank pressure gauge.

(e) With the aid of a spreadsheet determine the temperature from part d at which the safety gauge would release starting at 70°F and increasing in 10 degree increments.

$$P_v = RT$$

Pressure, P, lbf/ft²

Sp. Vol., v − ft³/lbm

Gas constant for air, (R_{air} = 53.33 ft · lbf/lbm°R)

Temperature, T − °R

7.25 A weir is used to measure flow rates in open channels. For a rectangular weir the expression can be written $Q = 288.8\,LH^{1.5}$ (See Fig. 7.2.)

where   Q = discharge rate, in gal/h

L = length of weir opening parallel to liquid, in inches

H = height of fluid above crest, in inches

(a) As the channel becomes larger, the weir opening can be expressed as follows: Q in cubic feet per second, with L and H in feet. Determine the new constant.

(b) Prepare a spreadsheet resulting in a table of values of Q in both ft³/s and gal/h, with L and H in inches. Use values of L that range from 1 to 16 in 1 inch increments and let H range from 2 to 32 in 2 inch increments. For example when L = 6, H = 12.

Figure 7.2

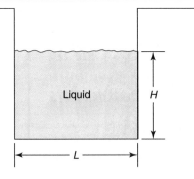

CHAPTER 8

Introduction to Engineering Economics

Chapter Objectives

When you complete your study of this chapter, you will able to:

- Understand that the value of money changes with time
- Distinguish between simple and compound interest
- Prepare a cash-flow diagram
- Compute present worth and future worth of multiple sums of money
- Calculate the equivalent uniform annual cost of a series of amounts
- Recognize and solve problems involving sinking funds and installment loans
- Solve problems with arithmetic and geometric gradients

8.1 Introduction

Engineers often serve as managers or executive officers of businesses and therefore are required to make financial as well as technical decisions. Even in companies where the managers are not engineers, engineers serve as advisers and provide reports and analyses that influence decisions. Also, the amount of capital investment (money spent for equipment, facilities, and so on) in many industries represents a significant part of the cost of doing business. Thus, estimates of the cost of new equipment, facilities, software, and processes must be carefully done if the business is to be successful and earn a profit on its products and services.

A couple of examples will be used to illustrate how an engineer might be involved in the financial decision-making process. Suppose a manufacturing company has decided to upgrade its computer network. The network will connect all parts of the company, such as engineering design, purchasing, marketing, manufacturing, field sales, and accounting. Ten different vendors are invited to submit bids based on the specifications prepared by engineers. The bids will include hardware, software, installation, and maintenance. As an engineer, you will then analyze the bids submitted and rank the 10 vendors' proposals based on predetermined criteria.

This assignment is possible once a method of comparison, such as the equivalent uniform annual cost (EUAC) or the present worth (PW) method is selected. These methods of analysis are discussed later in the chapter. However, they allow us to compare only tangible costs. Intangible items, such as safety or environmental concerns, must also be considered.

Another example of a common engineering task is justification of the purchase of a new machine to reduce the costs of manufacture. This justification is usually expressed as a rate of return on investment or *rate of return* (ROR). Rate of return is defined as the equivalent compound interest rate that must be earned on the investment to produce the same income as the proposed activity. Often the profit comes from the reduction of production costs as a result of the new machine.

Since any venture has some risk involved and the cost reductions expected are only estimated, companies will not choose to invest in new equipment unless there is a promise of a much greater return than could be realized by less risky investments, such as bank deposits or the purchase of treasury notes or government bonds.

In addition to the application of engineering economy methods in your professional life, you will also have applications in your personal life. Major purchases such as a vehicle or a home, as well as investments (e.g., money markets, bonds, treasury notes, company stock) you may make using your own funds, require understanding of the principles you will learn in this chapter. As an example, Table 8.1 is an Excel spreadsheet that shows an annual investment of $1 200 (nominally $100 per month) for 10 years beginning at age 22. This is left at 8% annual compounded interest until age 65. Compare that with starting the same yearly investment at age 32 and continuing until age 65. Of course you may not be able to earn 8% at all times, although over the long term, 8% is a reasonable number. Suppose, however, you could earn only 4%. Table 8.2 shows equal investments earning 8% and 4%, the difference being in one case the investment is made early and in the other the investment is made late. It is clear the 8% compound interest results in much greater accumulation than 4%, just as you would expect. Notice the difference in accumulation between investing early and investing late. Your conclusion should be that to optimize your retirement income you will need to invest (save) just as early as you possibly can.

8.2 Simple and Compound Interest

The idea of interest on an investment is certainly not new. The New Testament of the Christian Bible refers to banks, interest, and return. History records business dealings involving interest at least 40 centuries ago. Early business was largely barter in nature with repayment in kind. It was common during the early years of the development of the United States for people to borrow grain, salt, sugar, animal skins, and other products from each other to be repaid when the commodity was again available. Since most of these items depended on the harvest, annual repayment was the normal process. When it became impossible to repay the loan after a year, the interest was calculated by multiplying the

Table 8.1

Age	Annual Savings	Accumulation	Annual Savings	Accumulation
22	$1,200	$1,296		
23	$1,200	$2,696		
24	$1,200	$4,207		
25	$1,200	$5,840		
26	$1,200	$7,603		
27	$1,200	$9,507		
28	$1,200	$11,564		
29	$1,200	$13,785		
30	$1,200	$16,184		
31	$1,200	$18,775		
32		$20,277	$1,200	$1,296
33		$21,899	$1,200	$2,696
34		$23,651	$1,200	$4,207
35		$25,543	$1,200	$5,840
36		$27,586	$1,200	$7,603
37		$29,793	$1,200	$9,507
38		$32,176	$1,200	$11,564
39		$34,750	$1,200	$13,785
40		$37,530	$1,200	$16,184
41		$40,533	$1,200	$18,775
42		$43,776	$1,200	$21,573
43		$47,278	$1,200	$24,594
44		$51,060	$1,200	$27,858
45		$55,145	$1,200	$31,383
46		$59,556	$1,200	$35,189
47		$64,321	$1,200	$39,300
48		$69,466	$1,200	$43,740
49		$75,024	$1,200	$48,536
50		$81,025	$1,200	$53,714
51		$87,508	$1,200	$59,308
52		$94,508	$1,200	$65,348
53		$102,069	$1,200	$71,872
54		$110,234	$1,200	$78,918
55		$119,053	$1,200	$86,527
56		$128,577	$1,200	$94,745
57		$138,863	$1,200	$103,621
58		$149,973	$1,200	$113,207
59		$161,970	$1,200	$123,559
60		$174,928	$1,200	$134,740
61		$188,922	$1,200	$146,815
62		$204,036	$1,200	$159,856
63		$220,359	$1,200	$173,941
64		$237,988	$1,200	$189,152
65		$257,027	$1,200	$205,580

principal amount by the product of the interest rate and the number of periods (years), now called simple interest.

$$I = Pni \tag{8.1}$$

where

I = Interest accrued

P = Principal amount

n = Number of interest periods

i = Interest rate per period (as a decimal, not as a percent)

This is an example of a simple interest transaction where interest is calculated using the principal only, ignoring any interest accrued in preceding interest periods. Therefore, if $1 000 were to be loaned at 7% annual interest for five years, the interest would be

$$
\begin{aligned}
I &= Pni \\
&= (1\,000)(5)(0.07) \\
&= \$350
\end{aligned}
$$

and the total amount F to be repaid at the end of five years is

$$
\begin{aligned}
F &= P + I \\
&= 1\,000 + 350 \\
&= \$1\,350
\end{aligned}
\tag{8.2}
$$

It can be seen that

$$
\begin{aligned}
F &= P + I = P + Pni \\
&= P(1 + ni)
\end{aligned}
\tag{8.3}
$$

As time progressed and business developed, the practice of borrowing became more common, and the use of money replaced the barter system. It also became increasingly more common that money was loaned for longer periods of time. Simple interest was relegated to the single-interest period, and the practice of compounding developed. It can be shown by using Eq. (8.3), $n = 1$, that the amount owed at the end of one period is

$$P + Pi = P(1 + i)$$

The interest generated during the second period is then $(P + Pi)i$. It can be seen that interest is being calculated not only on the *principal* but on the previous interest as well. The sum F at the end of two periods becomes

P	principal amount
$+ Pi$	interest during first period
$+ Pi + Pi^2$	interest during second period
$P + 2Pi + Pi^2$	sum after two periods

This can be factored as follows:

$$P(1 + 2i + i^2) = P(1 + i)^2$$

Table 8.2

Age	Annual Savings	Accumulation (8%)	Accumulation (4%)	Annual Savings	Accumulation (8%)	Accumulation (4%)
22	$1,200	$1,296	$1,248			
23	$1,200	$2,696	$2,546			
24	$1,200	$4,207	$3,896			
25	$1,200	$5,840	$5,300			
26	$1,200	$7,603	$6,760			
27	$1,200	$9,507	$8,278			
28	$1,200	$11,564	$9,857			
29	$1,200	$13,785	$11,499			
30	$1,200	$16,184	$13,207			
31	$1,200	$18,775	$14,984			
32	$1,200	$21,573	$16,831			
33	$1,200	$24,594	$18,752			
34	$1,200	$27,858	$20,750			
35	$1,200	$31,383	$22,828			
36	$1,200	$35,189	$24,989			
37	$1,200	$39,300	$27,237			
38	$1,200	$43,740	$29,574			
39	$1,200	$48,536	$32,005			
40	$1,200	$53,714	$34,534			
41	$1,200	$59,308	$37,163			
42	$1,200	$65,348	$39,898			
43	$1,200	$71,872	$42,741			
44		$77,622	$44,451	$1,200	$1,296	$1,248
45		$83,831	$46,229	$1,200	$2,696	$2,546
46		$90,538	$48,078	$1,200	$4,207	$3,896
47		$97,781	$50,001	$1,200	$5,840	$5,300
48		$105,603	$52,002	$1,200	$7,603	$6,760
49		$114,052	$54,082	$1,200	$9,507	$8,278
50		$123,176	$56,245	$1,200	$11,564	$9,857
51		$133,030	$58,495	$1,200	$13,785	$11,499
52		$143,672	$60,834	$1,200	$16,184	$13,207
53		$155,166	$63,268	$1,200	$18,775	$14,984
54		$167,579	$65,799	$1,200	$21,573	$16,831
55		$180,986	$68,430	$1,200	$24,594	$18,752
56		$195,465	$71,168	$1,200	$27,858	$20,750
57		$211,102	$74,014	$1,200	$31,383	$22,828
58		$227,990	$76,975	$1,200	$35,189	$24,989
59		$246,229	$80,054	$1,200	$39,300	$27,237
60		$265,928	$83,256	$1,200	$43,740	$29,574
61		$287,202	$86,586	$1,200	$48,536	$32,005
62		$310,178	$90,050	$1,200	$53,714	$34,534
63		$334,992	$93,652	$1,200	$59,308	$37,163
64		$361,791	$97,398	$1,200	$65,348	$39,898
65		$390,735	$101,294	$1,200	$71,872	$42,741

The interest during the third period is

$$(P + 2Pi + Pi^2)i = Pi + 2Pi^2 + Pi^3$$

and the sum after three periods is

$P + 2Pi + Pi^2$	sum after second period
$+ Pi + 2Pi^2 + Pi^3$	interest during third period
$P + 3Pi + 3Pi^2 + Pi^3 = P(1 + i)^3$	sum after three periods

This procedure can be generalized to n periods of time and will result in

$$F = P(1 + i)^n \tag{8.4}$$

where F is future worth or the sum generated after n periods.

For compound interest, the interest accrued for each interest period is calculated on the principal plus the total interest accumulated in all previous periods. Thus, compound interest means interest on top of interest. Consider the sum or future worth at the end of five years on a $1 000 loan with 7% annual interest, compounded annually.

$$F = P(1 + i)^n$$
$$= (1\ 000)(1.07)^5$$
$$= \$1\ 402.55$$

Thus, the sum with annual compounding is $1 402.55, compared with the previous sum of $1 350 when simple interest was used.

Table 8.3 demonstrates the difference between simple and compound interest on a year-by-year basis for the problem just discussed. Care must be exercised in using interest rates and payment periods to make sure that the interest rate used is the rate for the period selected.

Consider calculating the sum after one year. If the annual interest rate is 12% compounded annually, then $i = 0.12$ and $n = 1$. However, when the annual rate is 12%, but it is to be compounded every six months (semiannually), then $i = 0.12/2$ and $n = 2$. This idea can be extended to a monthly compounding period, with $i = 0.12/12$ and $n = 12$, or a daily compounding period, with $i = 0.12/365$ and $n = 365$.

Table 8.3 Principal and Interest Paid on Money Borrowed for Five Years at 7% Annual Interest

Year	Simple Interest Principal	Simple Interest Interest	Compound Interest Principal	Compound Interest Interest
0 (today)	1000		1000	
1	1000	70.00	1000	70.00
2	1000	70.00	1000	74.90
3	1000	70.00	1000	80.14
4	1000	70.00	1000	85.75
5	1000	70.00	1000	91.76
	$1000 +	350.00	$1000 +	402.55
Total owed		**$1350.00**		**$1402.55**

Example Problem 8.1 What total amount (principal and interest) must be paid at the end of four years if $8 000 is borrowed from a bank at a 12% annual interest rate compounded (a) annually, (b) semiannually, (c) monthly, and (d) daily?

Solution

    (*a*) Compounded annually:
       $F = P(1 + i)^n$

where

       $i = 0.12$
       $n = 1 \times 4 = 4$ periods (years)
       $P = \$8\ 000$
       $F = 8\ 000(1.12)^4$
        $= \$12\ 588.15$

    (*b*) Compounded semiannually:
       $F = P(1 + i)^n$

where

       $i = 0.12/2$
       $n = 2 \times 4 = 8$ periods
       $P = \$8\ 000$
       $F = 8\ 000(1.06)^8$
        $= \$12\ 750.78$

    (*c*) Compounded monthly:
       $F = P(1 + i)^n$

where

       $i = 0.12/12$
       $n = 12 \times 4 = 48$ periods (months)
       $P = \$8\ 000$
       $F = 8\ 000(1.01)^{48}$
        $= \$12\ 897.81$

    (*d*) Compounded daily:
       $F = P(1 + i)^n$

where

       $i = 0.12/365$
       $n = 365 \times 4 = 1\ 460$ periods (days)
       $P = \$8\ 000$
       $F = 8\ 000(1 + 0.12/365)^{1460}$
        $= \$12\ 927.57$

Note: Always round the answer to the nearest penny.

As you can see from the example, even though the stated interest is the same, 12% in this case, the change in the compounding period changes the sum. Thus, to compare different alternatives, we must know the *stated* or *nominal* annual interest rate and the compounding period. We can also define an *effective annual rate,* often called *annual percentage rate* (APR), for comparison purposes. The annual percentage rate (APR) is then the interest rate that would have produced the final amount under annual (rather than semiannual, monthly, or other) compounding.

Then, continuing with Example Problem 8.1 part (b), with a nominal interest rate of 12% and semiannual compounding, the APR can be found as follows:

$$F = \$12{,}750.73 = 8{,}000(1 + APR)^4 = 8{,}000\left(1 + \frac{0.12}{2}\right)^8$$

or

$$(1 + APR)^4 = \left(1 + \frac{0.12}{2}\right)^8$$

then

$$APR = \left(1 + \frac{0.12}{2}\right)^2 - 1 = 0.123\,6 \qquad (12.36\% \ APR)$$

Considering part (c) with 12% nominal and monthly compounding, the APR is found from

$$\$12\,897.81 = 8\,000(1 + APR)^4 = 8\,000\left(1 + \frac{0.12}{12}\right)^{48}$$

$$APR = \left(1 + \frac{0.12}{12}\right)^{12} - 1 = 0.126\,8 \qquad (12.68\% \ APR)$$

Financial institutions sometimes "intentionally confuse" nominal and APR values in their advertising. They may state the nominal rate and simply call it the APR if this makes the rate appear to be a better deal. APR is always going to be larger than nominal interest if the compounding period is less than one year when the APR value is computed as previously defined. Since you know how to compute APR, you can always check it out.

8.3 Cash-Flow Diagram

The transaction described in Example Problem 8.1 for annual compounding can and should be graphically illustrated in a *cash-flow diagram* (Fig. 8.1). Since cash-flow diagrams are very useful in the visualization of any transaction, they will be used throughout this chapter. The following general rules apply:

1. The horizontal line is a time scale. The interval of time is normally given in years, though in some cases other periods may be more meaningful.
2. The arrows signify cash flow. A downward arrow means money out, and an upward arrow means money in.
3. The diagram is dependent on the point of view from which it is constructed—that is, on whether it is the lender's or the borrower's point of view (see Fig. 8.1).

Figure 8.1

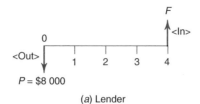

(a) Lender

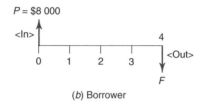

(b) Borrower

Cash-flow diagram: (a) as seen by the lender; (b) as seen by the borrower.

8.4 Present Worth and Future Worth

It is important to keep in mind that the value of any transaction (loan, invest-ment, and so on) changes with time because of interest. Thus, to express the value of a transaction, you must also give the point in time at which that value is computed. For example, the value of the loan described in Example Problem 8.1 (assuming annual compounding) is $8 000 at year zero but is $12 588.15 four years later. We will examine several methods of stating the value of a transaction.

Present worth (P) is the worth of a monetary transaction at the current time. It is the amount of money that must be invested now in order to produce a prescribed sum at another date.

Future worth (F) is the worth of a monetary transaction at some point in the future. It is an analysis of what the future amount of money will be if we take some particular course of action now.

To illustrate, if you were guaranteed an amount of money (F) four years from today, then P would be the present worth of F, where the interest is i and n = 4 (assuming annual compounding). Since this analysis is exactly the inverse of finding a future sum, we have

$$P = F(1 + i)^{-n} \tag{8.5}$$

As an example, if you can convince a lending institution that you will have a guaranteed amount of money available four years from today, it may be pos-sible to borrow the present worth of that amount. If the guaranteed sum (four years later) is equal to $12 588.15, the present worth at 12% annual interest (compounded annually) is $8 000. (See Fig. 8.2.)

$$P = F(1 + i)^{-n}$$

Figure 8.2

Banker's cash-flow diagram.

where
$$F = \$12\ 588.15$$
$$i = 0.12$$
$$n = 4$$
$$P = \frac{12\ 588.15}{(1.12)^4}$$
$$= \$8\ 000.00$$

In situations that involve economic decisions the following types of questions may arise:

1. Does it pay to make an investment now?
2. What is the current benefit of a payment that will be made at some other date?
3. How does the monthly price paid to lease a computer compare with the future cost of a new computer?
4. How much money do you need to invest annually, beginning at graduation, to accumulate $5 million by the time you retire?

In such cases the answer is found by calculating the present or future worth of the transaction.

Many businesses calculate their present worth each year since the change in their present worth is a measure of the growth of the company. The following example problem will help to demonstrate the concepts just described.

Example Problem 8.2   Listed below are five transactions. Determine their present worth if money is currently valued at 10% annual interest compounded annually. Determine the current net cash equivalent assuming no interest has been withdrawn or paid. Draw a cash-flow diagram for each.

Solution

(a) $1 000 deposited two years ago. (See Fig. 8.3.)
$$F = 1\ 000(1.10)^2 = \$1\ 210.00$$

(b) $2 000 deposited one year ago. (See Fig. 8.4.)
$$F = 2\ 000(1.10)^1 = \$2\ 200.00$$

Figure 8.3

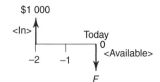

Owner's cash-flow diagram.

Figure 8.4

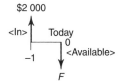

Owner's cash-flow diagram.

Note: For parts (a) and (b) we solve for F to bring the value of money deposited one or two years ago to today's equivalent amount.

 (c) $3 000 to be received one year from now. (See Fig. 8.5.)

$$P = 3\,000(1.10)^{-1} = \$2\,727.27$$

 (d) $4 000 to be paid two years from now (treated as negative for the owner since it must be paid). (See Fig. 8.6.)

$$P = -4000(1.10)^{-2}$$
$$= -3\,305.79$$

Figure 8.5

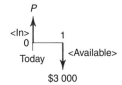

Owner's cash-flow diagram.

Figure 8.6

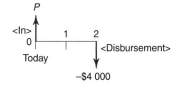

Owner's cash-flow diagram.

Figure 8.7

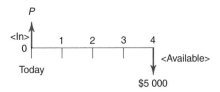

Owner's cash-flow diagram.

(*e*) $5 000 to be received four years from now. (See Fig. 8.7.)

$$P = 5\,000(1.10)^{-4} = \$3\,415.07$$

Present worth of the five transactions (*note:* the values can be added or subtracted because each value has been computed on the *same* date):

$$\$1\,210.00$$
$$2\,200.00$$
$$2\,727.27$$
$$-3\,305.79$$
$$\underline{3\,415.07}$$
$$\text{Present worth} = \$6\,246.55$$

Example Problem 8.3 A company can buy a vacant lot and have a new manufacturing plant constructed on the property. The timing and costs of various components for the factory are given in the cash-flow table below. If annual interest is 8% compounded annually, draw a cash-flow diagram and determine the future worth of the costs incurred when the firm begins production at the end of three years.

Year	Activity	Cost
0	Buy land	$ 75 000
1	Design and initial construction costs	150 000
2	Balance of construction costs	1 150 000
3	Setup production equipment	150 000

Solution See Figure 8.8.
Using Eq. (8.4):

$$F = P(1 + i)^n$$

$$F = \$75\,000(1 + 0.08)^3 + 150\,000(1 + 0.08)^2 + 1\,150\,000(1 + 0.08)^1 + 150\,000$$

$$F = \$94\,478.40 + 174\,960.00 + 1\,242\,000.00 + 150\,000.00$$

$$F = \$1\,661\,438.40$$

In this problem, one must examine the source of the money that is being spent. It is likely from one of two sources: (1) It is money that is borrowed

Figure 8.8

193
Annual Worth
and Gradients

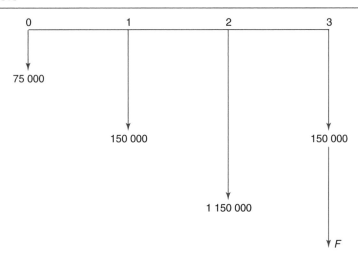

Future worth of the cost of a new manufacturing plant.

from a bank and the company is paying 8% annual interest, or (2) It is money that the company has on hand (profit). If the money is from the second source, then we must explore what other opportunities for investment are being forgone in order to build the new factory.

8.5 Annual Worth and Gradients

With present and future worth analysis, we resolved cash flows into single equivalent cash sums. But we can also state the value of a transaction on an equivalent annual basis.

Annual worth (A) is the worth of monetary transactions that have been converted to an equivalent uniform annual cost or benefit (EUAC or EUAB). An *annuity* involves a series of equal payments at regular intervals. The value of such a series will be developed in the following sections from the idea of compound interest. *Consideration of the point in time at which compounding begins will be of prime importance.*

Several forms of annuities will be discussed later in this chapter:

Annual Future Worth (Sinking Fund)
Annual Present Worth (Installment Loan)
Capitalized Cost (Infinite Life)

In many cases, however, annual payments do not occur in equal-amount payment series. For example, as your car ages, you may expect to pay more each year for automobile maintenance. If these costs increase (or decrease) in equal dollar amounts each year, they are referred to as arithmetic gradients *(G)*; a formula allows easy computation of equivalent present values for gradient series. Geometric gradients occur in cases where a uniform payment increases (or decreases) by a constant percentage. For example, you may expect your salary to increase by 8% per year for the first five years of your career. If you want to find the present worth of this series of cash flows, you may use one of the formulas derived and demonstrated later in this chapter:

Arithmetic Gradients
Geometric Gradients

8.5.1 Annual Future Worth (Sinking Fund)

A *sinking fund* is an annuity that is designed to produce an amount of money at some future time. It might be used to save for an expenditure that you know is going to occur—for instance, a Christmas gift fund or a new car fund. In business, the fund may be used to provide cash needed to replace obsolete equipment or to upgrade software. The cash-flow diagram for the sinking fund is shown in Figure 8.9.

If an amount A is deposited at the *end* of each period and interest is compounded each period at a rate of i, the sum F will be produced after n periods. Please note that the deposit period and the interest compounding period must be *equal* for the equations being developed to be valid.

It can be seen from Figure 8.9 that the last payment will produce no interest, the payment at period $n - 1$ will produce interest equal to A times i, the payment at period $n - 2$ will produce a sum (interest and principal) of $A(1 + i)^2$, and so on. Hence, the sum produced will be as follows:

Deposit at end of period	Interest generated	Sum due to this payment
n	None	$A(1)$
$n - 1$	$A(i)$	$A(1 + i)$
$n - 2$	$A(1 + i)i$	$A(1 + i)^2$
$n - 3$	$A(1 + i)^2 i$	$A(1 + i)^3$

Thus, for four payments

$$F = A(4 + 6i + 4i^2 + i^3)$$

Figure 8.9

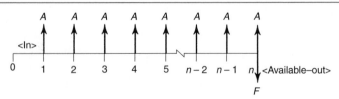

F

Saver's cash-flow diagram.

If we multiply and divide this expression by i, and then add and subtract 1 from the numerator, F becomes

$$F = A\left[\frac{(4i + 6i^2 + 4i^3 + i^4 + 1) - 1}{i}\right]$$

$$= A\left[\frac{(1 + i)^4 - 1}{i}\right]$$

It can be shown that the general term is

$$F = A\left[\frac{(1 + i)^n - 1}{i}\right] \tag{8.6}$$

Therefore the annual future worth equation, Eq. (8.6), should be used if you want to accumulate a future amount F over n periods, an amount A must be deposited at the end of each period at i interest rate compounded at each period.

Example Problem 8.4   How much money would be accumulated by a sinking fund if $90 is deposited at the end of each month for three years with an annual interest rate of 10% compounded monthly?

Solution   (See Fig. 8.10.)

$$F = A\left[\frac{(1 + i)^n - 1}{i}\right]$$

where

$$A = \$90$$
$$i = 0.10/12 \text{ (monthly compounding)}$$
$$n = 12 \times 3 = 36 \text{ (\# of periods)}$$
$$F = 90\left[\frac{(1 + 0.10/12)^{36} - 1}{0.10/12}\right]$$

$$= \$3\,760.36$$

Figure 8.10

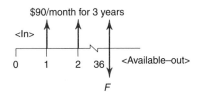

$90/month for 3 years

<In>

0   1   2   36   <Available–out>

F

Saver's cash-flow diagram.

Example Problem 8.5 $10 000 will be needed in eight years to replace a piece of equipment. How much money must be placed annually into a sinking fund that earns 7% interest? Assume the first payment is made today and the last one eight years from today with interest compounded annually.

Solution (See Fig. 8.11.)

$$F = A \left[\frac{(1 + i)^n - 1}{i} \right]$$

$$A = \frac{F(i)}{(1 + i)^n - 1}$$

$$= \frac{(10\,000)(0.07)}{(1.07)^9 - 1}$$

$$= \$834.86$$

Note that $n = 9$ in this example because nine payments will be made. For the sinking-fund formula to be valid, the time of the initial payment ("today," in the problem) must be considered to be the end of the first period.

8.5.2 Annual Present Worth (Installment Loan)

A second and very popular way that annuities are used to retire a debt is by making periodic payments instead of a single large payment at the end of a given time period. This time-payment plan, offered by most retail businesses and lending institutions, is called an *installment loan*. It is also used to amortize (pay off with a sinking-fund approach) bond issues. A cash-flow diagram for this scheme is illustrated in Figure 8.12.

Figure 8.11

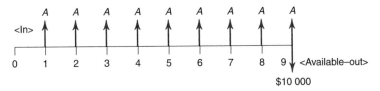

Company's cash-flow diagram.

Figure 8.12

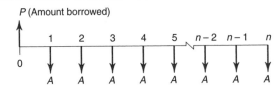

Buyer's cash-flow diagram.

In this case the principal amount P is the size of the debt and A is the amount of the periodic payment that must be made with interest compounded at the end of each period. It can be seen that if P were removed from the time line and F placed at the end of the nth period, the time line would represent a sinking fund. Furthermore, it can be shown that F would also be the value of P placed at compound interest for n periods $[F = P(1 + i)^n]$. Likewise, P can be termed the present worth of the sinking fund that would be accumulated by the deposits. Therefore, since

$$F = P(1 + i)^n \text{ and } F = A\left[\frac{(1 + i)^n - 1}{i}\right]$$

the present worth becomes

$$P = A\left[\frac{(1 + i)^n - 1}{i(1 + i)^n}\right] = A\left[\frac{1 - (1 + i)^{-n}}{i}\right] \tag{8.7}$$

The term within the brackets is known as the present worth of a sinking fund, or the *uniform annual payment present-worth factor*.

It follows that

$$A = P\left[\frac{i(1 + i)^n}{(1 + i)^n - 1}\right] \tag{8.8}$$

where the term in brackets is most commonly called the *capital recovery factor*, or the *uniform annual payment annuity factor*, and is the reciprocal of the uniform annual payment present-worth factor.

A third way use of annuities is when a sum of money is returned in monthly installments at retirement. The formula that applies is Eq. (8.7), and the cash-flow diagram is shown in Figure 8.13.

The problem could be stated as follows: How much money P must be available at retirement so that A dollars can be received for n periods, assuming i interest rate? Equation (8.7) can be solved for the amount of money P that must be accumulated by retirement if an amount A is to be withdrawn for n periods at a given interest rate.

To understand the concept of installment loans and retirement plans, consider the following example problems.

Figure 8.13

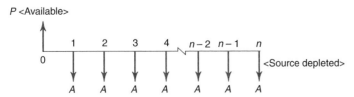

Series of monthly withdrawals.

Example Problem 8.6 Suppose you borrow $1 000 from a bank for one year at 9% annual interest compounded monthly. Consider paying the loan back by two different methods:

1. You keep the $1 000 for one year and pay back the bank at the end of the year in a lump sum. What would you owe? (See Fig. 8.14; note that the time line shown is in years but interest is compounded monthly.)

Solution

$$F = 1\,000(1 + 0.0075)^{12}$$

$$= \$1\,093.81$$

2. The second method is the installment loan. You borrow $1 000 from the bank and repay it in equal monthly payments. What is the amount of each payment? (See Fig. 8.15.)

Solution
From Eq. (8.8):

$$A = P\left[\frac{i(1 + i)^n}{(1 + i)^n - 1}\right]$$

$$= 1\,000\left[\frac{0.0075(1 + 0.0075)^{12}}{(1 + 0.0075)^{12} - 1}\right]$$

$$= \$87.45$$

Figure 8.14

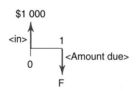

Borrower's cash-flow diagram.

Figure 8.15

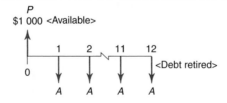

Borrower's cash-flow diagram.

Example Problem 8.7   A used automobile that has a total cost of $15 500 is to be purchased in part by trading in an older vehicle for which $6 250 is allowed. The balance will be financed over three years. If the interest rate is 8.5% per year, compounded monthly, what are the monthly payments? The first installment is to be paid at the end of the first month. (See Fig. 8.16.)

Solution   From Eq. (8.8):

$$A = (15\ 500 - 6\ 250) \left[\frac{\dfrac{(0.085)}{12} \left(1 + \dfrac{0.085}{12}\right)^{36}}{\left(1 + \dfrac{0.085}{12}\right)^{36} - 1} \right]$$

$$= \$292.00$$

Another way of expressing the relationship is by saying that $9 250 (that is, $15 500 − $6 250) is the present worth of 36 monthly payments of $292.00, beginning in one month at 8.5% annual interest compounded monthly.

Example Problem 8.8   Suppose that the auto purchase described in Example Problem 8.7 is modified so that no payment is made until six months after the purchase and then a total of 36 monthly payments are made. Now, what is the amount of each monthly payment? (See Fig. 8.17.)

Figure 8.16

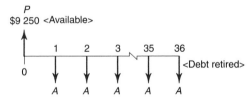

Purchaser's cash-flow diagram.

Figure 8.17

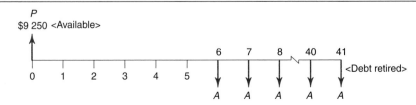

Purchaser's cash-flow diagram.

Solution Balance due after trade-in = $15 500 − $6 250 = $9 250. Balance due after five months = $9 250(1 + 0.085/12)^5 = $9 582.28. Note that the unpaid balance is compounded for only five months because the first payment marks the end of the first period of the annuity. This is called a *deferred annuity*. The monthly payment is then

$$A = 9\,582.28 \left[\frac{\dfrac{(0.085)}{12}\left(1 + \dfrac{0.085}{12}\right)^{36}}{\left(1 + \dfrac{0.085}{12}\right)^{36} - 1} \right]$$

$$= \$302.49$$

Example Problem 8.9 Amy and Kevin are purchasing their first home and have arranged for a mortgage of $200 000 at a fixed annual interest rate of 6.75% compounded monthly for a period of 30 years.

(a) What will be their monthly payment?
(b) At the end of year 10, what amount will be necessary to pay off the loan?
(c) If at the end of the first year they pay an additional sum of $50 000 on the principal, what is the remaining principal?
(d) If they continue with the monthly payments as found in part (a), how many payments will be necessary to retire the debt following the $50 000 payment described in part (c)?

Solution

(a) This problem describes a standard installment loan, so Eq. (8.8) applies. (See Fig. 8.18.)

$$A = P\left[\frac{i(1 + i)^n}{(1 + i)^n - 1} \right]$$

$$= 200\,000 \left[\frac{\dfrac{(0.0675)}{12}\left(1 + \dfrac{0.0675}{12}\right)^{360}}{\left(1 + \dfrac{0.0675}{12}\right)^{360} - 1} \right]$$

$$= \$1\,297.20$$

Figure 8.18

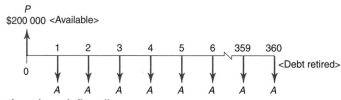

Home purchaser's cash-flow diagram.

Figure 8.19

201
*Annual Worth
and Gradients*

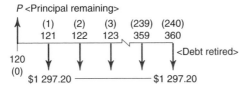

Purchaser's cash-flow diagram for last 240 months.

(b) The amount necessary to pay off the loan at year 10 is the principal remaining. This can be viewed as the present worth of the monthly payments following year 10. (See Fig. 8.19.) From Eq. (8.7):

$$P = A\left[\frac{(1+i)^n - 1}{i(1+i)^n}\right]$$

$$= 1\,297.20\left[\frac{\left(1 + \dfrac{0.0675}{12}\right)^{240} - 1}{\left(\dfrac{0.0675}{12}\right)\left(1 + \dfrac{0.0675}{12}\right)^{240}}\right]$$

$$= \$170\,602.50$$

(c) The principal remaining at the end of the first year can be found following the procedure in part (b). (See Fig. 8.20.)

$$P = 1\,297.20\left[\frac{\left(1 + \dfrac{0.0675}{12}\right)^{348} - 1}{\left(\dfrac{0.0675}{12}\right)\left(1 + \dfrac{0.0675}{12}\right)^{348}}\right]$$

$$= \$197\,869.08$$

After the lump-sum payment of $50 000 the principal remaining is $197 869.08 − $50 000 = $147 869.08.

(d) Equation (8.7) applies here with the present worth of $147 869.08, the monthly payment equal to $1 297.20, and the number of payments unknown. (See Fig. 8.21.)

$$147\,869.08 = 1\,297.20\left[\frac{\left(1 + \dfrac{0.0675}{12}\right)^n - 1}{\left(\dfrac{0.675}{12}\right)\left(1 + \dfrac{0.0675}{12}\right)^n - 1}\right]$$

Figure 8.20

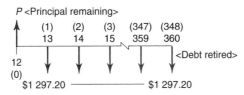

Purchaser's cash-flow diagram for last 29 years.

Figure 8.21

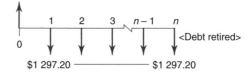

$147 869.08 <Principal remaining>

1   2   3   n − 1   n   <Debt retired>

0

$1 297.20 ——————— $1 297.20

Purchaser's cash-flow diagram.

Thus

$$(0.64120)\left(1 + \frac{0.0675}{12}\right)^n = \left(1 + \frac{0.0675}{12}\right)^n - 1$$

$$\left(1 + \frac{0.0675}{12}\right)^n = 2.7871$$

Taking the logarithm of both sides, we have

$$n \log\left(1 + \frac{0.0675}{12}\right) = \log 2.787\ 1$$

Then

$$n = \frac{\log 2.7871}{\log\left(1 + \frac{0.0675}{12}\right)}$$

$$= 182.73$$

As is often the case, the computed number of periods is not an integer value, meaning that there will be 182 full payments of $1 297.20 followed by a partial payment required to retire the mortgage.

Example Problem 8.10 Suppose following graduation you decide to purchase a vehicle and borrow $15 000 at 8.5% annual interest, compounded monthly, to finance the deal. The agreement requires monthly payments for a period of two years. Use a spreadsheet to calculate your payment schedule and prepare it so that the amount borrowed and the annual interest rate can be changed with simple cell modifications. Include the amount of each payment, the amount of each payment going toward principal and toward interest, the principal remaining after each payment, and a running total of the interest paid.

Solution *Note:* The spreadsheet application shown in this example is Microsoft Excel. Many other applications can be used with small coding changes.

Figure 8.22

203
*Annual Worth
and Gradients*

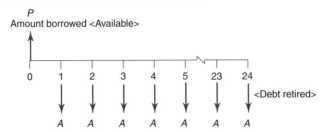

Purchaser's cash-flow diagram.

The cash-flow diagram for you as the purchaser is shown in Figure 8.22. As the problem is described, it fits the installment loan definition.

Table 8.4 shows the cell contents (values and formulas) used to compute the results. Please note the following points:

1. Cell C3 contains the principal amount and can be changed.
2. Cell C4 contains the annual interest rate in percent and can be changed. (Coded as 8.5%; formula view shows as 0.085.)
3. Cell C5 is the loan term in years and can be changed, but will not be changed in this example because the loan term affects the total number of payments and therefore the size of the spreadsheet.
4. A built-in function, PMT, is used in cell C6 to compute the payment based on the monthly interest rate, the number of payments, and the principal amount. Equation (8.7) could have been coded here rather than using the spreadsheet function.
5. Cells B12 through B35 all contain the value computed in C6.
6. The split of the payment into interest and principal is handled by first computing the interest for the first period (cell D12) and then subtracting it from cell B12 to obtain the portion of the payment applied to the principal. Note that cells D12 through D35 depend on the principal at the beginning of the period. The principal at the beginning of the first period is the original amount borrowed.
7. The principal remaining is found by subtracting the portion of the payment going to principal (column C) from the previous principal amount.
8. Finally, the total interest paid is simply a running sum.
9. You should note that this spreadsheet calculates values to greater precision than is displayed and thus there will be places where sums or differences will be off by a penny in the display. This can be prevented by using an option "precision as displayed" which actually changes the data at that point in the spreadsheet to the value displayed. The general result will be that the last payment will then be in error by a few cents.

Table 8.4 Spreadsheet Cell Contents (Values and Formulas)

	A	B	C	D	E	F
1			AUTO FINANCE SCHEDULE			
2						
3	Amount borrowed		15000			
4	Annual interest rate (percent)		0.085			
5	Loan term (years)		2			
6	Monthly payment		=PMT(C4/(12) .C5*12,-C3)			
7						
8			PAYMENT SCHEDULE			
9						
10	Number	Amount paid	Principal	Interest	Principal remaining	Total interest
11	0				=C3	
12	1	=$C$6	=B12-D12	=$C$4/(12)*E11	=E11-C12	=F11+D12
13	2	=$C$6	=B13-D13	=$C$4/(12)*E12	=E12-C13	=F12+D13
14	3	=$C$6	=B14-D14	=$C$4/(12)*E13	=E13-C14	=F13+D14
15	4	=$C$6	=B15-D15	=$C$4/(12)*E14	=E14-C15	=F14+D15
16	5	=$C$6	=B16-D16	=$C$4/(12)*E15	=E15-C16	=F15+D16
17	6	=$C$6	=B17-D17	=$C$4/(12)*E16	=E16-C17	=F16+D17
18	7	=$C$6	=B18-D18	=$C$4/(12)*E17	=E17-C18	=F17+D18
19	8	=$C$6	=B19-D19	=$C$4/(12)*E18	=E18-C19	=F18+D19
20	9	=$C$6	=B20-D20	=$C$4/(12)*E19	=E19-C20	=F19+D20
21	10	=$C$6	=B21-D21	=$C$4/(12)*E20	=E20-C21	=F20+D21
22	11	=$C$6	=B22-D22	=$C$4/(12)*E21	=E21-C22	=F21+D22
23	12	=$C$6	=B23-D23	=$C$4/(12)*E22	=E22-C23	=F22+D23
24	13	=$C$6	=B24-D24	=$C$4/(12)*E23	=E23-C24	=F23+D24
25	14	=$C$6	=B25-D25	=$C$4/(12)*E24	=E24-C25	=F24+D25
26	15	=$C$6	=B26-D26	=$C$4/(12)*E25	=E25-C26	=F25+D26
27	16	=$C$6	=B27-D27	=$C$4/(12)*E26	=E26-C27	=F26+D27
28	17	=$C$6	=B28-D28	=$C$4/(12)*E27	=E27-C28	=F27+D28
29	18	=$C$6	=B29-D29	=$C$4/(12)*E28	=E28-C29	=F28+D29
30	19	=$C$6	=B30-D30	=$C$4/(12)*E29	=E29-C30	=F29+D30
31	20	=$C$6	=B31-D31	=$C$4/(12)*E30	=E30-C31	=F30+D31
32	21	=$C$6	=B32-D32	=$C$4/(12)*E31	=E31-C32	=F31+D32
33	22	=$C$6	=B33-D33	=$C$4/(12)*E32	=E32-C33	=F32+D33
34	23	=$C$6	=B34-D34	=$C$4/(12)*E33	=E33-C34	=F33+D34
35	24	=$C$6	=B35-D35	=$C$4/(12)*E34	=E34-C35	=F34+D35

Table 8.5 shows the resulting values. Once the spreadsheet is prepared, you could readily change the amount borrowed and/or the interest rate to learn how the payment schedule would change.

Example Problem 8.11   Modify the spreadsheet developed in Example Problem 8.10 so that the following series of payments (always greater than or equal to the minimum payment for a two-year loan except for the final one) can be made:

Table 8.5 Spreadsheet Values for Example Problem 8.10

	A	B	C	D	E	F
1			AUTO FINANCE SCHEDULE			
2						
3	Amount borrowed		$15,000			
4	Annual interest rate(percent)		8.50%			
5	Loan term (years)		2			
6	Monthly payment		$681.84			
7						
8			PAYMENT SCHEDULE			
9						
10	Number	Amount paid	Principal	Interest	Principal remaining	Total interest
11	0				$15,000.00	
12	1	$681.84	$575.59	$106.25	$14,424.41	$106.25
13	2	$681.84	$579.66	$102.17	$13,844.75	$208.42
14	3	$681.84	$583.77	$98.07	$13,260.98	$306.49
15	4	$681.84	$587.90	$93.93	$12,673.08	$400.42
16	5	$681.84	$592.07	$89.77	$12,081.01	$490.19
17	6	$681.84	$596.26	$85.57	$11,484.75	$575.76
18	7	$681.84	$600.48	$81.35	$10,884.27	$657.11
19	8	$681.84	$604.74	$77.10	$10,279.53	$734.21
20	9	$681.84	$609.02	$72.81	$9,670.51	$807.02
21	10	$681.84	$613.34	$68.50	$9,057.17	$875.52
22	11	$681.84	$617.68	$64.15	$8,439.49	$939.68
23	12	$681.84	$622.06	$59.78	$7,817.44	$999.46
24	13	$681.84	$626.46	$55.37	$7,190.98	$1,054.83
25	14	$681.84	$630.90	$50.94	$6,560.08	$1,105.77
26	15	$681.84	$635.37	$46.47	$5,924.71	$1,152.23
27	16	$681.84	$639.87	$41.97	$5,284.84	$1,194.20
28	17	$681.84	$644.40	$37.43	$4,640.44	$1,231.64
29	18	$681.84	$648.97	$32.87	$3,991.47	$1,264.51
30	19	$681.84	$653.56	$28.27	$3,337.91	$1,292.78
31	20	$681.84	$658.19	$23.64	$2,679.72	$1,316.42
32	21	$681.84	$662.85	$18.98	$2,016.87	$1,335.40
33	22	$681.84	$667.55	$14.29	$1,349.32	$1,349.69
34	23	$681.84	$672.28	$9.56	$677.04	$1,359.25
35	24	$681.84	$677.04	$4.80	$0.00	$1,364.04

1. $681.84
2. $681.84
3. $700.00
4. $681.84
5. $1 000.00
6. $681.84
7. $1 000.00
8. $1 000.00
9. $1 500.00
10. $900.00
11. $700.00
12. $800.00
13. $1 000.00
14. $681.84
15. $1 000.00
16. $1 500.00
17. $1 000.00
18. $569.25

Assume that any amount paid that is greater than $681.84 will be applied toward the principal and that no prepayment penalty is added.

Solution   Table 8.6 shows the cell contents for this case. Column B is no longer constant and the specific value must be placed in the cells since no pattern of payments is evident. Column E is modified to test whether there is principal remaining. If not, a value of zero is entered in the cell. Finally, Table 8.7 provides the numerical solution to the problem.

Table 8.6   Spreadsheet Cell Contents for Example Problem 8.11

	A	B	C	D	E	F
1			AUTO FINANCE SCHEDULE			
2						
3	Amount borrowed		1500			
4	Annual interest rate (percent)		0.085			
5	Loan term (years)		2			
6	Monthly payment (minimum)		=PMT(C4/(12), C5*12,-C3)			
7						
8			PAYMENT SCHEDULE			
9						
10	Number	Amount paid	Principal	Interest	Principal remaining	Total interest
11	0				=C3	
12	1	681.84	=B12-D12	=$C$4/(12)*E11	=IF(B12>0, E11-C12, 0)	=F11+D12
13	2	681.84	=B13-D13	=$C$4/(12)*E12	=IF(B13>0, E14-C13, 0)	=F12+D13
14	3	700	=B14-D14	=$C$4/(12)*E13	=IF(B14>0, E15-C14, 0)	=F13+D14
15	4	681.84	=B15-D15	=$C$4/(12)*E14	=IF(B15>0, E16-C15, 0)	=F14+D15
16	5	1000	=B16-D16	=$C$4/(12)*E15	=IF(B16>0, E17-C16, 0)	=F15+D16
17	6	681.84	=B17-D17	=$C$4/(12)*E16	=IF(B17>0, E18-C17, 0)	=F16+D17
18	7	1000	=B18-D18	=$C$4/(12)*E17	=IF(B18>0, E19-C18, 0)	=F17+D18
19	8	1000	=B19-D19	=$C$4/(12)*E18	=IF(B19>0, E18-C19, 0)	=F18+D19
20	9	1500	=B20-D20	=$C$4/(12)*E19	=IF(B20>0, E19-C20, 0)	=F19+D20
21	10	900	=B21-D21	=$C$4/(12)*E20	=IF(B21>0, E22-C21, 0)	=F20+D21
22	11	700	=B22-D22	=$C$4/(12)*E21	=IF(B22>0, E23-C22, 0)	=F21+D22
23	12	800	=B23-D23	=$C$4/(12)*E22	=IF(B23>0, E24-C23, 0)	=F22+D23
24	13	1000	=B24-D24	=$C$4/(12)*E23	=IF(B24>0, E25-C24, 0)	=F23+D24
25	14	681.84	=B25-D25	=$C$4/(12)*E24	=IF(B25>0, E26-C25, 0)	=F24+D25
26	15	1000	=B26-D26	=$C$4/(12)*E25	=IF(B26>0, E27-C26, 0)	=F25+D26
27	16	1500	=B27-D27	=$C$4/(12)*E26	=IF(B27>0, E28-C27, 0)	=F26+D27
28	17	1000	=B28-D28	=$C$4/(12)*E27	=IF(B28>0, E29-C28, 0)	=F27+D28
29	18	569.25	=B29-D29	=$C$4/(12)*E28	=IF(B29>0, E28-C29, 0)	=F28+D29
30	19		=B30-D30	=$C$4/(12)*E29	=IF(B30>0, E29-C30, 0)	=F29+D30
31	20		=B31-D31	=$C$4/(12)*E30	=IF(B31>0, E30-C31, 0)	=F30+D31
32	21		=B32-D32	=$C$4/(12)*E31	=IF(B32>0, E31-C32, 0)	=F31+D32
33	22		=B33-D33	=$C$4/(12)*E32	=IF(B33>0, E32-C33, 0)	=F32+D33
34	23		=B34-D34	=$C$4/(12)*E33	=IF(B34>0, E33-C34, 0)	=F33+D34
35	24		=B35-D35	=$C$4/(12)*E34	=IF(B35>0, E34-C35, 0)	=F34+D35

Table 8.7 Spreadsheet Values for Example Problem 8.11

	A	B	C	D	E	F
1			AUTO FINANCE SCHEDULE			
2						
3	Amount borrowed		$15,000			
4	Annual interest rate (percent)		8.50%			
5	Loan term (years)		2			
6	Monthly payment		$681.84			
7						
8			PAYMENT SCHEDULE			
9						
10	Number	Amount paid	Principal	Interest	Principal remaining	Total interest
11	0				$15,000.00	
12	1	$681.84	$575.59	$106.25	$14,424.41	$106.25
13	2	$681.84	$579.66	$102.17	$13,844.75	$208.42
14	3	$700.00	$601.93	$98.07	$13,242.81	$306.49
15	4	$681.84	$588.04	$93.80	$12,654.77	$400.29
16	5	$1000.00	$910.36	$89.64	$11,744.41	$489.93
17	6	$681.84	$598.65	$83.19	$11,145.76	$573.12
18	7	$1000.00	$921.05	$78.95	$10,224.71	$652.07
19	8	$1000.00	$927.57	$72.43	$9,297.13	$724.49
20	9	$1500.00	$1,434.15	$65.85	$7,862.99	$790.35
21	10	$900.00	$844.30	$55.70	$7,018.69	$846.05
22	11	$700.00	$650.28	$49.72	$6,368.40	$895.76
23	12	$800.00	$754.89	$45.11	$5,613.51	$940.87
24	13	$1000.00	$960.24	$39.76	$4,653.27	$980.63
25	14	$681.84	$648.88	$32.96	$4,004.39	$1,013.59
26	15	$1000.00	$971.64	$28.36	$3,032.76	$1,041.96
27	16	$1500.00	$1,478.52	$21.48	$1,554.24	$1,063.44
28	17	$1000.00	$988.99	$11.01	$565.25	$1,074.45
29	18	$569.25	$565.25	$4.00	$0.00	$1,078.45
30	19		$0.00	$0.00	$0.00	$1,078.45
31	20		$0.00	$0.00	$0.00	$1,078.45
32	21		$0.00	$0.00	$0.00	$1,078.45
33	22		$0.00	$0.00	$0.00	$1,078.45
34	23		$0.00	$0.00	$0.00	$1,078.45
35	24		$0.00	$0.00	$0.00	$1,078.45

Example Problem 8.12   Estimate the amount of money you will need to save for retirement. Assume that you want to continue to receive 80% of your current (just prior to retirement) take-home pay ($8 200/month) for 30 years and that you can earn 12% interest on your retirement funds.

Solution   The cash-flow diagram is shown in Figure 8.23.

Eighty percent of $8 200 = $6 560/month

Number of periods = 30 years × 12 months per year = 360 months

Interest (i) = 0.12/12 = 0.01

$$P = 6\,560\left[\frac{1 - (1 + 0.01)^{-360}}{0.01}\right]$$

$P = \$637\,752.23$

Note: This calculation does not include taxes collected by federal and state governments. You should plan on saving an additional 35 to 50% to cover tax payments.

8.5.3 Capitalized Cost (Infinite Life Analysis)

Capitalized cost (CC) refers to the present worth (P) of a project or investment that is assumed to last forever. Public works projects like dams, railroads, and irrigation systems are typical capitalized-cost calculations. Investments requiring perpetual (infinite) payments such as endowed scholarships or trusts also rely on capitalized-cost calculations.

Capitalized cost can be used to extend the retirement planning problem in Example Prob. 8.12. Suppose that you want the monthly amount (A) to continue after your death to be disbursed to your heirs. Begin with Eq. 8.7:

$$P = A\left[\frac{1 - (1 + i)^{-n}}{i}\right]$$

and consider that, as n approaches infinity (∞), the numerator becomes 1, yielding

$$P = \frac{A}{i} \tag{8.9}$$

We can restate the problem by asking what amount of money P must be available at retirement so that an amount A can be withdrawn each month and never affect the principal amount P.

Figure 8.23

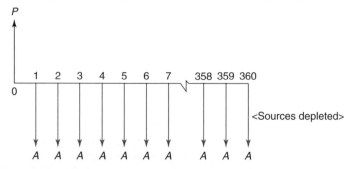

Monthly withdrawals of $6 560 for 360 months.

If we use the compound interest formula, that is, $F = P(1 + i)^n$, and find the amount of interest generated for one month (here that is equal to A), then $F = P + A = P(1 + i)^1$. If we solve this equation for P, then

$$P = \frac{A}{i}$$

Given an interest rate earned by the annuity (say, $i = 0.12/12$) and a fixed monthly income (say, $A = \$6\,560$), $P = \$656\,000$.

Example Problem 8.13   Recalculate Example Problem 8.12 assuming annual interest of 7% and desired perpetual monthly income of $6 560.

Solution (See Fig. 8.24.)

$$P_{CC} = \frac{A}{i} = \frac{6\,560}{(0.07/12)} = \$1\,124\,571.43$$

Example Problem 8.14   A wealthy alum from your institution wishes to provide ten $5 000 scholarships to deserving engineering students starting next year and to continue giving the same number of scholarships and the same dollar amount of scholarship money every year forever. Assuming that 8% interest can be earned annually, how much money would this alum need to turn over to the university today?

Solution Ten scholarships at $5 000 each = $50 000 per year = A

$$P = \frac{A}{i} = \frac{50\,000}{(0.08)} = \$625\,000$$

Note in this problem the $50 000 amount is spent only once a year (not every month as in previous examples), thus the interest rate is the annual 8%.

8.5.4 Arithmetic Gradients

Now consider the case where annual cash flows do not occur in equal amounts for every period. An arithmetic gradient is a cash-flow series which either

---

Figure 8.24

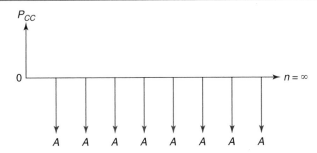

Capitalized cost of a $6 560 monthly income.

Figure 8.25

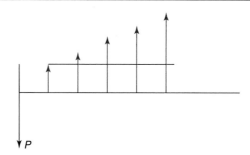

increases or decreases uniformly. That is, income or payments change by the same amount each interest period. The amount of the increase or decrease is the gradient (G). Consider the cash-flow series shown in Figure 8.25.

Cash flows can be resolved into two components with present-worth values of P' and P'' as shown in Figure 8.26.

We already have an equation for P' (Equation 8.7) and we can use the present worth equation to derive an equation for P''. The result is

$$P'' = G\left[\frac{(1 + i)^n - in - 1}{i^2(1 + i)^n}\right] \tag{8.10}$$

So, the overall present worth for the arithmetic gradient is $P = P' + P''$ or

$$P = A\left[\frac{1 - (1 + i)^{-n}}{i}\right] + G\left[\frac{(1 + i)^n - in - 1}{i^2(1 + i)^n}\right]$$

Note in Figure 8.26 that the gradient factor begins in period 2 and P'' is located in period zero. Equation 8.10 takes into account that there are $(n - 1)$ terms containing G.

Figure 8.26

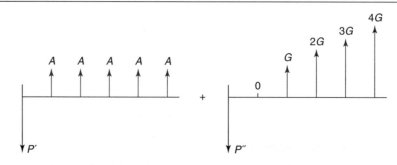

Example Problem 8.15 The annual receipts from operation of a gravel pit are expected to decrease until the pit closes. If next year's receipts are $11 200, and the second year's receipts are $9 800, determine the following:

(a) How many years will it be before the income stream is zero?
(b) Draw a cash-flow diagram to depict the situation.
(c) What is the present worth of the income assuming an annual interest rate of 11%?

Solution

(a) Number of years remaining = [First-year receipts/annual decrease (gradient)] = [$11 200/($11 200 − $9 800)] = 8 years

Thus, there would be zero income in year 9.

(b) See Figure 8.27.

(c) $P = A\left[\dfrac{1 - (1 + i)^{-n}}{i}\right] - G\left[\dfrac{(1 + i)^n - in - 1}{i^2(1 + i)^n}\right]$

$P = 11\ 200\left[\dfrac{1 - (1 + 0.11)^{-8}}{0.11}\right]$

$- 1\ 400\left[\dfrac{(1 + 0.11)^8 - 0.11(8) - 1}{(0.11)^2(1 + 0.11)^8}\right]$

$P = 57\ 636.57 - 21\ 314.50 = \$36\ 323.97$

Notice that the value of n in the gradient factor is 8, not 7. The gradient factor is derived based on $(n - 1)$ terms containing G. In the case, there are seven terms containing G, thus $(n - 1) = 7$, so $n = 8$.

Example Problem 8.16 A manufacturing plant installed a new machining cell. It is expected that initial tooling, adjustments, and repair costs will be high but that the costs will decline for several years. The project costs are shown below:

Year	Costs
1	$2 400
2	$1 800
3	$1 200
4	$600

Figure 8.27

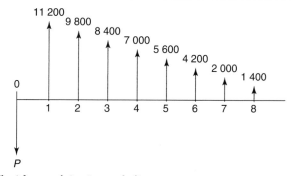

Arithmetic gradient for receipts at gravel pit.

(a) Draw a cash-flow diagram to depict this situation.

(b) What is the present worth of these projected costs if annual interest is 10%?

Solution

(a) See Figure 8.28.

(b) $P = A\left[\dfrac{1 - (1 + i)^{-n}}{i}\right] - G\left[\dfrac{(1 + i)^n - in - 1}{i^2(1 + i)^n}\right]$

$P = 2\,400\left[\dfrac{1 - (1 + 0.10)^{-4}}{0.10}\right] - 600\left[\dfrac{(1 + 0.10)^4 - 0.10(4) - 1}{(0.10)^2(1 + 0.10)^4}\right]$

$P = 7\,607.68 - 2\,626.87 = \$4\,980.81$

8.5.5 Geometric Gradients

Oftentimes, cash flows change by a constant percentage or uniform rate, g, in consecutive payment periods. This type of cash flow is called a geometric gradient series. An example of this is the maintenance costs for an automobile that begin at $150 the first year and are expected to increase at a uniform rate of 10% per year for the next four years. The general cash-flow diagram for geometric gradients is shown in Figure 8.29.

Just as we did for the arithmetic gradient, we can use the present-worth equation to derive two equations to find the present worth of this unique series of annual payments. Care must be taken to apply the appropriate formula based on whether the interest rate is equal (or not equal) to the annual rate of increase of the gradient (g).

$$P = A\left[\frac{1 - (1 + g)^n(1 + i)^{-n}}{i - g}\right] \text{ where } i \neq g \qquad (8.11)$$

$$P = A[n(1 + i)^{-1}] \text{ where } i = g \qquad (8.12)$$

Figure 8.28

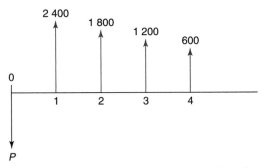

Arithmetic gradient for machining-cell tooling, adjustment, and repair costs.

Figure 8.29

213
*Annual Worth
and Gradients*

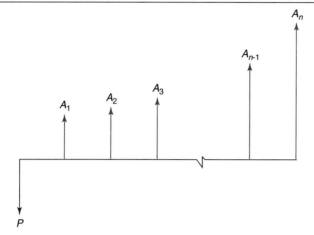

Geometric gradient cash flow.

Example Problem 8.17 The maintenance for an automobile is estimated to be $150 in the first year and is expected to increase at a uniform rate of 10% per year. What is the present worth of the cost of the first five years of maintenance if an 8% annual interest rate is assumed?

Solution See Figure 8.30 for the cash-flow diagram.
See Table 8.8 for a year-by-year calculation of the present value (P).

Figure 8.30

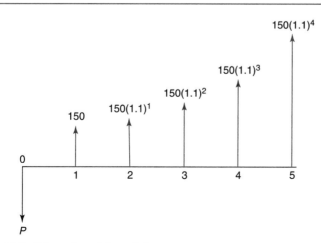

Geometric gradient for automobile maintenance.

Table 8.8 Automobile Maintenance Costs (Geometric Gradient)

Year		Cash Flow, $	Convert to P	P, $ (Maintenance)
1	$150	150.00	$150(1 + 0.08)^{-1}$	138.89
2	$150.00 + 10\%(150.00) = 150(1 + 0.10)^1$	165.00	$165(1 + 0.08)^{-2}$	141.46
3	$165.00 + 10\%(165.00) = 150(1 + 0.10)^2$	181.50	$181.50(1.08)^{-3}$	144.08
4	$181.50 + 10\%(181.50) = 150(1 + 1.10)^3$	199.65	$199.65(1.08)^{-4}$	146.75
5	$199.65 + 10\%(199.65) = 150(1 + 0.10)^4$	219.62	$219.62(1.08)^{-5}$	149.47
				720.65

$$P = A\left[\frac{1 - (1 + g)^n(1 + i)^{-n}}{i - g}\right]$$

$$P = 150\left[\frac{1 - (1 + 0.10)^5(1 + 0.08)^{-5}}{0.08 - 0.10}\right]$$

$$P = 150(4.8043) = \$720.65$$

Example Problem 8.18 Recalculate Example Problem 8.17 using a 10% annual interest rate.

Solution (See Fig. 8.30.) *Note:* The cash-flow diagram is identical to the one for Example Problem 8.17; only the interest rate has changed.

Using Eq. (8.12):

$$P = A[n(1 + i)^{-1}]$$
$$P = 150[5(1 + 0.10)^{-1}]$$
$$P = \$681.82$$

Example Problem 8.19 The utility bill for a small paper recycling center is expected to increase by $528 per year. If the utility cost in year 1 was $4 000, what is the equivalent uniform annual worth through year 8 if the interest rate is 15% per year?

Solution (See Fig. 8.31.)
Since we are to solve for the annual worth, we can first find the present worth of the gradient:

$$P = G\left[\frac{(1 + i)^n - in - 1}{i^2(1 + i)n}\right]$$

$$P = 528\left[\frac{(1 + 0.15)^8 - (0.15)(8) - 1}{(0.15)^2(1 + 0.15)^8}\right]$$

$$P = 528(12.4807) = \$6\ 589.82$$

Figure 8.31

215

Summary Table

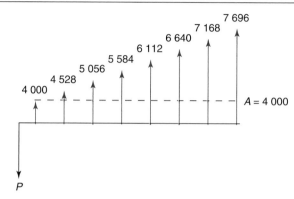

Arithmetic gradient cash flow for utility bill.

Then we can find the annual equivalent of this present worth:

$$A = 6\,589.82\left[\frac{0.15(1.15)^8}{(1.15)^8 - 1}\right]$$

$$A = \$1\,468.54$$

Finally, we add the initial $4\,000 annual cost to the value just calculated to find the equivalent annual worth:

$$A = 4\,000 + 1\,468.54 = \$5\,468.54$$

Thus, for the eight years we are interested in, the variable cost of utilities can be resolved into an equivalent uniform annual cost (EUAC = \$5\,468.54).

8.6   Summary Table

Table 8.9 summarizes the equations used in this chapter.

Problems

8.1   Determine the difference in interest earned on $5\,000 for 15 years at 6% simple interest to that earned when the interest is compounded annually.

8.2   What is the present worth of $1\,000 payable in five years, if money is thought to be worth (*a*) 5%, (*b*) 15%, (*c*) 25%?

8.3   How much must be invested now to grow to $30\,000 in seven years if the annual interest rate is 8.0% compounded (*a*) annually, (*b*) semiannually, (*c*) monthly?

8.4   Compute the unknown values for each of the following cash-flow diagrams (time shown in years and interest compounded annually).

(*a*) Figure 8.32   (*b*) Figure 8.33

(*c*) Figure 8.34   (*d*) Figure 8.35

Table 8.9 Summary Table

Find/Given	Sample Cash-Flow Diagram	Formula Name/Eq. No.	Formula
To find F Given P		Compound Amount Eq. (8.2)	$F = P(1 + i)^n$
To find P Given F		Present Worth Eq. (8.3)	$P = F(1 + i)^{-n}$
To find F Given A		Future Compound Amount Eq. (8.6)	$F = A\left[\dfrac{(1 + i)^n - 1}{i}\right]$
To find A Given F		Sinking Fund Eq. (8.5)	$A = F\left[\dfrac{i}{(1 + i)^n - 1}\right]$
To find A Given P		Capital Recovery Eq. (8.8)	$A = P\left[\dfrac{i(1 + i)^n}{(1 + i)^n - 1}\right]$
To find P Given A		Present Compound Amount Eq. (8.7)	$P = A\left[\dfrac{(1 + i)^n - 1}{i(1 + i)^n}\right]$
To find P Given G (Arithmetic)		Arithmetic Gradient Present Worth Eq. (8.10)	$P = G\left[\dfrac{(1 + i)^n - in - 1}{i^2(1 + i)^n}\right]$
To find P Given G (Geometric)		Geometric Gradient Present Worth Eq. (8.11) when $i \neq g$ Eq. (8.12) when $i = g$	$P = A\left[\dfrac{1 - (1 + g)^n(1 + i)^{-n}}{i - g}\right]$ $P = A\left[n(1 + i)^{-1}\right]$

8.5 Your real estate taxes are $3 600 per year with one-half due April 1 and the remainder due October 1 each year. What single sum of money must you place in an account earning 5.25% interest 15 months before the first payment is due in order to accumulate enough money to pay each tax bill? Assume monthly compounding.

Figure 8.32

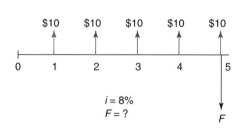

Figure 8.33

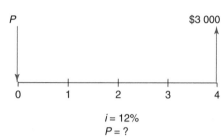

Figure 8.34

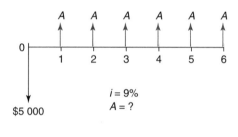

Figure 8.35

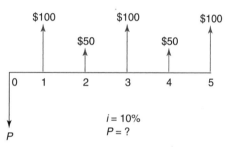

8.6 Your hometown has been given $1 000 000 from the estate of a citizen. The gift stipulates that the money cannot be used for five full years but must be invested. If the money is invested at 6.7% annual interest, how much will be available in five years if it is compounded (*a*) annually, (*b*) semiannually, (*c*) daily?

8.7 If the interest rate is 6.8% per year, how long will it take for an investment to double in value with semiannual, monthly, and daily compounding? Is the time to double a linear function of the compounding period?

8.8 You have just made an investment that will repay $976.12 at the end of each month; the first payment is one month from today and the last one is six years from today (thereby depleting the account).
  (*a*) If the interest rate is 6.5% per year compounded monthly, what amount did you invest?
  (*b*) If you can refuse all the payments and allow the money to earn at the same rate as stated, how much will you have at the end of six years?

8.9 You have made an investment that will yield $5 000 exactly 10 years from today. If the current interest rate is 6.2% compounded quarterly, what is your investment worth today?

8.10 Your parents have agreed to lend you $30 000 to help with your college expenses. They do not expect you to repay the loan or any interest until you have finished school (five years) and worked for 10 years. At the end of 15 years, they will require a lump-sum payment of $60 000.

(*a*) What equivalent annual interest rate are your parents charging for the use of their money?

(*b*) If your parents cannot lend you money and you have to borrow $30 000 at 6.6% annual interest, what would the lump-sum payment be at the end of 15 years?

8.11 You just borrowed $2 500 and have agreed to repay the bank $560 at the end of each of the next five years. What is the annual interest rate of the loan?

8.12 A firm purchased some equipment at a price of $50 000. The equipment resulted in an annual net savings of $2 000 per year during the 10 years it was used. At the end of 10 years, the equipment was sold for $40 000. Draw a cash-flow diagram that depicts this situation. Assuming $i = 8\%$, what was the equivalent cost to the company of this transaction on the purchase date?

8.13 Today you have a savings account of $15 620. Based on an annual interest rate of 4.2%, what equal amount can you withdraw from the account at the end of each month for three years and leave $4 000 in the account? If you could earn 6.0% interest, what would be the value of your monthly withdrawals?

8.14 On March 1 of this year you borrowed $75 000 toward materials for a product you hope to have on the market in November of this year. You have agreed upon an annual interest rate of 8.2% compounded monthly. You also have agreed to begin repaying the debt on December 1, making equal monthly payments until the loan is repaid on April 1 next year. What are your monthly payments?

8.15 Referring to Problem 8.14, suppose that you are unable to make the monthly payments and your creditor agrees to allow you to make a single payment on April 1 next year. How much will you owe at that time?

8.16 What uniform annual payment is equivalent to the following payment schedule if the interest rate is 7.5 percent, compounded annually?

(*a*) $600 at the end of the first year

(*b*) $800 at the end of the second year

(*c*) $1 200 at the end of the third year

(*d*) $2 000 at the end of the fourth year

(*e*) $2 400 at the end of the fifth year

8.17 If sales at your company are doubling every five years, what is the annual rate of increase? What annual rate of increase would be necessary for sales to double in four years? Three years?

8.18 You have $200 000 to invest, and you have decided to purchase bonds that will mature in six years. You have narrowed your choices to two types of bonds. The first class pays 8.75% annual interest. The second class pays 5.4% annual interest but is tax-free, both federal and state. Your income bracket is such that your highest income tax rate is 31% federal and the state income tax is 9%. Which is the better investment for you at this time? Assume that all conditions remain unchanged for the six-year period.

8.19 The average age of engineering students when they graduate is a little over 23 years. Let's assume that the working career of most engineers is exactly 40 years (retiring at 63). How much would an engineer need to save each month to have $3 million saved by the end of his or her working career? Assume $i = 7.0\%$ compounded monthly.

8.20 Assume that when you graduate you will owe a total of $29 500 in student loans. Assume that the interest rate is 7.5%, compounded monthly, and that the entire amount must be repaid within 10 years. Draw the cash-flow diagram that describes this situation. Determine what your minimum monthly payment will be.

8.21 On your 23rd birthday you open a 401K account (retirement account). At that time and on each succeeding birthday up to and including your 60th birthday, you deposit $2 000. During this period the interest rate on the account remains constant at 6.8%. No further payments are made, and beginning one month after your 65th birthday you begin withdrawing equal payments (the annual interest rate is the same as before). How much will you withdraw each month if the account is to be depleted with the last check on your 85th birthday? Assume annual compounding up to your 65th birthday and monthly compounding thereafter.

8.22 You wish to retire at age 66 and at the end of each month thereafter, for 30 years, to receive $5 000. Assume that you begin making monthly payments into an account at age 24. You continue these payments until age 66. If the interest rate is constant at 7.5%, how much must you deposit monthly between ages 24 and 66?

8.23 Your aunt has decided to give most of her wealth to charity and to retain for herself only enough money to provide for her living. She feels that $4 000 per month will provide for her needs. She will establish a trust fund at a bank that will pay 6% interest, compounded monthly. She has also arranged that upon her death, the balance in the account is to be paid to you. If she opens the trust fund and deposits enough money to withdraw her $4 000 a month forever, how much will you receive when your aunt dies?

8.24 You have borrowed as follows:

January 1, 2008	$52 000
July 1, 2009	$35 000
January 1, 2010	$30 000

The agreed-upon annual interest rate was 7.50% compounded semiannually. How much did you owe on July 1, 2010?
You agreed to make the first of 15 monthly payments on October 1, 2010. Assume the interest rate was still 7.50% but was compounded monthly. How much was each of the 15 payments?

8.25 Compute the unknown values for each of the following cash-flow diagrams (time shown in years and interest compounded annually).
(*a*) Figure 8.36 (*b*) Figure 8.37
(*c*) Figure 8.38 (*d*) Figure 8.39

Figure 8.36

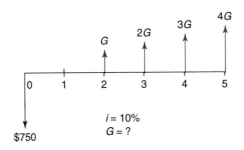

Figure 8.37

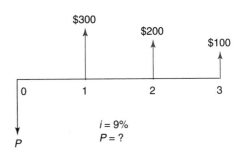

Figure 8.38

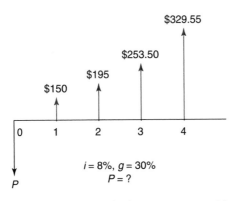

Figure 8.39

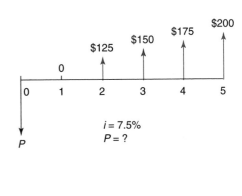

8.26 You have reached an agreement with an auto dealer regarding a new car. She has offered you a trade-in allowance of $8 500 on your old car for a new one that she has "reluctantly" reduced to only $19 995 (before trade-in). She has further agreed upon a contract that requires you to pay $315.22 each of the next 48 months, beginning one month from today. What is your interest rate, expressed as an annual percentage? Give your answer to the nearest 0.01%. With your instructor's approval, write a computer program or use a spreadsheet to solve the problem.

8.27 You have been assigned the task of estimating the annual cost of operating and maintaining a new assembly line in your plant. Your calculations indicate that during the first four years the cost will be $400 000 per year; the next five years will cost $520 000 per year; and the following six years will cost $600 000 per year. If the interest rate is constant at 8.2% over the next 15 years, what will be the equivalent uniform annual cost (EUAC) of operation and maintenance?

8.28 You are buying a new home for $385 000. You have an agreement with the savings and loan company to borrow the needed money if you pay 15% in cash and monthly payments for 30 years at an interest rate of 6.25% compounded monthly.

(a) What monthly payments will be required?

(b) How much principal reduction will occur in the first payment?

(c) Prepare a spreadsheet that will show each payment, how much of each will go to principal and how much to interest, the current balance, and the cumulative interest paid.

(d) Repeat steps (a), (b), and (c) for interest rates of 6.50, 6.75, 7.00, 7.25, and 7.50%. Work as a team if approved by your instructor.

8.29 If you had started a savings account that paid 4.5%, compounded monthly, and your payments into the account were the monthly payments for 6.25% interest as in Problem 8.28, how long would you have had to make payments in order to purchase the home for cash? (Assume the same down-payment amount was available as in Problem 8.28.)

8.30 Your bank pays 2.25% on Christmas Club accounts. How much must you put into an account weekly beginning on January 2 in order to accumulate $1 000 on December 4? Assume weekly compounding and a non-leap year.

8.31 You can purchase a treasury note today for 94.2% of its face value of $20 000. Every six months you will receive an interest payment at the annual rate of 4.88% of face value. You can then invest your interest payments at the annual rate of 5.0% compounded semiannually. If the note matures six years from today, how much money will you receive from all the investments? Express this also as an annual rate of return.

8.32 What is the present worth of each of the following assets and liabilities and your net present worth?
(a) You deposited $2 000 exactly four years ago.
(b) You have a checking account with a current balance of $1 427.22.
(c) You must pay $3 500 exactly four years from now.
(d) Today you just made the 29th of 36 monthly payments of $142.60.
(e) You will receive $12 000 exactly six years from now.
Assume all annual interest rates are 5.5%. Assume monthly compounding in figuring part (d) and annual compounding for the rest.

8.33 Smith Fabrication has estimated that the purchase of a milling machine costing $160 000 will reduce the firm's fabrication expenses by $13 000 per month during a two-year period. If the milling machine has zero salvage value in two years, what is the firm's expected annual rate of return on investment? (*Rate of return* is the equivalent interest rate that must be earned on the investment to produce the same income as the proposed project. What would be the rate of return if the salvage value for 50% of the purchase price? If approved by your instructor, write a computer program or use a spreadsheet to solve the problem. Repeat for expense reductions of $13 500, $14 000, $14 500 $15 000 and $15 500 per month.

8.34 Engineers at Specialty Manufacturing are writing a justification report to support the purchase of a DNC milling center (mill, controller, microcomputer, installation, etc.). They have learned that the total initial cost will be $90 000. The labor savings and improved product quality will result in an estimated benefit to the company of $2 400 each month over a 10-year time period. If the salvage value of the center is about $12 000 in 10 years, what annual rate of return (annual percentage) on investment did the engineers calculate? Write a computer program or use a spreadsheet if assigned by your instructor. Repeat for estimated benefits of $2 600, $2 800, $3 000, $3 200 and $3 400 per month.

8.35 In payment for engineering services rendered, you have been offered the choice of (a) a lump sum payment of $8 000 to be paid five years from now, or (b) five yearly payments of $1 000 that begin one year from now and increase by $300 per year. Draw the cash-flow diagrams for parts (a) and (b). If $i = 10\%$ compounded annually, which option should you select?

8.36 What present expenditure is warranted for business that is expected to produce a savings of $8 000 per year that will decrease by $800 per year for nine years with an interest rate of 10%? Draw a cash-flow diagram that depicts this situation.

8.37 You have decided to invest a fixed percentage of your salary in the stock market at the end of each year. This year (today) you will invest $3 000. For the next five years, you expect that your salary will increase at an 8% annual rate and you will increase your savings at 8% per year. Thus, there will be a total of six investments ($3 000 today plus five more).

(a) What is the present worth of your investment if the stock market yields a 15% annual rate?

(b) What is the present worth of your investment if the market yields only 8%?

8.38 Many new engineering graduates purchase and finance new cars. Automobiles are typically financed for four years with monthly payments made to the lending agency. Assume you will need to borrow $18 000 with 48 monthly payments at 6.5% annual interest.

(a) Write a computer program or prepare a spreadsheet to produce the mortgage table below:

Payment number	Monthly payment	Amount to principal	Amount to interest	Cumulative interest	Current balance
1	$xxx.xx	$xxx.xx	$xxx.xx	$xxx.xx	$xxx.xx
2	$xxx.xx	$xxx.xx	$xxx.xx	$xxx.xx	$xxx.xx
3	.	.	.	.	.
.	.	.	.	.	.
.	.	.	.	.	.
.	.	.	.	.	.

(b) If you decided to pay the loan off at the end of 10 months, what amount is needed? At the end of 20 months? At the end of 40 months?

(c) What is the cumulative interest paid in the first 12 payments? Second 12? Third 12? Last 12?

(d) Repeat parts (a), (b), and (c) assuming the interest rate is 8.5% instead of 6.5%. The amount borrowed remains the same.

(e) Repeat parts (a), (b), and (c) assuming you find it necessary to borrow $20 000. The interest rate is still 6.5%.

(f) What is the result if you borrow $20 000 and the interest rate is 8.5%?

8.39 Many of you will eventually purchase a house. Few will have the total cash on hand, so it will be necessary to borrow money from a home loan agency. Often you can borrow the money at a fixed annual interest for, say, 15 or 30 years. Monthly payments are made to the lending agency. Write a computer program or use a spreadsheet to prepare mortgage tables similar to the one described in Problem 8.38 for the following situations:

(a) $80 000 at 7% interest for 15 years

(b) $80 000 at 6.25% interest for 15 years

(c) $200 000 at 7% interest for 15 years

(d) $200 000 at 7% interest for 30 years

(e) Other cases as may be assigned

(f) Critically examine the monthly payments and the cumulative interest amount produced by changing from a 30-year loan to a 15-year loan, all other parameters being constant.

CHAPTER 9

Economics: Decision Making

Chapter Objectives

When you complete your study of this chapter, you will able to:

■ Make considered economical decisions based on comparing alternatives

9.1 Economic Decision Making

An engineer makes use of engineering economy principals in a very practical way. They are used to analyze a situation so that an intelligent decision can be made. Normally, several alternatives are available, each having some strong attributes. The task is to compare each alternative and to select the one that appears superior, all things considered.

The most obvious method of comparing costs is to determine the total cost of each alternative. An immediate problem arises in that the various costs occur at different times, so the *total number of dollars spent is not a valid method of comparison*. You have seen that the present worth of an expenditure can be calculated. If this is done for all costs, the present worth of buying, operating, and maintaining two or more alternatives can then be compared. Simply stated, the present worth is the sum of money needed now to buy, maintain, and operate a facility for a given interest rate. The alternatives must obviously be compared for the same length of time, and replacements due to short-life expectancies must be considered.

A second method, preferred by those who work with annual budgets, is to calculate the *equivalent uniform annual cost* of each. The approach is similar to the present worth method, but the numerical value is in essence the annual contribution to a sinking fund that would produce a sum identical to the present worth placed at compound interest.

Many investors approach decisions on the basis of the profit that a venture will produce in terms of percent per year. The purchase of a piece of equipment, a parcel of land, or a new product line is thus viewed favorably only if it appears that it will produce an annual profit greater than the money could earn if invested elsewhere. The acceptable return fluctuates with the money market. Since there is doubt about the amount of the profit, and certainly there is a chance of a loss, it would not be prudent to proceed if the prediction of return was not considerably above "safe" investments such as bonds.

The example that follows illustrates the use of these two methods (present worth and average annual cost) and includes a third technique called *future worth* that provides a check.

Each method compares money at the same point in time or over the same time period. Each method is different yet each yields the same conclusion.

Example Problem 9.1   Consider the purchase of two computer-aided design (CAD) systems. Assume the annual interest rate is 12%.

	System 1	System 2
Initial cost	100 000	65 000
Maintenance & operating cost	4 000/year	8 000/year
Salvage	18 000 after 5 years	5 000 after 5 years

Using each of the three methods below, compare the two CAD systems and offer a recommendation:

1. Annual cost
2. Present worth
3. Future worth

Solution
1. Annual cost
(See Fig. 9.1.)

System 1	System 2
(a) Initial cost	(a) Initial cost
$A = P\left[\dfrac{i(1 + i)^n}{(1 + i)^n - 1}\right]$	$A = P\left[\dfrac{i(1 + i)^n}{(1 + i)^n - 1}\right]$
$P = 100\ 000$	$P = 65\ 000$
$i = 0.12$	$i = 0.12$
$n = 5$	$n = 5$
$A = \$27\ 740.97/\text{year}$	$A = \$18\ 031.63/\text{year}$
(b) Maintenance and operating costs $MC = 4\ 000/\text{year}$	(b) Maintenance and operating costs $MC = 8\ 000/\text{year}$

Figure 9.1

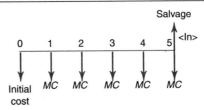

Company's cash-flow diagram (*MC* = *maintenance cost*).

(*c*) Salvage

$$F = A \left[\frac{(1 + i)^n - 1}{i} \right]$$

$$A = \left[\frac{F(i)}{(1 + i)^n - 1} \right]$$

$F = 18\ 000$
$i = 0.12$
$n = 5$
$A = \$(-)2\ 833.38/\text{year}$

(*c*) Salvage

$$F = A \left[\frac{(1 + i)^n - 1}{i} \right]$$

$$A = \left[\frac{F(i)}{(1 + i)^n - 1} \right]$$

$F = 5\ 000$
$i = 0.12$
$n = 5$
$A = \$(-)787.05/\text{year}$

System 1 (annual-cost analysis)	System 2 (annual-cost analysis)
+27 740.97	+18 031.63
+4 000.00	+8 000.00
(−)2 833.38	(−) 787.05
$28 907.59	$25 244.58

Conclusion: System 2 is less expensive.

2. Present worth

System 1	System 2

(*a*) Initial cost = $100 000

(*a*) Initial cost = $65 000

(*b*) Maintenance and operating costs

$$P = A \left[\frac{(1 + i)^n - 1}{i(1 + i)^n} \right]$$

$A = \$4\ 000$
$i = 0.12$
$n = 5$
$P = \$14\ 419.11$

(*b*) Maintenance and operating costs

$$P = A \left[\frac{(1 + i)^n - 1}{i(1 + i)^n} \right]$$

$A = \$8\ 000$
$i = 0.12$
$n = 5$
$P = \$28\ 838.21$

(*c*) Salvage

$P = F(1 + i)^{-n}$
$F = 18\ 000$
$i = 0.12$
$n = 5$
$P = \$(-)10\ 213.68$

(*c*) Salvage

$P = F(1 + i)^{-n}$
$F = 5\ 000$
$i = 0.12$
$n = 5$
$P = \$(-)2\ 837.13$

System 1 (present-worth analysis)	System 2 (present-worth analysis)
$100 000.00	$ 65 000.00
14 419.11	28 838.21
(−)10 213.68	(−)2 837.13
$104 205.43	$ 91 001.08

Conclusion: System 2 is less expensive.

3. Future worth

System 1	System 2

(a) Initial cost = $100 000

$$F = P(1 + i)^n$$

$P = \$100\ 000$

$i = 0.12$

$n = 5$

$F = \$176\ 234.17$

(a) Initial cost = $65 000

$$F = P(1 + i)^n$$

$P = \$65\ 000$

$i = 0.12$

$n = 5$

$F = \$114\ 552.21$

(b) Maintenance and operating costs

$$F = A\left[\frac{(1 + i)^n - 1}{i}\right]$$

$A = \$4\ 000$

$F = \$25\ 411.39$

(b) Maintenance and operating costs

$$F = A\left[\frac{(1 + i)^n - 1}{i}\right]$$

$A = \$8\ 000$

$F = \$50\ 822.78$

(c) Salvage = $(−)18 000

(c) Salvage = $(−)5 000

System 1 (future-cost analysis)	System 2 (future-cost analysis)
$176 234.17	$114,552.21
25 411.39	50 822.78
(−)18 000.00	(−)5 000.00
$183 645.56	$160 374.99

Conclusion: System 2 is less expensive.

Example Problem 9.2   A major potentiometer manufacturer is considering two alternatives for new production machines with capacity to produce 20 000 units per day. One alternative is for a high-capacity automated production machine capable of producing 20 000 units per day when operated for three shifts per day. A quarter-time employee would be assigned to monitor the machine (employee would monitor other machines at the same time). With the three-shift schedule this would be equivalent to a three-quarter-time employee.

A second alternative would be to use two manually operated machines, each capable of 10 000 units per day assuming three-shift operation. Here, a total of six employees (2 per shift, 3 shifts) would be needed.
The following data have been estimated:

	Alternative 1	Alternative 2
Cost to purchase	$500 000	$100 000
Number of machines required	1	2
Number of employees required	0.75	6
Expected life of machine	10 yr	10 yr
Interest rate	8%	8%
Annual maintenance cost per machine	$30 000	$10 000
Salvage value at 10 years per machine	$100 000	$20 000

Figure 9.2

227
*Economic
Decision Making*

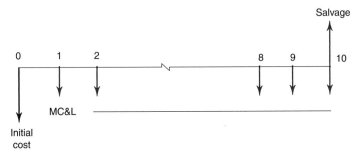

Company's cash-flow diagram
(MC&L = maintenance cost and labor)

If labor costs (including wages, benefits, etc.) are \$40 000 per employee per year, recommend which alternative is best using the equivalent uniform annual cost method.

Solution: Refer to Fig. 9.2

Alternative 1	**Alternative 2**

(a) Initial cost (annualized)

$$A = P\left[\frac{i(1 + i)^n}{(1 + i)^n - 1}\right]$$

$P = \$500\ 000$
$i = 0.08$
$n = 10$
$A = \$74\ 514.74$

(b) Maintenance \$30 000

(c) Labor costs \$30 000

(d) Salvage value

$$A = \frac{F(i)}{[(1 + i)^n - 1]}$$

$F = \$100\ 000$
$i = 0.08$
$n = 10$
$A = -6\ 902.95$

(a) Initial cost (annualized)

$$A = P\left[\frac{i(1 + i)^n}{(1 + i)^n - 1}\right]$$

$P = \$200\ 000$
$i = 0.08$
$n = 10$
$A = \$29\ 805.90$

(b) Maintenance \$20 000

(c) Labor costs \$240 000

(d) Salvage value

$$A = \frac{F(i)}{[(1 + i)^n - 1]}$$

$F = \$40\ 000$
$i = 0.08$
$n = 10$
$A = -2\ 761.18$

Alternative 1 (EUAC)	Alternative 2 (EUAC)
$74 514.74	$29 805.90
$30 000	$20 000
$30 000	$240 000
−$6 902.95	−$2 761.18
$127 611.79	$287 044.72

Clearly alternative 1 results in the lowest EUAC and would be recommended for implementation. The labor costs dominated all other costs in this example. If the EUAC for the two alternatives differed by only a few thousand dollars, and remembering that the numbers in the analysis are necessarily estimates, then other factors would have to be included in order to make a final recommendation.

9.2 Depreciation and Taxes

As demonstrated in Example Problem 9.1, economic decision analysis is concerned with judging the economic desirability of alternative investment proposals and policies. The desirability of a venture is measured in terms of the difference between income and costs, receipts and disbursements, or some other measure of profit. One additional topic we have not yet discussed is taxes. Income taxes represent additional costs and are therefore levies on a company's profit. In fact, businesses can expect to pay federal, state, and local governments 35–50% of the company's net income.

The U.S. government taxes individuals and businesses to support its processes, such as lawmaking, domestic and foreign economic policymaking, infrastructure (roads, dams, etc.), even the making and issuing of money itself.

The tax codes are a complex set of rules that outline appropriate deductions, calculations, and acceptable depreciation. Depreciation is defined as a reduction in value of a property such as a machine, building, or a vehicle because, with the passage of time, the value of most physical property suffers a reduction. Depreciation, while not a cash-flow item itself, results in a positive cash flow (savings) of income tax payments by decreasing the net income on which taxes are based.

For our purposes in this chapter, we define taxes as simply another disbursement similar to operating costs, maintenance, labor and materials, etc. While income taxes are important to the bottom line of a company, the details go beyond the scope of this text. Further study in this area is advised, as the after-tax consequences of an economic decision are critical.

Problems

9.1 Your company is trying to reduce energy costs for one of its warehouses by improving its insulation. Two options are being considered and it is up to you to recommend either urethane foam or fiberglass insulation. Use a 12-year analysis period and an interest rate of 14%. The initial cost of the foam will be $37 000 and it will have to be painted every three years at a cost of $3 000. The energy savings is expected to be $6 500 per year. Alternatively, fiberglass batts can be installed for $14 000 with no maintenance costs. Fiberglass batts will likely save $2 600 per year in energy costs.

9.2 Two machines are being considered to do a certain task. Machine A costs $24 000 new and $2 600 to operate and maintain each year. Machine B costs $32 000 new and $1 200 to operate and maintain each year. Assume that both will be worthless after eight years and that the interest rate is 10.0%. Determine by the equivalent uniform annual cost method which alternative is the better buy.

9.3 Two workstations are being considered by your company. Workstation 1 costs $12 000 new and $1 300 to operate and maintain each year. Workstation 2 costs $15 000 new and $600 to operate and maintain each year. Assume both will be worthless after six years and that the interest rate is 9.0%. Determine by the equivalent uniform annual cost method which alternative is the better buy.

9.4 Assume you needed $10 000 on April 1, 2006, and two options were available:

(a) Your banker would lend you the money at an annual interest rate of 7.0%, compounded monthly, to be repaid on September 1, 2006.

(b) You could cash in a certificate of deposit (CD) that was purchased earlier. The cost of the CD purchased September 1, 2005, was $10 000. If left in the savings and loan company until September 1, 2006, the CD's annual interest is 3.8% compounded monthly. If the CD is cashed in before September 1, 2006, you lose all interest for the first three months and the interest rate is reduced to 1.9%, compounded monthly, after the first three months.

Which option is better and by how much? (Assume an annual rate of 3.6%, compounded monthly, for any funds for which an interest rate is not specified.)

9.5 Two machines are being considered for purchase. The Sande 10 costs $36 000 new and is estimated to last five years. The cost to replace the Sande 10 will increase by 4% each year. Annual operation and maintenance costs are $2 400. It will have a trade-in (salvage) value of $3 000. The Sande 20 costs $76 000 to buy, but will last 10 years and will have a trade-in (salvage) value of $4 000. The cost of operation and maintenance is $1 400 per year.

Compare the two machines and state the basis of your comparison. Include a cash-flow diagram for each alternative. Assume all interest rates at 6% per year unless otherwise stated.

9.6 Two systems are being considered for the same task. System 1 costs $63 000 new and is estimated to last four years. It will then have a salvage or trade-in value of $4 500. The cost to replace system 1 will be 3.5% more each year than it was the year before. It will cost $4 200 per year to operate and maintain system 1, payable at the end of each year. System 2 costs $120 000 to buy and will last eight years. It will have a salvage or trade-in value of $6 000. The cost to operate and maintain system 2 will be $2 400 per year, payable at the end of each year. Assume the task will be performed for eight years. Compare the two systems, state your basis of comparison, and include a cash-flow diagram. All interest rates are 7.0% per year unless otherwise stated.

9.7 Compare two units, A and B. A has a new cost of $42 000, a life expectancy of 14 years, a salvage value of $4 000, and an annual operating cost of $3 000. B has a new cost of $21 000, a life expectancy of 7 years, a salvage value of $2 000, and an operating cost of $5 000. Assume an annual interest rate of 7 percent. Which of the two units would you recommend? What initial cost of machine A would make the two machines identical in overall cost?

9.8 One of two machines, alpha and beta, is to be purchased to provide for a new production operation in a factory. Machine alpha costs $10 000 and machine beta, $15 000. However, machine beta will result in an annual savings in operating costs of $800 over machine alpha. Which machine would you recommend purchasing and why, if each has a useful life of 10 years and money is worth 9 percent? Assume that both machines will be worthless at the end of 10 years. What value of annual savings of machine beta over machine alpha would result in each being equally desirable?

9.9 Your small company is considering whether to buy a new automobile or to lease it. You have determined that to purchase a new vehicle it will cost $21 000. After 8 years of use, the vehicle can be sold for $4 500. The cost to lease the same vehicle is found to be $3 000 per year for a 4-yr lease after a delivery payment of $2 000. It is expected that the annual operating and maintenance will cost about $1 800 whether the vehicle is purchased or leased. Should you buy or lease? Use present worth analysis to justify your answer. Assume the interest rate is 8% compounded annually.

9.10 A chemical plant is considering three different pieces of equipment to perform a process within the plant. You have gathered the following data:

	Machine 1	Machine 2	Machine 3
Initial cost	$150 000	$70 000	$39 000
Annual O/M	$1 900	$8 500	$11 000
Salvage value	$18 000	$9 000	$5 500
Life expectancy, years	6	4	3

Which machine would you recommend be purchased using an equivalent equal annual cost method? Interest rate is expected to be 8 percent. (O/M means operating and maintenance costs.)

CHAPTER 10

Statistics

Chapter Objectives

When you complete your study of this chapter, you will able to:

- Analyze a wide variety of data sets using descriptive techniques (mean, mode, variance, standard deviation, and correlation)
- Learn to apply the appropriate descriptive statistical techniques in a variety of situations
- Create graphical representations of individual and grouped data points with graphs and histograms
- Make inferences about the relationship between two variables via linear regression analysis
- Determine the strength of linear relationships by calculating and interpreting the correlation coefficient

10.1   Introduction

Statistics, as used by the engineer, can most logically be called a branch of applied mathematics. It constitutes what some call the science of decision making in a world full of uncertainty. In fact, some degree of uncertainty exists in most day-to-day activities, from a simple coin toss or the outcome of a ball game to the results of an election or the relative efficiency of various production processes.

It would be virtually impossible to understand a great deal of the work done in engineering without having a thorough knowledge of statistics. Numerical data derived from surveys and experiments constitute the raw material upon which interpretations, analyses, and decisions are based; it is essential that engineers learn how to properly use the information derived from such data. Everything concerned even remotely with the collection, processing, analysis, interpretation, and presentation of numerical data belongs to the domain of statistics.

There exist today a number of different and interesting stories about the origin of statistics, but most historians believe it can be traced to two dissimilar areas: games of chance and political science. Figure 10.1 illustrates but one of a multitude of familiar games that are associated with a statistical probability.

Figure 10.1

During the eighteenth century, various games of chance involving the mathematical treatment of errors led to the study of probability and, ultimately, to the foundation of statistics. At approximately the same time, an interest in the description and analysis of the voting of political parties led to the development of methods that today fall under the category of *descriptive statistics*, which is basically designed to summarize or describe important features of a set of data without attempting to infer conclusions that go beyond the data.

Descriptive statistics is an important sector of the entire subject area; it is used whenever a person wishes to represent data derived from observation.

10.2 Frequency Distribution

Imagine that you've been given a very large set of data. Unless the information can be appropriately summarized, it will be very difficult to interpret. That is precisely why statistics is so important. Even though raw data contain a lot of information, it is not very meaningful because people have a limited capacity to absorb, remember, sort, and interpret a collection of disparate information. In order to convey meaning, it is necessary to summarize the data.

A frequency distribution is a systematic collection of data illustrating the number of times a given value or collection of values occurs. Frequency distributions can be graphically represented as data curves, bar graphs, scattergrams, histograms, frequency polygrams, and so forth.

Table 10.1

27.2	13.9	40.1	25.9	17.7
4.1	32.3	32.2	22.9	15.6
15.4	36.4	24.1	5.3	12.1
32.9	24.2	19.2	14.5	28.9
21.3	28.8	27.1	8.7	25.2
15.2	19.1	16.5	17.9	12.3
20.9	22.1	21.2	10.8	11.9
35.2	30.2	30.7	37.1	16.3
19.9	20.8	25.6	30.3	21.6

Table 10.2

Range	Tally	Frequency
4.0–8.9	III	3
9.0–13.9	ﬀﬀ	5
14.0–18.9	ﬀﬀ III	8
19.0–23.9	ﬀﬀ ﬀﬀ	10
24.0–28.9	ﬀﬀ IIII	9
29.0–33.9	ﬀﬀ I	6
34.0–38.9	III	3
39.0–43.9	I	1

Various ways of describing measurements and observations, such as the grouping and classifying of data, are a fundamental part of statistics. In fact, when dealing with a large set of collected numbers, a good overall picture of the data can often be conveyed by proper grouping into classes. The following examples will serve to illustrate this point.

Consider the percentage of body fat in 45 men under the age of 50 listed in Table 10.1. Table 10.2 is one possible numerical arrangement showing percentage of body fat distributed among selected classes. Some information such as the highest and lowest values will be lost once the raw data have been sorted and grouped.

The construction of numerical distributions as in this example normally consists of the following series of steps: select classes into which the data are to be grouped, distribute data into appropriate classes, and count the number of items in each class. Since the last two steps are essentially mechanical processes, our attention will be directed primarily toward the *classification* of data.

Two things must be considered when arranging data into classes: the number of classes into which the data are to be grouped and the range of values each class is to cover. Both these areas are somewhat arbitrary, but they do depend on the nature of the data and the ultimate purpose the distribution is to serve.

The following are guidelines that should be followed when constructing a frequency distribution.

1. Use no fewer than six and no more than 15 classes. The square root of n, where n is the number of data points, provides an approximate number of classes to consider.

Figure 10.2

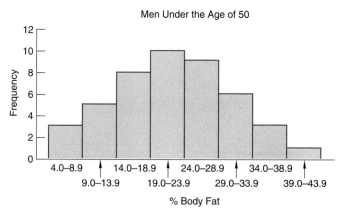

Men Under the Age of 50

% Body Fat

2. Select classes that will accommodate all the data points.
3. Make sure that each data point fits into only one class.
4. Whenever possible, make the class intervals of equal length.

The numbers in the right-hand column of Table 10.2 are called the *class frequencies*, which denote the number of items that are in each class. Since frequency distributions are constructed primarily to condense large sets of data into more easily understood forms, it is logical to display or present that data graphically. The most common form of graphical presentation is called the *histogram*. It is constructed by representing measurements or grouped observations on the horizontal axis and class frequencies along the graduated and calibrated vertical axis. This representation affords a graphical picture of the distribution with vertical bars whose bases equal the class intervals and whose heights are determined by the corresponding class frequencies. Figure 10.2 demonstrates a histogram of the percentage of body fat measures tabulated in Table 10.2.

10.3 Measures of Central Tendency

The solution of many engineering problems in which a large set of data is collected can be facilitated by the determination of single numbers that describe unique characteristics about the data. The most popular measure of this type is called the arithmetic mean or average.

The arithmetic mean, or *mean* of a set of n numbers, is defined as the sum of the numbers divided by n. In order to develop a notation and a simple formula for arithmetic mean, it is helpful to use an example.

Suppose that the mean height of a starting basketball team is to be determined. Let the height in general be represented by the letter x and the height of each individual player be represented by $x_1, x_2, x_3, x_4,$ and $x_5.$ More generally, there are n measurements that are designated $x_1, x_2, \ldots, x_n.$ From this notation, the mean can be written as follows:

$$\text{Mean} = \frac{x_1 + x_2 + x_3 + \cdots + x_n}{n}$$

A mathematical notation that indicates the summation of a series of numbers is normally written

$$\sum_{i=1}^{n} x_i$$

which represents $x_1 + x_2 + x_3 + \ldots + x_n$. This notation will be written in the remainder of the chapter as Σx_i but the intended summation will be from 1 to n.

To distinguish between descriptive measures for a population and for a sample we will use different symbols. When a set of all possible observations is used, it is referred to as the *population*. For the mean or average, we will use mu (μ) to represent the mean of a population.

When a portion or subset of that population is used, it is referred to as a *sample*. The notation for arithmetic mean will be $\bar{x}$ (read xbar) when the x values are representative of a random sample and not an entire population.

The standard notations discussed above provide the following common expressions for the arithmetic mean of the population and of a sample:

$$\mu = \frac{\Sigma x_i}{n} \text{ (population)} \tag{10.1a}$$

$$\bar{x} = \frac{\Sigma x_i}{n} \text{ (sample)} \tag{10.1b}$$

where the sum for the population is over all members of the population, whereas for the sample the sum is just the members of the sample.

The mean is a popular measure of central tendency because (1) it is familiar to most people, (2) it takes into account every item, (3) it always exists, (4) it is always unique, and (5) it lends itself to further statistical manipulations.

One disadvantage of the arithmetic mean, however, is that any outlier or gross error in a number can have a pronounced effect on the value of the mean. To avoid this difficulty, it is possible to describe the "center" of a set of data with other kinds of statistical descriptions. One of these is called the *median*, which can be defined as the value of the middle item of data arranged in increasing or decreasing order of magnitude. For example, the median of the five numbers 15, 27, 10, 18, and 22 can be determined by first arranging them in increasing order: 10, 15, 18, 22, and 27. The median or middle number for this series is 18.

If a data set contains an even number of items, there is never a specific middle item, so the median is defined as the mean of the values of the two middle items. For example, the median of the six numbers 5, 9, 11, 13, 16, and 19 is (11 + 13)/2, or 12.

The mean and median of a set of data rarely coincide. Both terms describe the center of a set of data, but in different ways. The median divides the data so that half of all entries are greater than or equal to the median; the mean may be thought of as the center of gravity of the data.

The median, like the mean, has certain desirable properties. It always exists and is always unique. Unlike the mean, the median is not affected by extreme values. If the exclusion of the highest and lowest values causes a significant change in the mean, then the median should be considered as the indicator of central tendency of that data.

In addition to the mean and the median, there is one other average, or center, of a set of data, which we call the *mode*. It is simply the value that occurs with the highest frequency. In the set of numbers 18, 19, 15, 17, 18, 14, 17, 18, 20, 19, 21, and 14, the number 18 is the mode because it appears more often than any of the other values. There may exist more than one mode, for example, if there were an equal number of 18's and 19's in the above data set, there would be two modes; therefore the data set would be bimodal.

An important point for a practicing engineer to remember is that there are any number of ways to suggest the middle, center, or average value of a data set. If comparisons are to be made, it is essential that similar methods be compared. It is only logical to compare the mean of brand A with the mean of brand B, not the mean of one with the median of the other. If one particular item, brand, or process is to be compared with another, the same measures must be used. If the average grade in one section of college calculus is to be compared with the average grade in other sections, the mean of each section would be the important statistic.

10.4 Measures of Variation

It is possible, but not likely, that the mean values of the course grades of different sections of college calculus will be of equal magnitude. However, mean values are only one measure of importance; another is variation.

Measures of variation indicate the degree to which data are dispersed—that is, how much they are spread out or bunched together. Suppose that by coincidence two sections of a college calculus course have exactly the same mean grade values on the first hour exam. It would be of interest to know how far individual scores varied from the mean. Perhaps one class was bunched very closely around the mean, while the other class demonstrated a wide variation, with some very high scores and some very low scores. This situation is typical and is often of interest to the engineer.

It is reasonable to define this variation in terms of how much each number in the sample deviates from the mean value of the sample, that is, $x_1 - \bar{x}$, $x_2 - \bar{x}$, ... $x_n - \bar{x}$. If you wanted an average deviation from the mean you might try adding $x_1 - \bar{x}$ through $x_n - \bar{x}$ and dividing by n. But this does not give a useful result, since the sum of the deviations is always zero. The procedure generally followed is to square each deviation, sum the resulting squares, divide the sum by n, and take the square root. By definition, the formula for the standard deviation of the entire population is

$$\sigma = \left[\frac{\Sigma(x_i - \mu)^2}{n} \right]^{1/2} \text{(population standard deviation)} \tag{10.2}$$

Statistical procedures used in most engineering applications are concerned with the standard deviation of a sample (s), which can then be used to estimate the standard deviation in the whole population (σ).

To compute the standard deviation of a sample (s), we must alter formula 10.2 in two ways. First we use the mean of the sample ($\bar{x}$) instead of the population mean (μ) and second we replace n in the denominator with $n - 1$. It has been determined that when using just a sample of the population the resulting value of (s) represents a "better" estimate of the true standard deviation of the entire population. To clarify, when a sample size is less than 30, dividing by $n - 1$ has more influence on the calculated value of s. The difference in the calculated value of s when using n or $n - 1$ in the denominator decreases when the sample size is larger.

$$s = \left[\frac{\Sigma(x_i - \bar{x})^2}{n - 1}\right]^{1/2} \text{(sample standard deviation)} \tag{10.3}$$

Throughout this book, we will use n in the denominator for population standard deviations and $(n - 1)$ in the denominator for all sample standard deviations. Although you may encounter either n or $n - 1$ in the denominator in various textbooks, most calculator and spreadsheet software will use $n - 1$ for sample standard deviations.

An alternate form of Eq. (10.3) that is sometimes easier to use is derived by expanding $(x_i - \bar{x})^2$, substituting for $\bar{x}$ from Eq. (10.1b), and reducing terms. The result is as follows.

$$s = \left[\frac{n(\Sigma x_i^2) - (\Sigma x_i)^2}{n(n - 1)}\right]^{1/2} \text{(alternate sample standard deviation)} \tag{10.4}$$

Another common measure of variation is actually called the *variance*; it is the square of the standard deviation. Therefore, the sample variance is given by

$$s^2 = \frac{\Sigma(x_i - \bar{x})^2}{n - 1} \tag{10.5}$$

Example Problem 10.1 A Midwestern university campus has 10 540 male students. Using a random selection process, 50 of these students were chosen and weighed to the nearest pound (pound-mass); the raw data were as recorded in Table 10.3. The data were then grouped (Table 10.4), and the histogram in Figure 10.3 was constructed. Calculate the sample mean, sample standard deviation, and sample variance of the data.

Solution From Eq. (10.1b) the sample mean can be calculated (summary of computations shown in Table 10.5):

$$\bar{x} = \frac{\Sigma x_i}{n} = \frac{8037}{50}$$
$$= 160.74 \text{ lbm} = 161 \text{ lbm}$$

From Eq. (10.3) the sample standard deviation can be determined:

Table 10.3

164	171	154	160	158	150	159	185	168	158
143	159	162	165	160	167	166	164	152	172
177	165	170	155	155	163	180	157	145	160
149	153	137	173	157	175	163	147	156	156
162	167	165	166	162	136	158	170	162	159

Table 10.4

Range	Frequency
136–140	2
141–145	2
146–150	3
151–155	5
156–160	13
161–165	11
166–170	7
171–175	4
176–180	2
181–185	1
	50

$$s = \left[\frac{\Sigma(x_i - \bar{x})^2}{n - 1}\right]^{1/2} = \left[\frac{4\,793.73}{49}\right]^{1/2} = 9.89\,\text{lbm}$$

The sample standard deviation can also be determined from Eq. (10.4):

$$s = \left[\frac{n(\Sigma x_i^2) - (\Sigma x_i)^2}{n(n - 1)}\right]^{1/2} = \left[\frac{50(1\,296\,661) - (8037)^2}{50(49)}\right]^{1/2} = 9.89\,\text{lbm}$$

The sample variance can be calculated by squaring Eq. (10.3):

$$s^2 = \frac{\Sigma(x_i - \bar{x})^2}{n - 1}$$

$$s^2 = \frac{4\,793.74}{49} = 97.8\,\text{lbm}^2$$

By examining the raw data in Example Problem 10.1 we can see the range in variation of values that occurs from a random sample. Certainly we would expect to find both larger and smaller values if all males at the university— that is, the entire population—were weighed. If we were to select additional

Figure 10.3

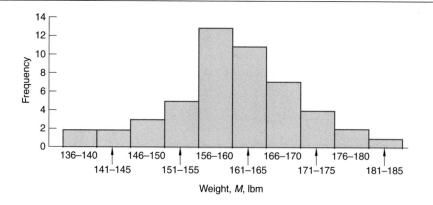

Weight, *M*, lbm

Table 10.5

Mass, x_i, lbm	x_i^2	$x_i - \bar{x}$	$(x_i - \bar{x})^2$
164	26 896	3.26	10.63
143	20 449	−17.74	314.71
177	31 329	16.26	264.39
149	22 201	−11.74	137.83
. . .	. . .	. . .	. . .
156	24 336	−4.74	22.47
159	25 281	−1.74	3.03
8 037	1 296 661	0.00	4 793.73

random values and develop a second sample from the population, we would expect to find a different sample mean and a different sample standard deviation. We would not expect, however, the differences in these measures of central tendency and variation to be significant if the two samples were truly random in nature.

Example Problem 10.2 Interstate Safety Corridors are established on certain roadways with a propensity for strong cross winds, blowing dust, and frequent fatal accidents. A driver is expected to turn on the headlights and pay special attention to the posted speed limit in these corridors. In one such Safety Corridor in northern New Mexico, the posted speed limit is 75 miles per hour. The Department of Public Safety set up a radar checkpoint and the actual speed of 36 vehicles that passed the checkpoint is shown in Table 10.6.

For this data set, use a hand calculation or spreadsheet software to solve parts c, d, and e.
  (a) Make a frequency distribution table using 5 as a class width (e.g., 60.0-64.9)
  (b) Construct a histogram from the frequency distribution
  (c) The sample mean

Table 10.6

70	78	70	80	86	76
85	69	68	61	81	80
71	82	69	71	62	71
75	76	85	72	63	72
65	90	77	89	76	70
66	78	91	69	80	92

Table 10.7

Interval	Frequency
60-64.9	3
65-69.9	6
70-74.9	8
75-79.9	7
80-84.9	5
85-89.9	4
90-94.5	3

(d) The sample standard deviation
(e) The sample variance
(f) If speeding tickets are issued to drivers exceeding 84 mph, what percentage of drivers in this sample would receive tickets?

Solution A spreadsheet can be used to solve this problem; however, the equations presented in the text material are representative of the theory necessary to determine mean, standard deviation, and variance. A critical step in the use of packaged software is to thoroughly understand how the numbers provided by these packages are calculated.

(a) The data from Table 10.6 was grouped into the frequency distribution shown in Table 10.7.
(b) A histogram was developed using Excel and is shown in Figure 10.4. To determine the sample mean, standard deviation, and variance an Excel spreadsheet was developed in Figure 10.5. This spreadsheet quickly provides the required sums for the equations below. In addition, you may verify that your calculations are correct using the statistical formulas within Excel.

Figure 10.4

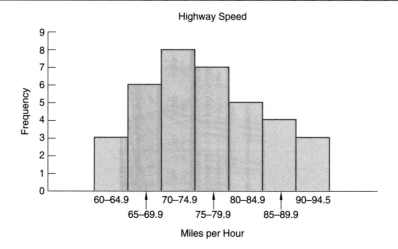

Figure 10.5

241

*Measures
of Variation*

	A	B	C
1		Speed	Speed squared
2		70	4 900
3		85	7 225
4		71	5 041
5		75	5 625
6		65	4 225
7		66	4 356
8		78	6 084
9		69	4 761
10		82	6 724
11		76	5 776
12		90	8 100
13		78	6 084
14		70	4 900
15		68	4 624
16		69	4 761
17		85	7 225
18		77	5 929
19		91	8 281
20		80	6 400
21		61	3 721
22		71	5 041
23		72	5 184
24		89	7 921
25		69	4 761
26		86	7 396
27		81	6 561
28		62	3 844
29		63	3 969
30		76	5 776
31		80	6 400
32		76	5 776
33		80	6 400
34		71	5 041
35		72	5 184
36		70	4 900
37		92	8 464
38	Sum	2 716	20 7360
39	Average	75.4444444	Excel software
40	Std Dev	8.3715315	Excel software
41	Variance	70.0825397	Excel software

Excel Spreadsheet—Highway Safety Corridors

(c) The sample mean

$$\bar{x} = \frac{\Sigma x_i}{n} = \frac{2\,716}{36} = 75.44$$

(d) The sample standard deviation

$$s = \left[\frac{n(\Sigma x_i^2) - (\Sigma x_i)^2}{n(n-1)}\right]^{1/2} = \left[\frac{36(207\,360) - (2\,716)^2}{36(35)}\right]^{1/2} = 8.371\,5$$

(e) The sample variance

$$s^2 = 70.082\,5$$

(f) Speeding tickets issued
7/36 drivers are exceeding 84 mph, therefore 19% would be issued tickets

10.5 Linear Regression

There are many occasions in engineering analysis when the ability to predict or forecast the outcome of a certain event is extremely valuable. The difficulty with most practical applications is the large number of variables that may influence the analysis process. Regression analysis is a study of the relationships among variables. If the situation results in a relationship among three or more variables, the study is called *multiple regression.* There are many problems, however, that can be reduced to a relationship between an independent and a dependent variable. This introduction will limit the subject and treat only two-variable regression analyses.

Of the many equations that can be used for the purposes of prediction, the simplest and most widely used is a linear equation of the form $y = mx + b$, where m and b are constants. Once the constants have been determined, it is possible to calculate a predicted value of y (dependent variable) for any value of x (independent variable).

Before investigating the regression concept in more detail, we must examine how the regression equation is established.

If there is a reason to believe that a relationship exists between two variables, the first step is to collect data. For example, suppose x denotes the age of an automobile in years and y denotes the annual maintenance cost. Thus, a sample of n cars would reveal the age $x_1, x_2, x_3, \ldots, x_n$ and the corresponding annual maintenance cost $y_1, y_2, y_3, \ldots, y_n$.

The next step would be to plot the data on rectangular coordinate paper or by using a spreadsheet. The resulting graph is called a *scatter diagram.*

From the scatter diagram shown in Figure 10.6, it may be possible to construct a straight line that adequately represents the data, in which case a linear relationship exists between the variables. In other cases, the line may be curved, and the relationship between variables will therefore be nonlinear in nature.

Ideally, we would hope to determine the best possible line (straight or curved) through the points. A standard approach to this problem is called the *method of least squares.*

To demonstrate how the process works, as well as to explain the concept of the method of least squares, consider the following situation. A class of 20 students is given a math test and the resulting scores are recorded. Each student's IQ score is also available. Both scores for the 20 students are shown in Table 10.8.

First, the data must be plotted on rectangular coordinate paper (see Figure 10.7). As you can see by observing the plotted data, there is no limit to the number of straight lines that could be drawn through the points. In order to find the line of best fit, it is necessary to state what is meant by "best." The method of least squares requires that the sum of the squares $\Sigma(y - y')^2$ of the vertical deviations from the data points (y) to the straight line (y') be minimized. This vertical deviation $(y - y')^2$ is also called a residual (see Figure 10.8).

Figure 10.6

243
Linear Regression

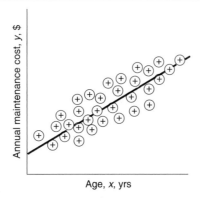

To demonstrate how a least-squared line is fit to data, let us consider this problem further. There are n pairs of numbers $(x_1, y_1), (x_2, y_2), \ldots, (x_n, y_n)$, where $n = 20$, with x and y being IQ and math scores, respectively. Suppose that the equation of the line that best fits the data is of the form

$$y' = mx + b \tag{10.6}$$

where the symbol y' (y prime) is used to differentiate between the observed values of y and the corresponding values calculated by means of the equation of the line. (Note that although y' is sometimes used to represent a derivative in calculus, it is not a derivative here, rather a computed value on the line being sought.) In other words, for each value of x, there exist an observed value (y) and a calculated value (y') obtained by substituting x into the equation $y' = mx + b$.

Table 10.8

Student No.	Math Score	IQ	Student No.	Math Score	IQ
1	85	120	11	100	130
2	62	115	12	85	130
3	60	100	13	77	118
4	95	140	14	63	112
5	80	130	15	70	122
6	75	120	16	90	128
7	90	130	17	80	125
8	60	108	18	100	140
9	70	115	19	95	135
10	80	118	20	75	130

Figure 10.7

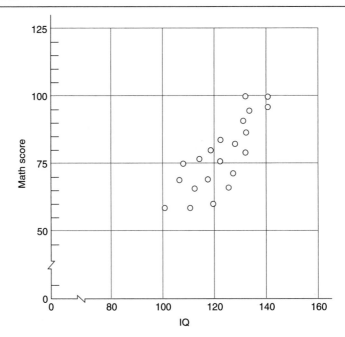

Figure 10.8

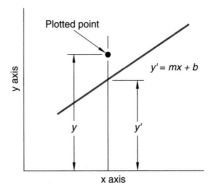

The least-squares criterion requires that the sum of all $(y - y')^2$ terms, as illustrated in Figure 10.8, be the smallest possible. One must determine the constants m and b so that the difference between the observed and the predicted values of y will be minimized.

When this analysis is applied to the linear equation $y = mx + b$, it follows that we wish to minimize the summation of all deviations (residuals):

$$SUM = \Sigma[y_i - (mx_i + b)]^2 \tag{10.7}$$

From the calculus, to minimize *SUM*, the partial derivatives with respect to m and b must be zero as follows:

$$\frac{\partial(SUM)}{\partial m} = \frac{\partial}{\partial m}\{\Sigma[y_i - (mx_i + b)]^2\} = 0$$

$$\frac{\partial(SUM)}{\partial b} = \frac{\partial}{\partial b}\{\Sigma[y_i - (mx_i + b)]^2\} = 0$$

Performing these partial derivatives gives

$$nb + m\Sigma x_i = \Sigma y_i$$

$$b\Sigma x_i + m\,\Sigma x_i2 = \Sigma x_i y_i$$

Solving these two equations simultaneously for m and b gives

$$m = \frac{n(\Sigma x_i y_i) - (\Sigma x_i)(\Sigma y_i)}{n(\Sigma x_i^2) - (\Sigma x_i)^2} \qquad (10.8)$$

$$b = \frac{\Sigma y_i - m(\Sigma x_i)}{n} \qquad (10.9)$$

Table 10.9 is a tabulation of the values necessary to determine the constants m and b for the math score–IQ problem. The independent variable is the IQ, and the dependent variable is the math score.

Substituting the values from Table 10.9 into Eqs. (10.8) and (10.9), we get the following values for the two constants:

$$m = \frac{20(198\,527) - (2\,466)(1\,592)}{20(306\,124) - (2\,466)^2}$$

$$= 1.080\,9$$

$$b = \frac{1\,592 - (1.080\,9)(2\,466)}{20}$$

$$= -53.67$$

The equation of the line relating math score and IQ using the method of least squares becomes

$$\text{Math score} = 1.08\,(\text{IQ}) - 53.67$$

Interesting questions arise from this problem. Can IQ be used to predict success on a math exam and, if so, how well? Regression analysis or estimation of one variable (dependent) from one or more related variables (independent) does not provide information about the strength of the relationship. Later in this chapter we will suggest a method to determine how well an equation developed from the method of least squares describes the strength of the relationship between variables.

The method of least squares as just explained is a most appropriate technique for determining the best-fit line. You should clearly understand that this method as presented is *linear regression* and is valid only for *linear* relationships. The technique of least squares can, however, be applied to power ($y = bx^m$) and exponential ($y = be^{mx}$) relationships if done correctly. The power function can be handled by noting that there is a linear relationship between log y and

Table 10.9

IQ	(IQ)²	Math Score	(Math Score)²	IQ (Math Score)
Independent Variable		**Dependent Variable**		
120	14 400	85	7 225	10 200
115	13 225	62	3 844	7 130
100	10 000	60	3 600	6 000
140	19 600	95	9 025	13 300
130	16 900	80	6 400	10 400
120	14 400	75	5 625	9 000
130	16 900	90	8 100	11 700
108	11 664	60	3 600	6 480
115	13 225	70	4 900	8 050
118	13 924	80	6 400	9 440
130	16 900	100	10 000	13 000
130	16 900	85	7 225	11 050
118	13 924	77	5 929	9 086
112	12 544	63	3 969	7 056
122	14 884	70	4 900	8 540
128	16 384	90	8 100	11 520
125	15 625	80	6 400	10 000
140	19 600	100	10 000	14 000
135	18 225	95	9 025	12 825
130	16 900	75	5 625	9 750
2 466	306 124	1 592	129 892	198 527

log x (log $y = m$ log x + log b, which plots as a straight line on log-log paper). Thus we can apply the method of least squares to the variables log y and log x to obtain parameters m and log b.

The exponential function written in natural logarithm form is ln $y = mx$ + ln b. Therefore, there exists a linear relationship between ln y and x (this plots as a straight line on semilog paper). The next examples will demonstrate the use of least squares method for power and experimental curves.

Example Problem 10.3 The data in Table 10.10 were obtained from measuring the distance dropped by a falling body with time. We would expect that the distance should be related to the square of the time according to theory (neglecting air friction) s = 1/2 gt². Find the equation of the line of best fit.

Solution The expected form of the equation is $s = bt^m$ or, in logarithmic form, log $s = m$log t + log b. Therefore, if we use log t in place of x and log s in place of y in Eqs. (10.8) and (10.9) we can solve for m and log b. (Note carefully that the parameters are m and log b, not m and b.) Refer to Table 10.11.

Substitute into Eq. (10.8):

$$m = \frac{6(5.521\,9) - (2.857\,4)(9.855\,8)}{6(1.774\,9) - (2.857\,4)^2}$$

$$= 2.000\,0$$

Table 10.10

Time t, s	Distance s, m
0	0
1	4.9
2	19.6
3	44.1
4	78.4
5	122.5
6	176.4

Table 10.11

t	s	Independent Variable		Dependent Variable	
		log t	(log t)²	log s	(log t)(log s)
1	4.9	0.000 0	0.000 0	0.609 2	0.000 0
2	19.6	0.301 0	0.090 6	1.292 3	0.389 0
3	44.1	0.477 1	0.227 6	1.644 4	0.784 5
4	78.4	0.602 1	0.362 5	1.894 3	1.140 6
5	122.5	0.699 0	0.488 6	2.088 1	1.459 6
6	176.4	0.778 2	0.605 6	2.246 5	1.748 2
		2.857 4	1.774 9	9.855 8	5.521 9

Substitute into Eq. (10.9), using log *b*, not *b*:

$$\log b = \frac{9.855\ 8 - (2.000\ 0)(2.857\ 4)}{6}$$

$$\log b = 0.690\ 17$$

$$b = 4.899\ 7$$

The equation is then $s = 4.9t^2$

Example Problem 10.4   Using the method of least squares, find the equation that best fits the data shown in Table 10.12.

Solution   This data set produces a straight line when plotted on semilog graph paper; therefore, its equation will be of the form $Q = be^{mV}$ or $\ln Q = mV + \ln b$.

An examination of the equation and the graph paper leads us to the following:

1. Since the line is straight, the method of least squares can be used.
2. Since the abscissa is a uniform scale, the independent variable values (velocity in this problem) may be used without adjustment.
3. Since the ordinate is a log scale, the dependent variable values (fuel consumption) must be the logarithms of the data, not the raw data.

Table 10.12

Fuel Consumption, *Q*, mm³/s	Velocity, *V*, m/s
25.2	10.0
44.6	20.0
71.7	30.0
115	40.0
202	50.0
367	60.0
608	70.0

Table 10.13

Independent Variable		Dependent Variable	
v	*v²*	ln *Q*	*v*(ln *Q*)
10	100	3.226 8	32.27
20	400	3.797 7	75.95
30	900	4.272 5	128.17
40	1 600	4.744 9	189.80
50	2 500	5.308 3	265.41
60	3 600	5.905 4	354.32
70	4 900	6.410 2	448.71
280	14 000	33.665 8	1 494.63

Table 10.13 provides us with the needed values to substitute into Eqs. (10.8) and (10.9). Substitute into Eq. (10.8):

$$m = \frac{7(1\,494.63) - (280)(33.665\,8)}{7(14\,000) - (280)^2}$$

$$= 0.052\,86$$

Substitute into Eq. (10.9), using ln *b* rather than *b*:

$$\ln b = \frac{33.665\,8 - (0.052\,86)(280)}{7}$$

$$= 2.695\,0$$

$$b = 14.805$$

The equation becomes $Q = 14.8\,e^{0.053V}$

10.6 Coefficient of Correlation

The technique of finding the best possible straight line to fit experimentally collected data is certainly useful, as previously discussed. The next logical and interesting question is how well such a line actually fits. It stands to reason that if the differences between the observed *y*'s and the calculated *y*'s are small, the sum of squares $\Sigma(y - y')^2$ will be small; and if the differences are large, the sum of squares will tend to be large.

Figure 10.9

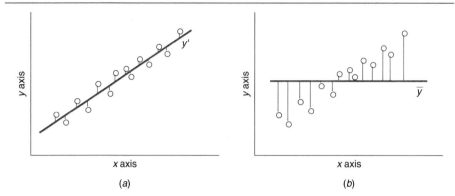

(a)   (b)

Although $\Sigma(y - y')^2$ provides an indication of how well a least-squares line fits particular data, it has the disadvantage that it depends on the units of y. For example, if the units of y are changed from dollars to cents, it will be like multiplying $\Sigma(y - y')^2$ by a factor of 10 000. To avoid this difficulty, the magnitude of $\Sigma(y - y')^2$ is normally compared with $\Sigma(y - y')^2$. This allows the sum of the squares of the vertical deviations from the least-squares line to be compared with the sum of squares of the deviations of the y's from the mean.

To illustrate, Figure 10.9a shows the vertical deviation of the y's from the least-squares line, while Figure 10.9b shows the deviations of the y's from their collective mean. It is apparent that where there is a close fit, $\Sigma(y - y')^2$ is much smaller than $\Sigma(y - y')^2$.

In contrast, consider Figure 10.10. Again, Figure 10.10a shows the vertical deviation of the y's from the least-squares line, and Figure 10.10b shows the deviation of the y's from their mean. In this case, $\Sigma(y - y')^2$ is approximately the same as $\Sigma(y - y')^2$. This would seem to indicate that if the fit is good, as in Figure 10.9, $\Sigma(y - y')^2$ is much less than $\Sigma(y - y')^2$; and if the fit is as poor as in Figure 10.10, the two sums of squares are approximately equal.

Figure 10.10

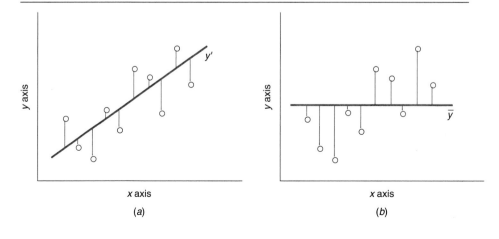

(a)   (b)

The coefficient of correlation puts this comparison on a precise basis:

$$r = \pm \sqrt{1 - \frac{\Sigma(y_i - y')^2}{\Sigma(y_i - \bar{y})^2}} \qquad (10.10)$$

If the fit is poor, the ratio of the two sums is close to 1 and r is close to zero. However, if the fit is good, the ratio is close to zero and r is close to +1 or −1. From the equation, it is obvious that the ratio can never exceed 1. Hence, r cannot be less than −1 or greater than +1.

The statistic is used to measure the strength of a linear relationship between any two variables. It indicates the goodness of fit of a line determined by the method of least squares, and this in turn indicates whether a relationship exists between x and y.

Although Eq. (10.10) serves to define the coefficient of correlation, it is seldom used in practice. An alternative form of the formula is

$$r = \frac{n(\Sigma x_i y_i) - (\Sigma x_i)(\Sigma y_i)}{\sqrt{n(\Sigma x_i^2) - (\Sigma x_i)^2}\sqrt{n(\Sigma y_i^2) - (\Sigma y_i)^2}} \qquad (10.11)$$

The interpretation of r is not difficult if it is ±1 or zero: When it is zero, the points are scattered and the fit of the regression line is so poor that a knowledge of x does not help in the prediction of y; when it is +1 or −1, all the points actually lie on the straight line, so an excellent prediction of y can be made by using x values. The problem arises when r falls between zero and +1 or zero and −1.

General guidelines for interpreting the correlation coefficient are as follows:

Correlation Coefficient	Correlation Interpretation
0.9 to 1.0	Very high positive
0.7 to 0.9	High positive
0.5 to 0.7	Moderate positive
0.3 to 0.5	Low positive
−0.3 to 0.3	Little, if any
−0.5 to −0.3	Low negative
−0.7 to −0.5	Moderate negative
−0.9 to −0.7	High negative
−1.0 to −0.9	Very high negative

The physical interpretation of r can be explained in the following manner. If the coefficient of correlation is known for a given set of data, then $100r^2\%$ of the variation of the y's can be attributed to differences in x, namely, to the relationship of y with x. If $r = 0.6$ in a given problem, then 36%—that is, $100(0.6^2)$—of the variation of the y's is accounted for (perhaps caused) by differences in X values. The square of the correlation coefficient (r^2) is called the *coefficient of determination*.

Again consider the problem of IQ and math scores, substituting values from Table 10.9 into the equation for the correlation coefficient.

$$r = \frac{(20)(198\,527) - (2\,466)(1\,592)}{\sqrt{(20)(306\,124) - (2\,466)^2}\sqrt{(20)(129\,892) - (1\,592)^2}}$$

$$= 0.87$$

Computing the coefficient of determination and multiplying by 100 to obtain percent yields

$$100r^2 = 76\%$$

This would indicate that 76% of the variations in math scores can be accounted for by differences in IQ.

One word of caution when using or considering results from linear regression and coefficients of correlation and determination. There is a fallacy in interpreting high values of r or r^2 as implying cause-effect relations. If the increase in television coverage of professional football is plotted against the increase in traffic accidents at a certain intersection over the past 3 years, an almost perfect positive correlation (+1.0) can be shown to exist. This is obviously not a cause-effect relation, so it is wise to interpret the correlation coefficient and coefficient of determination carefully. The variables must have a measure of association if the results are to be meaningful.

The following example problem is intended to illustrate how much of the material presented in this chapter could be used in the solution to a practical engineering problem.

Example Problem 10.5   Water conservation can be a critical matter. Many regions of the country can have a serious lack of water for many years and an excess in other years. One solution to this problem is construction of dams that allows water storage in large reservoirs. The amount of water flowing into a region is a function of the watershed area, annual rainfall, snow melt, and so on. Assume the dam height, discharge area, and spillways must be designed to accommodate a 100-year flood.

Data showing the flow rate (discharge) from a river that is under consideration for the construction of a dam as a function of the recovery time needed to refill the reservoir is given in Table 10.14. These are measurements from the primary watershed of about 5 540 mi² located in a southwestern state.

1. Using the least-squares method, obtain an equation relating the discharge Q to the recovery period P.
2. Compute the correlation coefficient and interpret the results.
3. From the derived equation, compute the projected discharge for a 100-year flood.

Solution   A spreadsheet is a convenient tool for this problem. The data were entered in columns A and B as shown in Figure 10.11. Test plots were then done to determine visually whether linear, semilog, or log-log techniques would be most likely to produce a straight line. As shown in Figure 10.12, the log-log plot, although not a perfect straight line, is more nearly so than either the linear plot or the semilog plot. Thus we will use the method of least squares to find an equation of the form $Q = bP^m$ ($\log Q = m\log P + \log b$).

The spreadsheet was then modified (see Fig. 10.11) by adding columns to compute $\log P$, $\log Q$, $(\log P)(\log Q)$, $(\log P)^2$, and $(\log Q)^2$. The sums of each

Table 10.14

Period (years)	Discharge (cfs)	Period (years)	Discharge (cfs)
1	2 480	2.66	5 360
1.09	2 690	2.87	5 600
1.15	2 730	3.11	5 950
1.2	2 780	3.48	6 300
1.41	3 300	3.75	7 300
1.45	3 320	4.18	7 530
1.49	3 350	4.73	8 650
1.55	3 420	5.43	9 000
1.62	3 750	6.39	9 150
1.68	3 800	7.75	10 700
1.72	4 200	9.84	12 000
1.85	4 330	13.5	15 250
1.95	4 470	21.8	21 100
2.06	4 600	33.7	26 870
2.18	4 620	41.5	33 250
2.32	5 110	50.5	34 750
2.48	5 320	55.8	44 800

of columns A through G were also computed. From these sums, the parameter m could be found as follows,

$$m = \frac{n(\Sigma x_i y_i) - (\Sigma x_i)(\Sigma y_i)}{n(\Sigma x_i^2) - (\Sigma x_i)^2}$$

$$= \frac{34(81.329) - (19.765)(130.032)}{34(19.969) - (19.765)^2} = 0.676\,8$$

$$\log b = \frac{\Sigma y_i - m(\Sigma x_i)}{n} = \frac{130.032 - 0.676\,8(19.765)}{34} = 3.431\,1$$

$$b = 2\,698.1$$

and once m has been determined, $\log b$ can be calculated. The parameter b is obtained from the expression $b = 10^{\log b}$.

The least-squares equation is then

$$Q' = 2.7 \times 10^3\, P^{0.677}$$

after rounding the coefficients.

The correlation coefficient was then computed and found to be about 0.997. Squaring this value (coefficient of determination) and multiplying by 100% gives a result of 99.4%, which is an indication of a very good agreement between the data and the prediction line. It suggests that nearly 99% of the variation in the discharge can be accounted for by variation in the return period.

Spreadsheets also provide a mechanism for computing a least-squares curve fit (Excel calls this a trendline) to a data set. Figure 10.13 shows the result of this approach for Example Problem 10.5. The data were plotted as a scatter chart (log-log) and a trendline was added.

Figure 10.11

	A	B	C	D	E	F	G	H
1	P	Q	log P	log Q	(log P)(log Q)	(log P)²	(log Q)²	Q'
2	1	2480	0	3.394452	0	0	11.5223022	2698.6
3	1.09	2690	0.037426	3.429752	0.128363617	0.001401	11.7632007	2860.627
4	1.15	2730	0.060698	3.436163	0.208567652	0.003684	11.8072137	2966.243
5	1.2	2780	0.079181	3.444045	0.272703758	0.00627	11.8614446	3052.9
6	1.41	3300	0.149219	3.518514	0.525029528	0.022266	12.3799404	3404.867
7	1.45	3320	0.161368	3.521138	0.568199018	0.02604	12.3984134	3469.925
8	1.49	3350	0.173186	3.525045	0.610489356	0.029993	12.4259409	3534.405
9	1.55	3420	0.190332	3.534026	0.67263719	0.036226	12.4893405	3630.086
10	1.62	3750	0.209515	3.574031	0.748813213	0.043897	12.7736995	3740.213
11	1.68	3800	0.225309	3.579784	0.806558471	0.050764	12.8148506	3833.388
12	1.72	4200	0.235528	3.623249	0.853378278	0.055474	13.1279354	3894.907
13	1.85	4330	0.267172	3.636488	0.971566757	0.071381	13.2240442	4091.729
14	1.95	4470	0.290035	3.650308	1.058715524	0.08412	13.324745	4240.098
15	2.06	4600	0.313867	3.662758	1.14961962	0.098513	13.4157949	4400.49
16	2.18	4620	0.338456	3.664642	1.240321873	0.114553	13.4296008	4572.336
17	2.32	5110	0.365488	3.708421	1.355383282	0.133581	13.7523856	4769.004
18	2.48	5320	0.394452	3.725912	1.469692106	0.155592	13.8824175	4989.127
19	2.66	5360	0.424882	3.729165	1.584453639	0.180524	13.90667	5231.346
20	2.87	5600	0.457882	3.748188	1.716227443	0.209656	14.0489135	5507.334
21	3.11	5950	0.49276	3.774517	1.859932448	0.242813	14.2469783	5814.872
22	3.48	6300	0.541579	3.799341	2.057643982	0.293308	14.4349886	6274.383
23	3.75	7300	0.574031	3.863323	2.217668119	0.329512	14.9252635	6599.759
24	4.18	7530	0.621176	3.876795	2.408173089	0.38586	15.0295393	7102.748
25	4.73	8650	0.674861	3.937016	2.656939181	0.455438	15.5000958	7722.351
26	5.43	9000	0.7348	3.954243	2.905576722	0.539931	15.6360338	8478.208
27	6.39	9150	0.805501	3.961421	3.190928091	0.648832	15.6928571	9465.431
28	7.75	10700	0.889302	4.029384	3.583337854	0.790858	16.2359336	10785.49
29	9.84	12000	0.992995	4.079181	4.050606983	0.986039	16.6397196	12676.49
30	13.5	15250	1.130334	4.18327	4.728491167	1.277654	17.4997466	15700.82
31	21.8	21100	1.338456	4.324282	5.787863932	1.791466	18.6994188	21713.92
32	33.7	26870	1.52763	4.429268	6.766281726	2.333653	19.6184121	29156.33
33	41.5	33250	1.618048	4.521792	7.316476372	2.61808	20.4465997	33566.79
34	50.5	34750	1.703291	4.540955	7.734569174	2.901202	20.6202706	38334.12
35	55.8	44800	1.746634	4.651278	8.124081248	3.050731	21.6343872	41012.03
36	Sum P	Sum Q	Sum log P	Sum log Q	Sum (log P)(log Q)	Sum (log P)²	Sum (log Q)²	
37	299.19	327830	19.7654	130.0321	81.32929041	19.96931	501.209098	

Column H was added to compute values of the predicted discharge, Q', for each return period P. The original data ($R^2 = 0.994\,2$) as well as the predicted curve ($R^2 = 1$) are plotted in Figure 10.13.

To determine the discharge for a 100-year flood, use the least-squares equation.

$$Q' = 2.7 \times 10^3\, P^{0.677}$$

$$Q' = 2.7 \times 10^3 (100)^{0.677}$$

$$= 61 \times 10^3 \text{ cfs}$$

Figure 10.12

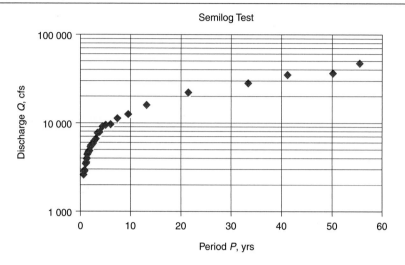

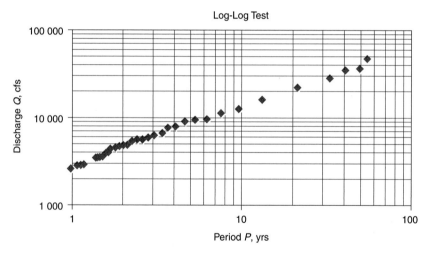

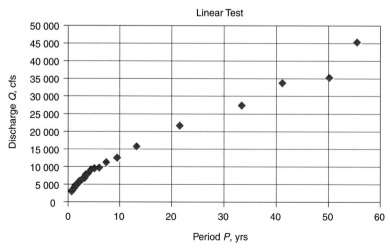

Figure 10.13

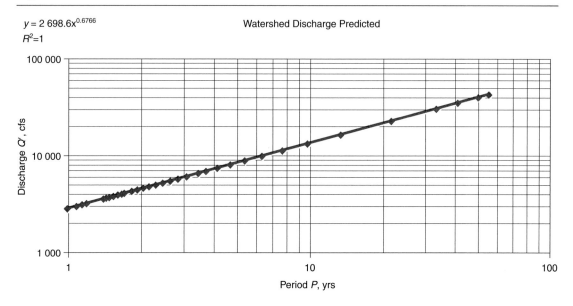

$y = 2\,698.6x^{0.6766}$
$R^2 = 1$

Watershed Discharge Predicted

Discharge Q', cfs

Period P, yrs

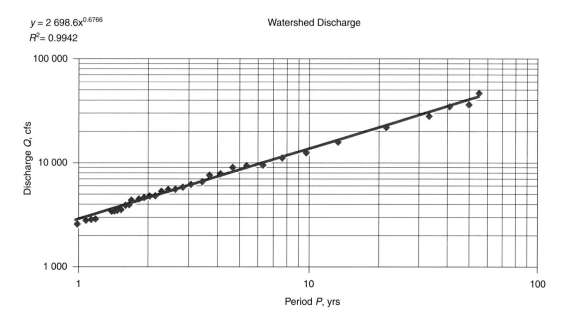

$y = 2\,698.6x^{0.6766}$
$R^2 = 0.9942$

Watershed Discharge

Discharge Q, cfs

Period P, yrs

Problems

10.1 The number of home runs hit per game for the Millard girls' softball team are: 1, 2, 4, 3, 2, 4, 3, 0, 1, 2, 3, 5, 2, 1, and 5.
  (*a*) What is the mean number of home runs hit?
  (*b*) What is the median?
  (*c*) What is the mode?

10.2 The temperature, in degrees Celsius, each day over a three-week period was recorded as follows:

17, 18, 20, 22, 21, 19, 16, 15, 18, 20, 21, 21, 22, 21, 19, 20, 19, 17, 16, 16, 17

  (a) Compute the mean, median, and mode.
  (b) Using two-degree intervals starting with 15-16, draw a frequency diagram.

10.3 The following table gives the life expectancy of males and females at birth in the United States, 1979 to 2004.

Year	Male	Female	Year	Male	Female
1979	70.0	77.4	1992	72.1	78.9
1980	70.0	77.5	1993	72.1	78.9
1981	70.4	77.8	1994	72.4	79.0
1982	70.9	78.1	1995	72.5	78.9
1983	71.0	78.1	1996	73.1	79.1
1984	71.2	78.2	1997	73.6	79.4
1985	71.2	78.2	1998	73.8	79.5
1986	71.3	78.3	1999	73.9	79.4
1987	71.5	78.4	2000	74.3	79.7
1988	71.5	78.3	2001	74.4	79.8
1989	71.7	78.5	2002	74.5	79.9
1990	71.8	78.8	2003	74.7	80.0
1991	72.0	78.9	2004	75.2	80.4

Source: NCHS (National Center for Health Statistics)

  (*a*) Calculate the mean and median, of the life expectancies for men and for women.
  (*b*) Calculate the standard deviation for the data for men and for women.

10.4 The exam scores of 50 students in a class follow:

92	71	91	53	99	93	88	95	65	67
98	76	65	68	82	91	93	44	77	100
88	87	56	85	60	98	89	70	78	82
78	70	78	89	34	95	67	88	89	78
77	65	88	78	59	50	66	76	91	87

  (*a*) What is the average (mean) score in the class?
  (*b*) More students got a _____ score than any other score. (mode)
  (*c*) What is the median score?
  (*d*) Compute the standard deviation of the data using equation 10.4 and a calculator or software.

10.5 Survey at least 30 engineering students to obtain each student's total investment in calculators and computers in the past 18 months. The investment figure should include all equipment and software, whether bought personally or received as a gift.
  (*a*) Find the mean, mode, and median investment.
  (*b*) Find the standard deviation.

10.6 The output of a gas furnace has large quantities of carbon dioxide, which must be monitored carefully. This table gives the percentage of CO_2 in the output, with samples taken every 9 seconds.

53.8	56	48.4	48.3	51.5	54.4	57.7
53.6	56.8	47.9	47	51.6	56	57
53.5	56.8	47.6	45.8	51.2	56.9	56
53.5	56.4	47.5	45.6	50.5	57.5	54.7
53.4	55.7	47.5	46	50.1	57.3	53.2
53.1	55	47.6	46.9	49.8	56.6	52.1
52.7	54.3	48.1	47.8	49.6	56	51.6
52.4	53.2	49	48.2	49.4	55.4	51
52.2	52.3	50	48.3	49.3	55.4	50.5
52	51.6	51.1	47.9	49.2	56.4	50.4
52	51.2	51.8	47.2	49.3	57.2	
52.4	50.8	51.9	47.2	49.7	58	
53	50.5	51.7	48.1	50.3	58.4	
54	50	51.2	49.4	51.3	58.4	
54.9	49.2	50	50.6	52.8	58.1	

Source: Time Series Data Library

(a) Group these measurements into equal classes, and construct a frequency distribution table for the data. (Use a spreadsheet to help solve this problem.)
(b) Compute the mean, median, and mode of the data.
(c) Compute the standard deviation of the data.

10.7 A farm-implement manufacturer in the Midwest purchases castings from the Omaha Steel foundry. Thirty castings were selected at random and weighed, and their masses were recorded to the nearest kilogram, as shown below:

235	232	228	228	240	231
225	220	218	230	222	229
217	233	222	221	228	228
244	241	238	219	242	222
227	227	229	229	224	227

(a) Group the measurements into a frequency distribution table having six equal classes from 215 to 244.
(b) Construct a histogram of the distribution.
(c) Determine the median, mode, and mean of the data.

10.8 The number of users that were logged into an Internet server was monitored every minute, over a period of 100 minutes. Here are the results:

88	138	140	171	112	91	193
84	146	134	172	104	91	204
85	151	131	172	102	94	208
85	150	131	174	99	101	210
84	148	129	175	99	110	215
85	147	126	172	88	121	222
83	149	126	172	88	135	228
85	143	132	174	84	145	226
88	132	137	174	84	149	222
88	131	140	169	88	156	220
91	139	142	165	89	165	
99	147	150	156	88	171	
104	150	159	142	85	175	
112	148	167	131	86	177	
126	145	170	121	89	182	

Source: Time Series Data Library

(*a*) Group these measurements into 10 equal classes, (81–95, 96–110, etc.) and construct a frequency distribution table for the data.

(*b*) Draw a histogram of the distribution.

(*c*) Find the average number of users during the sample period.

(*d*) Using the histogram, estimate the probability of 141–165 users being logged in at any given time.

10.9 The team earned-run averages for the American and National League baseball teams in 2009 are shown below. Calculate the median, mean, and standard deviation for each league individually and for Major League Baseball as a whole.

American League Team	ERA	National League Team	ERA
Seattle Mariners	3.87	Los Angeles Dodgers	3.41
Chicago White Sox	4.14	San Francisco Giants	3.55
Oakland Athletics	4.26	Atlanta Braves	3.57
New York Yankees	4.26	St Louis Cardinals	3.66
Detroit Tigers	4.29	Chicago Cubs	3.84
Tampa Bay Rays	4.33	Philadelphia Phillies	4.16
Boston Red Sox	4.35	Cincinnati Reds	4.18
Texas Rangers	4.38	Colorado Rockies	4.22
Los Angeles Angels	4.45	Florida Marlins	4.29
Toronto Blue Jays	4.47	San Diego Padres	4.37
Minnesota Twins	4.50	Arizona Diamondbacks	4.42
Kansas City Royals	4.83	New York Mets	4.45
Cleveland Indians	5.06	Houston Astros	4.54
Baltimore Orioles	5.15	Pittsburgh Pirates	4.59
		Milwaukee Brewers	4.83
		Washington Nationals	5.00

10.10 A land-grant university located in the Midwest has an entering freshman class of 1 522 students. A random sample of 150, or approximately 10%, of the entering first-year students' cumulative grade points were recorded. Academic standards require each student to have a first-year grade average of 1.5 or better to return for the sophomore year and 3.5 or better to be eligible for the Dean's list. The grade point averages (GPA) below have been partially grouped.

(a) Determine the mean grade point average and standard deviation for the sample.

(b) Group the data into equal classes and construct a frequency distribution table.

(c) Draw a histogram of the distribution.

GPA	# Students	GPA	# Students	GPA	# Students	GPA	# Students
0	0	1	3	2	7	3	5
0.1	0	1.1	2	2.1	7	3.1	2
0.2	1	1.2	3	2.2	10	3.2	1
0.3	0	1.3	2	2.3	8	3.3	3
0.4	1	1.4	4	2.4	12	3.4	2
0.5	1	1.5	4	2.5	14	3.5	4
0.6	0	1.6	3	2.6	6	3.6	2
0.7	2	1.7	5	2.7	7	3.7	1
0.8	1	1.8	6	2.8	5	3.8	0
0.9	2	1.9	6	2.9	6	3.9	1
						4	1

10.11 The WHO (Wyatt Hiring Organization) would like to predict how the trainees in its sales force will perform. At the beginning of their two-month training course, they are given an aptitude test. Sales records are kept for each trainee over the first year, the results are shown below.

Aptitude Score	Number of Sales
18	54
26	64
28	54
34	62
36	68
42	70
48	76
52	66
54	76
60	74

(a) Plot the data on linear graph paper.
(b) Using the method of least squares, determine the equation of the line.
(c) Draw the line on the curve.
(d) Calculate and interpret the coefficient of correlation.
(e) What level of sales would you expect from an aptitude score of 40, 50, 70?

10.12 All materials are elastic to some degree. It is desirable that certain parts of some designs compress when a load is applied to assist in making the part air- or watertight. The test results in the following table resulted from a test on a material known as Silon Q-177.

Pressure P (MPa)	Relative Compression R (%)
0.1	15
0.2	17
0.3	18
0.4	19
0.5	20
0.6	21
0.7	22
0.8	23
0.9	24
1.0	25.5
1.1	27
1.2	28
1.3	29.7
1.4	31.2
1.5	32.8
2.0	42
2.5	54
3.0	70

(a) Plot the data on semilog paper.
(b) Using the method of least squares, find the equation of the line of best fit.
(c) Draw the line on the curve.
(d) Calculate the coefficient of correlation.
(e) What pressure should be applied to achieve a relative compression of 60%?

10.13 An insurance company is interested in the relationship between the number of licensed vehicles in a state and the number of accidents per year in that state. It collects a random sample of 10 counties within the state:

X, Number of Licensed Vehicles (in thousands)	Y, Number of Accidents (in hundreds)
4	1
10	4
15	5
12	4
8	3
16	4
5	2
7	1
9	4
10	2

(a) Plot the points on linear graph paper.
(b) Find the equation that relates X and Y.
(c) Find the correlation coefficient.
(d) Estimate from the derived equation the number of accidents in the largest county, with 35 000 vehicles.

10.14 The annual flows (lowest one-day flow rate within a calendar year) of two tributaries of a river are expected to be linearly related. Data for a 12-year period are shown below.

Year	North Tributary Flow (cfs)	South Tributary Flow (cfs)
1	225	232
2	354	315
3	201	174
4	372	402
5	246	204
6	324	324
7	216	189
8	210	224
9	195	210
10	264	281
11	276	235
12	183	174

Use a spreadsheet to do the following:
(a) Compute the best-fit line using the method of least squares. Use data for the north tributary as the independent variable.
(b) Plot the data and the prediction equation on the same graph.
(c) Compute the correlation coefficient and interpret it.
(d) If the low flow for the north tributary is 150 cfs, what would you expect the flow for the south tributary to be?

10.15 The capacity of a screw conveyor that is moving dry ground corn is expressed in liters per second and the conveyor speed in revolutions per minute. The results of tests conducted on a new model conveyor are given below:

Capacity C (L/s)	Angular Velocity V (r/min)
3.01	10.0
6.07	21.0
15.0	58.2
30.0	140.6
50.0	245.0
80.0	410.0
110.0	521.0

(a) Plot the data on log-log graph paper.
(b) Using the method of least squares, find the equation of the line of best fit.
(c) Draw the line on the graph.
(d) Calculate the correlation coefficient.
(e) If the angular velocity of the auger was accelerated to 1 000 r/min, how many liters of corn could be moved each second?

10.16 The following table shows the U.S. consumption and production of oil per day in barrels × 1 000 from 1994 to 2008:

Year	U.S. Daily Oil Consumption Barrels × 1 000	U.S. Daily Oil Production Barrels × 1 000
1994	17 719	8 389
1995	17 725	8 322
1996	18 309	8 295
1997	18 621	8 269
1998	18 917	8 011
1999	19 519	7 731
2000	19 701	7 733
2001	19 649	7 669
2002	19 761	7 626
2003	20 033	7 400
2004	20 517	7 241

Source: BP Statistical Review of World Energy June 2005.

(a) Plot the data on linear graph paper.
(b) Find the equation that best describes the data for production and consumption.
(c) Use the equation found in part (b) to predict the expected U.S. oil consumption for the year 2015.
(d) Use the equation found in part (b) to predict the expected U.S. oil production for the year 2015.
(e) Estimate the number of dollars per day the United States will spend on imported oil at $80.00/barrel in 2015.

10.17 A mechanized "swinging hammer" is used to drive bolts into a wall. It is known that at the point of contact, the energy dissipated as heat is proportional to the square of the terminal velocity of the hammer. We have collected the following data:

Terminal Velocity (m/s)	Energy Dissipated (J)
26	2 800
35	4 220
40	5 380
49	8 250
53	10 097

(a) Manually plot the points on a suitable graph.

(b) Determine the mathematical relationship between the variables using the equations from the chapter.

(c) Using a spreadsheet, plot the points on a suitable graph.

(d) Using a spreadsheet, construct a trendline and determine the equation of the curve and the coefficient of determination.

(e) Compare and discuss the results between the two methods.

10.18 Experimental aircraft for the military are to be ranked relative to their Mach number, Ma. Mach number is a ratio of the velocity, V, of the aircraft to the speed of sound, c. The speed of sound, however, is a function of temperature. The theoretical equations for the speed of sound can be expressed as $c = (kRT)^{0.5}$, where

$k =$ specific heat ratio that varies with temperature but has no dimension. However, for this problem assume that it is a constant with a value of 1.4.

$R =$ Gas constant for air (53.33 ft · lbf/lbm · °R)

$T =$ Temperature, °R

Let us imagine that we did not know the equation for the speed of sound but had access to data relating the speed of sound in air to temperature. That set of data is recorded in the table below.

(a) Develop a table using spreadsheet software that will provide the necessary variables for Eqs. (10.16), (10.17), and (10.19).

From the data in part (a) above and the equations presented in the chapter material, determine:

(b) The equation of the line $c = bT^m$

(c) The coefficient of correlation

(d) Using the spreadsheet package, determine the equation of the line and the coefficient of determination.

(e) Does the equation obtained from the data verify the theoretical equation presented for the speed of sound?

(f) What is the numerical value and what are the units on the constant b? Recall

$$g_c = \frac{32.2 \text{ lbm} \cdot \text{ft}}{\text{lbf} \cdot \text{s}^2}$$

(g) Three aircraft were measured at the following velocities and temperatures. Determine the Mach number for each from the equation of the curve in part (b):

1. X101   855 mph at −100 degrees Fahrenheit
2. X215   548 m/s at −95 degrees Celsius
3. X912   2 120 ft/s at 520 degrees Rankine

Speed (ft/s)	Temperature Rankine	Speed (ft/s)	Temperature Rankine
100	490	550	1149
150	600	600	1201
200	693	650	1250
250	775	700	1297
300	849	750	1342
350	917	800	1387
400	980	850	1429
450	1040	900	1471
500	1096	950	1511
		1000	1550

Figure 11.1

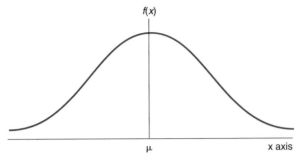

Normal distribution.

The location and shape of the normal curve can be specified by two parameters: (1) the population mean (μ), which locates the center of the distribution, and (2) the population standard deviation (σ), which describes the amount of variability or dispersion of the data.

Mathematically, the normal distribution is represented by Eq. (11.1):

$$f(x) = \frac{1}{\sigma\sqrt{2\pi}}\,e^{-(1/2)[x-\mu/\sigma]^2} \tag{11.1}$$

This expression can be used to determine the area under the curve between any two locations in the x-axis as long as we know the mean and standard deviation of the data:

$$\text{Area} = \int_{x_1}^{x_2} f(x)\,dx \tag{11.2}$$

Since the evaluation of this expression is difficult, in practice we obtain areas under the curve either from a special table of values that was developed from this equation, from functions built into a spreadsheet, or from a mathematical analysis program. What follows guides you through the use of the table.

As indicated previously, a normal curve is symmetrical about the mean; however, the specific shape of the distribution depends on the deviation of the data about the mean. As can be seen in Figure 11.2, when the data are bunched around the mean, the curve drops off rapidly toward the x-axis. However,

Figure 11.2

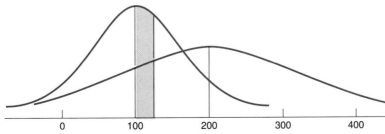

Normal curves having standard deviations and means of different magnitudes.

Figure 11.3

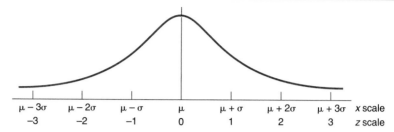

Normal curve with normalized distribution.

when the data have a wide deviation about the mean, the curve approaches the *x*-axis more slowly. The problem this presents can be seen in Figure 11.2, where the area under the curves between *x* values of 100 and 125 is not the same for the two distributions.

Thus with each different mean and standard deviation we would have to construct a separate table depicting the normal-curve areas. To avoid having to use many tables, a transformation can be applied converting all curves to a standard form that has $\mu = 0$ and $\sigma = 1$ (see Fig. 11.3). Thus we can normalize the distribution by performing a change of scale that converts the units of measurement into standard units by means of the following equation:

$$z = \frac{x - \mu}{\sigma} \tag{11.3}$$

In order to determine areas under a standardized normal curve, we must convert *x* values into *z* values and then use the Table in Appendix C, Areas under the Standard Normal Curve. Note that numbers in this table have no negative values. The reason for this is due to the symmetry of the normal curve about the mean; however, this does not reduce the utility of the table.

Before looking at example problems that demonstrate the practical application of this concept, we will demonstrate the use of the normal-curve table. The total area under the curve to the left of $z = 0$ as well as the area to the right of $z = 0$ are both equal to 0.500 0 because for this standardized normal curve, the area beneath the entire curve is equal to 1.000.

Referring to Figure 11.4, we shall calculate the probability of getting a *z* value less than +0.85 (the shaded portion of the curve). The area left of $z = 0$ is 0.5. The area from $z = 0$ to $z = +0.85$ can be determined from the table as 0.302 3. Adding these values gives us a probability of 0.802 3.

Figure 11.4

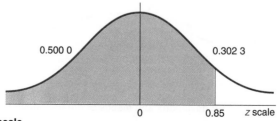

Normal curve, z scale.

Figure 11.5

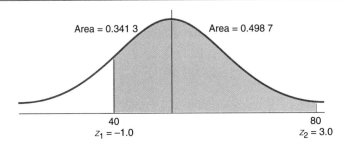

When working problems with this table, we must remember that the data should closely approximate a normal distribution, and we must know the population mean (μ) and the standard deviation (σ).

Example Problem 11.1 A random variable has a normal distribution with $\mu = 5.0 \times 10^1$ and $\sigma = 1.0 \times 10^1$. What is the probability of the variable assuming a value between 4.0×10^1 and 8.0×10^1? (See Fig. 11.5.)

Solution Normalize the value of μ and σ.

$$z_1 = \frac{40 - 50}{10} = -1.0$$
$$z_2 = \frac{80 - 50}{10} = +3.0$$

From the table in the appendix (Areas under the Standard Normal Curve), we obtain a value for $z_1 = -1.0$ (0.341 3). Note that the value obtained from the table is always the area from any given value of z to $z = 0$. Therefore we determine $z_2 = +3$ (0.498 7), so by adding these together:

Area = probability = 0.341 3 + 0.498 7 = 0.840 0

You may also solve this problem using a spreadsheet (NORMDIST in Excel). This is left as an exercise for you to try.

In most statistical applications we do not know the population parameters; instead we collect data in the form of a random sample from that population. It is possible to substitute sample mean and sample standard deviation, provided the sample size is sufficiently large ($n \geq 30$). A different theory called the *t distribution* is applicable to smaller sample sizes; this distribution is discussed in the next section but calculations using this distribution are left to more advanced statistics textbooks.

Example Problem 11.2 Assume that a normal distribution is a good representation of the data provided in Table 11.1.

(*a*) Determine the probability of a male student weighing more than 1.7×10^2 lbm.

Table 11.1

164	171	154	160	158	150	159	185	168	158
143	159	162	165	160	167	166	164	152	172
177	165	170	155	155	163	180	157	145	160
149	153	137	173	157	175	163	147	156	156
162	167	165	166	162	136	158	170	162	159

(*b*) Determine the percentage of male students who weigh between 1.4×10^2 and 1.5×10^2 lbm.

(*Note:* You may find it easier to visualize the following solution if you sketch a normal curve with appropriate z values; see Figs. 11.6 and 11.7.)

Solution

(*a*) $z_1 = \dfrac{170 - 160.74}{9.89}$

$= 0.94$

From the normal-curve table, area $= 0.326\ 4$. Since the area under the curve to the right of $z = 0$ is 0.500 0, the probability of a male student weighing more than 170 is determined by subtracting the table value from 0.500 0.

Probability $= 0.500\ 0 - 0.3264 = 0.173\ 6$

Figure 11.6

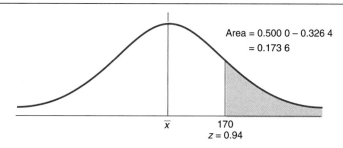

Area $= 0.500\ 0 - 0.326\ 4$
$= 0.173\ 6$

$\bar{x}$ 170
$z = 0.94$

Figure 11.7

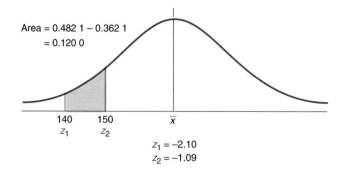

Area $= 0.482\ 1 - 0.362\ 1$
$= 0.120\ 0$

140   150 $\bar{x}$
z_1 z_2

$z_1 = -2.10$
$z_2 = -1.09$

(b) $z_1 = \dfrac{140 - 160.74}{9.89}$

$\quad = -2.10$

$z_2 = \dfrac{150 - 160.74}{9.89}$

$\quad = -1.09$

For $z_1 = 2.10$

Area = 0.482 1 (area between $z = -2.10$ and $z = 0$)

For $z_2 = 1.09$,

Area = 0.362 1 (area between $z = -1.09$ and $z = 0$)

The desired area is the difference, that is,

Probability = area = $0.4821 - 0.362\ 1 = 0.120\ 0$

Therefore, we would expect 12% of the males to weigh between 140 and 150 lbm. Try solving this problem with a spreadsheet function.

11.4.2 Student's *t* Distribution

William Sealy Gosset (1876–1937), a chemist working on quality control at a brewery in Dublin, Ireland, wrote many statistical papers under the pseudonym "Student." He recognized that, for small samples (generally for samples containing 30 or fewer observations), distributions departed substantially from the normal distribution which was described in the previous section. Gosset also noted that as the sample sizes changed, the distributions changed. This gave rise to not one distribution but a family of distributions. One may also note that as sample size increases and approaches infinity, the distributions increasingly approximate the normal distribution. The statistic for small samples from a normal distribution is universally known as *Student's t.*

The *t* distributions are a family of distributions that, like the normal distribution, are symmetrical, bell shaped, and centered on the mean. Because the *t* distribution changes as the sample size changes, there is a specific *t* distribution for every sample of a given size. Figure 11.8 demonstrates a family of *Student's t* curves.

Figure 11.8

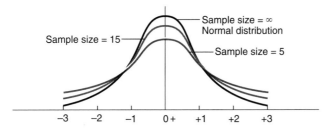

Student's *t* distribution for sample size of α, 15, and 5.

11.4.3 *F*–Distribution

In some instances we are asked to evaluate two separate data sets and deter-mine whether they are equal with respect to their population variances. If we hypothesize that they are equal, it is equivalent to hypothesizing that the ratio of the variances equals 1.00. The corresponding test statistic, called the *F* ratio, is the ratio of the two sample variances.

The sampling distribution for this test statistic cannot be approximated by either the normal or *Student's t* distributions. Therefore, a different sampling distribution must be used. It is called the *F* distribution, named for R. A. Fisher (1890–1962) who is widely credited with the modern development of statisti-cal methods. Like the *Student's t* distribution, the *F* distribution is a family of distributions and determined by the size of the two samples being compared. Unlike the normal and *t* distributions, the *F* distributions are not symmetrical, and go from zero to + infinity. In addition, the shapes of the *F* distributions vary drastically, especially when the sample sizes are small. These characteris-tics make determination of critical values for *F* distributions more complicated than for the normal and *t* distributions. Advanced statistics texts will provide a table of values from which critical values for the *F* distribution will be found.

11.4.4 *Chi-Square* Distribution

Lastly, we discuss the *chi-square* distribution. As with the use of the normal, *Student's t* and *F* distributions, we use the *chi-square* to test various hypotheses concerning sample statistics and population parameters. In each of the previ-ous distributions, certain assumptions were made about the parameters of the population from which the samples were drawn, namely, assumptions of normality and homogeneity of variance. But what do we do when the data for a research project are not normally distributed and not homogeneous? Luckily, there are a number of tests which have been developed for such data with the *chi-square* test being the most common and useful.

Often in statistical experimentation, data is simply classified into catego-ries, then the number of objects in each category is counted. For instance, if we make a nominal measurement of automobiles, we might classify them into various makes and then count the number of cars of each make. Gender, color

Figure 11.9

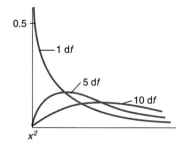

Chi-square **distribution for varying sample sizes (degrees of freedom).**

of eyes, outcome of a coin flip (heads or tails), and the number which occurs as a result of rolling a dice (1, 2, 3, 4, 5, or 6) are additional examples of variables measured on a nominal scale.

The most frequent use of the *chi-square* (X^2) distribution is in the analysis of nominal data. In such analyses, we compare observed frequencies with theoretical or expected frequencies. For example, in 100 flips of a coin, we would expect 50 heads and 50 tails. But what if we observed 46 heads and 54 tails? Would we suspect that the coin is unfair or would we attribute the difference between observed and expected frequencies to random fluctuation? The *chi-square* test statistic would allow us to decipher whether an outcome is or is not probable given what we know about population parameters.

A family of *chi-square* distributions have been generated which are asymmetrical and positively skewed with values ranging between zero and infinity. The *chi-square* distributions for 1, 3, 5, and 10 degrees of freedom are illustrated in Figure 11.9.

11.5 Level of Significance and Confidence Intervals

Referring back to the five steps for successfully utilizing inferential statistics outlined in Section 11.1, the last step asks us to make informed inferences from a sample to the population and it allows us to assign degree of certainty in our results. We know that variability in sample data is to be expected and it is typical in all experimentation. In inferential statistics, the problem of drawing conclusions from the results is addressed with statistical methods like levels of significance and confidence intervals that have a known and controllable probability of being correct.

A test of significance is a probability that represents the chance of our experimental results being valid. Normally, tests of significance have probabilities of 95% or 99%, which are simply common, useful conventions to indicate the experimenters' level of certainty in the results. A test of significance is useful in that it exerts a sobering influence on an experimenter who tends to jump to conclusions based on inadequate data.

Confidence intervals on the other hand may give the experimenter lower and upper bounds on the acceptable outcome of a study. We may be able to identify a range of values that we are confident (but not certain) contains the population parameter. For example we may be able to state that we are 95% confident that some population mean is between 82 and 85. This also means that there is a 5% chance that the mean is not between 82 and 85. As the experimenter, you must decide whether a 95% or 99% confidence interval is necessary and design your experiment accordingly.

To summarize, the power of inferential statistics is significant. However, as an experimenter you are required to carefully plan, closely monitor, and adhere to the prescribed methods of analysis in order to gain insight about populations based on a collected sample.

It is important to add that much additional reading and study in the area of inferential statistics is required to take full advantage of its potential. The information provided in this chapter and specifically on Student's *t*, *F*, and *chi-square* distributions is purely a fundamental introduction to these areas.

PROBLEMS

11.1 The quality control department at Alexander Fasteners measured the length of 100 bolts randomly selected from a specified order. The mean length was found to be 9.75 cm, and the standard deviation was 0.01 cm. If the bolt lengths are normally distributed, find
(a) The percentage of bolts shorter than 9.74 cm
(b) The percentage of bolts longer than 9.78 cm
(c) The percentage of bolts that meet the length specification of 9.75 ± 0.02 cm
(d) The percentage of bolts that are longer than the nominal length of 9.75 cm

11.2 Australian Koala bears are thought to have a mean height of 21 inches and a standard deviation of 3 inches. You plan to choose a sample of 55 bears at random. What is the probability of a sample mean between 20 and 22? How many bears in the sample are likely to be taller than 25 inches?

11.3 A random sample of 150 students attending Iowa State University has a mean age of 20.3 years and a standard deviation of 1.2 years, while a random sample of 110 students attending the University of Iowa has a mean age of 21.1 years and a standard deviation of 1.5 years. Discuss whether we can conclude that the average age of students at the two universities are not the same. (Calculations not required.)

11.4 The Karine Light Company manufactures inexpensive light-emitting diode (LED) bulbs whose estimated lifetime is normally distributed with a mean of 9 200 hours and a standard deviation of 600 hours. Assume that they are lighted continuously:
(a) Calculate the probability that the LED will fail within one year.
(b) Calculate the probability that the LED will still be operating 400 days after installation.
(c) Out of 100 LEDs, about how many will have lifetimes between 8 500 and 10 200 hours?

11.5 A machine in the JLK bottling plant is designed to fill 16 oz cans with coffee. The average weight of the coffee in the can is 16.08 oz, and the standard deviation is estimated to be 0.10 oz. If a batch consists of 100 000 cans, how many cans in a batch can be expected to weigh less than 16 oz?

11.6 The average height of a male ROTC cadet is 70 inches. Twelve percent of the male cadets are taller than 72 inches. Assuming that the heights of the cadets are normally distributed, what is the standard deviation?

11.7 The JMH Company manufactures resistors with an ideal resistance value of 400 ohms. A sample of 2 000 resistors were tested and found to have a mean resistance of 391.5 ohms with a standard deviation of 22.5 ohms. Assume a normal distribution.
(a) If the ideal resistor (400 ohms) has a manufacturing tolerance of ± 10% (±40 ohms), find the number of resistors from the sample that are expected to be satisfactory.
(b) Repeat part (a) for a tolerance of 1%.
(c) Given a tolerance of ±10%, what is the probability that a particular resistor will be unusable?
(d) Find the number of sample resistors with values exceeding 425 ohms.

11.8 Suppose that you have your choice between two jobs. Your annual earnings from a manufacturing job will have a normal distribution with a mean of $40 000 and standard deviation of $5 000. Your annual earnings from an engineering sales job will have a normal distribution with a mean of $35 000 and standard deviation of $15 000. What is the probability that you would earn more than $45 000 per year if you select one or the other?

11.9 Select one of the six following distributions and using the Internet or other sources, learn what type of data it is likely to represent and explain how it is used in statistical inference.

Geometric, Poisson, Exponential, Binominal, Gamma, Hypergeometric

11.10 There are circumstances for which cluster sampling methods provide access to specific groups within a population. Do your own research on this method as well as multi-stage sampling techniques and write a one–page paper on your findings.

11.11 Investigate the Level of Significance in hypothesis testing. Explain what Type I and Type II errors are and what they mean to a researcher.

11.12 Investigate the term "Six Sigma." Write a paper which outlines the method and its significance to quality control. Interview an engineer who has earned the "Six Sigma Certified" recognition to learn how this relates to inferential statistical methods.

CHAPTER 12

Mechanics: Statics

Chapter Objectives

When you complete your study of this chapter, you will able to:

- Define force, transmissibility, moment, couple, and equilibrium
- Identify collinear, concurrent, coplanar, and concurrent coplanar force systems
- Write a force vector in two dimensions
- Combine forces into a resultant and resolve forces into components
- Calculate the moment of a force and a couple
- Construct a free-body diagram of a rigid body at rest
- Write and solve the equations of equilibrium for a rigid body at rest

12.1 Introduction

Mechanics is the study of the effects of forces acting on bodies. The principles of mechanics have application in the study of machines and structures utilized in several engineering disciplines. Mechanics is divided into three general areas of application: rigid bodies, deformable bodies, and fluids.

Mechanics of rigid bodies is conveniently divided into two branches: statics and dynamics. When a body is acted upon by a balanced force system, the body will remain at rest or move with a constant velocity, creating a condition called equilibrium. This branch of mechanics is called *statics* and will be the focus of discussion in this chapter. The study of unbalanced forces on a body, creating an acceleration, is called *dynamics*. The relationship between forces and acceleration is governed by $F \propto ma$, an expression of Newton's second law of motion.

Strength of materials, or *mechanics of materials,* is the branch of mechanics dealing with the deformation of a body due to the internal distribution of a system of forces applied to the body. A brief introduction to this branch of mechanics is found in Chapter 13. The cranes in Figure 12.1 were designed using the principles of statics and mechanics of materials.

Before developing concepts from statics, a brief review of coordinate systems and units is appropriate.

Figure 12.1

Design of tower cranes requires analysis of the loads that are to be safely handled.

12.1.1 Coordinate Systems

The Cartesian coordinate system (shown as the xyz-axes in Figure 12.2) is the basis of reference for describing the body and force system for mechanics. In this introduction to statics, we restrict our study to two-dimensional systems (bodies and forces lying in the xy plane).

12.1.2 Units

As discussed in Chapter 7, three systems of units are commonly used in the United States. First, the International System of Units (SI) has the fundamental units meter (m) for length, kilogram (kg) for mass, and second (s) for time. In the SI system, force is a derived unit and is called a newton (N). A newton has units of $1.0 \ \text{kg} \cdot \text{m/s}^2$.

This system is officially accepted in the United States but is not mandatory, and two other systems are still widely practiced: the U.S. Customary System

Figure 12.2

279

Introduction

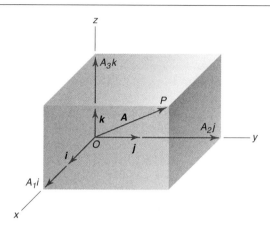

Graphical representation of a vector.

and the Engineering System. The U.S. Customary System (formerly known as the British gravitational system) has the fundamental units foot (ft) for length, pound (lb) for force, and second (s) for time. In this system mass (m) is a derived unit and is called the slug. Since it is a derived unit its magnitude can be established. A slug is defined as the amount of mass that would be accelerated to one foot per second squared when acted upon by a force of one pound. A slug, therefore, is coherent and has units of 1.0 lb·s^2/ft. The other system of units commonly used in the United States, the Engineering System, also selects length, time, and force as the fundamental dimensions. The pound (lb) is used to denote both mass and force; to avoid confusion (lbm) is used for pound-mass and (lbf) for pound-force. In the Engineering System force is specified as a fundamental dimension, and its units (lbf) are adopted independent of other base units. By convention, one pound-force is the effort required to hold a one pound-mass elevated in a gravitational field where the local acceleration of gravity, g_L, is 32.174 0 ft/s^2. We see that a force of 1.0 lbf is quite large. In fact it is large enough to provide an acceleration of 1.0 ft/s^2 to a mass of 32.174 0 lbm, or an acceleration of 32.174 0 ft/s^2 to a mass of 1.0 lbm.

In order to write Newton's equation as an equality in the Engineering System, i.e., $F = ma/g_c$, we must include the constant of proportionality (g_c). This is not the local gravitational constant, g_L, but rather a constant of proportionality. Therefore, when solving problems where the local acceleration of gravity is 32.174 0 ft/s^2, the magnitudes of lbm and lbf are equal.

$$F = ma/g_c \text{ or } g_c = ma/F \text{ where } g_c = 32.174 \text{ 0 lbm·ft/lbf·s}^2$$

Be sure when solving a problem that all units involved are in the same system, usually the system in which the given data is specified. The examples in this chapter and the problems at the end of the chapter include all the unit systems.

12.2 Scalars and Vectors

A **scalar** is a physical quantity having magnitude but no direction. Examples of scalar quantities are mass, length, time, and temperature.

A **vector**, on the other hand, is a physical quantity having both a direction and a magnitude. Some familiar examples of vector quantities are displacement, velocity, acceleration, and force. Vector quantities may be represented by either graphical or analytical methods. Graphically, a vector is represented by an arrow, such as *CD* in Figure 12.3. Point *C* identifies the origin and *D* signifies the terminal point. The magnitude of the vector quantity is represented by selecting a scale and constructing the arrow to the appropriate length. The arrow positioned in either direction determines the *line of action* of the vector.

Analytically a vector can be represented in different ways. First, consider the familiar Cartesian coordinate system shown in Figure 12.2. Graphically the vector is shown from *O* to *P*. Analytically the point *O* is at the (0, 0, 0) coordinate position, and the point *P* is at the (A_1, A_2, A_3) position. The vectors (**i**, **j**, **k**) are called *unit vectors* in the *x*, *y*, *z* directions, respectively, and the vectors A_1**i**, A_2**j**, and A_3**k** are called *rectangular component vectors*. The unit vectors **i**, **j**, **k** establish the direction, while the magnitude of the rectangular components of the vector is given by A_1, A_2, and A_3, for the *x*, *y*, *z* directions respectively.

A typical notation for representing a vector quantity in printed material is a boldface type, for example, the vector **A**. However, since engineers generally solve problems using pencil and paper, the boldface type is difficult to duplicate. Hence, the convention of placing a bar or arrow over the vector symbol has been adopted, for example, $\overline{A}$ or $\overrightarrow{A}$.

The notation we will use in this text is to represent the vector as boldface **A** and its magnitude as $|A|$ or *A*. The unit vectors will be identified as **i**, **j**, **k**. The Cartesian vector **A** and its magnitude can then be written

$$\mathbf{A} = A_1\mathbf{i} + A_2\mathbf{j} + A_3\mathbf{k} \tag{12.1}$$

$$|\mathbf{A}| = A = \sqrt{A_1^2 + A_2^2 + A_3^2} \tag{12.2}$$

Figure 12.3

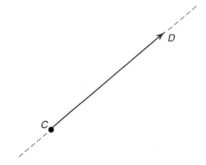

Description of a vector

Since our discussion of statics is limited to two-dimensional systems, vectors will have x and y components only.

$$\mathbf{B} = B_1\mathbf{i} + B_2\mathbf{j} \tag{12.3}$$

Care must be exercised when working with vectors. Vector quantities normally add or subtract by the laws of algebra. For the most part we will work with the scalar components of the vectors. Remember, if you are representing a vector, both magnitude and direction must be specified. If you wish to represent the scalar component of a vector, then only magnitude need be specified.

12.3 Forces

The action of one body acting upon another body tends to change the motion of the body acted upon. This action is called a *force*. Because a force has both magnitude and direction, it is a vector quantity, and the previous discussion on vector notation applies.

Newton's third law states that if a body P acts upon another body Q with a force of a given magnitude and direction, then body Q will *react* upon P with a force of equal magnitude but opposite direction. Therefore, as indicated earlier, to describe a force you must give its magnitude, direction, and the location of at least one point along the line of action. Figure 12.4 indicates common ways of depicting a force. Although the reference line for the angle may be selected arbitrarily, Figure 12.4a shows the standard procedure of measuring the angle counterclockwise from the positive horizontal axis. Instead of specifying the angle, one may indicate the slope, as shown in Figure 12.4b.

An alternative method used to describe a force is illustrated in Figure 12.4c. When two points along the line of action are known, the location and direction of the force can be identified. The vector is then represented as **AB**. Incidentally, the reverse notation **BA** would specify a vector of equal magnitude but of opposite direction.

12.4 Types of Force Systems

Forces acting along the same line of action, as illustrated in Figure 12.5a, are called *collinear forces*. The magnitude of collinear forces can be added and

Figure 12.4

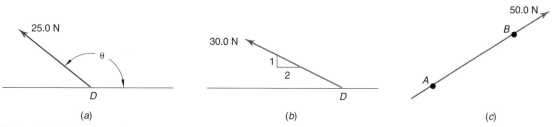

(a)   (b)   (c)

Methods of depicting a force vector.

Figure 12.5

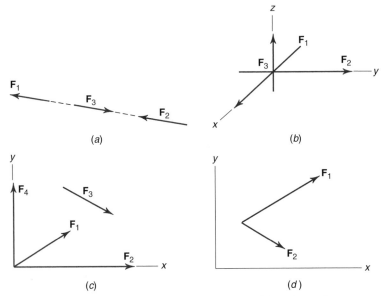

Force systems: (a) collinear forces; (b) concurrent forces; (c) coplanar forces; (d) concurrent, coplanar forces.

subtracted algebraically. The three forces in Figure 12.5a—F_1, F_2, and F_3—can be replaced by a single resultant force F_4.

$$|F_4| = |F_1| + |F_2| - |F_3| \tag{12.4}$$

The convenience of this addition and subtraction will be used to advantage when we resolve vectors into x and y components.

Forces whose lines of action pass through the same point in space are called *concurrent forces*. The illustration in Figure 12.5b represents a three-dimensional concurrent force system.

Forces that lie in the same plane, as seen in Figure 12.5c, are known as *coplanar*. Combinations of these force systems may also occur in problems. For example, the force system in Figure 12.5d is both concurrent and coplanar.

12.5 Transmissibility

The principle of transmissibility, as illustrated in Figure 12.6, is an extension of the concept associated with collinear forces. Two equal and opposite collinear forces, **R** and **S**, are applied to the object at points A and B respectively

Figure 12.6

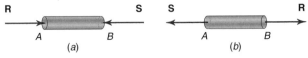

A force may be applied at different points along its line of action.

in Figure 12.6a. The object is in compression. If the force **R** is moved along its line of action to B, and **S** to A, as shown in Figure 12.6b, the object is now in tension. Thus the internal effect has been changed but since the forces are equal the object is in equilibrium and the external effect remains the same.

Thus a force can be moved along its line of action without altering the external effect. The principle of transmissibility is applied frequently to simplify computations in statics problems.

12.6 Resolution of Forces

If a force is determined, then its magnitude, direction, and one point on the line of action are known. When calculating the effect of a force, quite often it is convenient to divide or resolve the force into two or more components. Consider the force **F** given in Figure 12.7. It is acting through the point A and forms an angle of θ with the positive x-axis.

In this chapter we will concentrate on two-dimensional force systems; thus we will be working in the xy-coordinate system shown in Figure 12.7. In most problems it will be convenient to work with the scalar components of forces in the x and y directions.

The force **F** can be replaced by its two vector components $\mathbf{F}_x$ and $\mathbf{F}_y$. This means that if A is a point on a rigid body, the net effect of a force **F** applied to that body is identical to the combined effect of its components $\mathbf{F}_x$ and $\mathbf{F}_y$ applied to that point. Had $\mathbf{F}_x$ and $\mathbf{F}_y$ been known initially, it would follow that **F** is the resultant of $\mathbf{F}_x$ and $\mathbf{F}_y$. The *resultant* of a force system acting on a body is the simplest system (normally a single force) that can replace the original system without changing the external effect on the body.

Expressed in mathematical terms, the scalar relationship of the quantities in Figure 12.7 are

$$F_x = |\mathbf{F}| \cos \theta \tag{12.5}$$
$$F_y = |\mathbf{F}| \sin \theta$$

$$|\mathbf{F}| = \sqrt{F_x^2 + F_y^2} \tag{12.6}$$
$$\theta = \tan^{-1} \frac{F_y}{F_x}$$

Figure 12.7

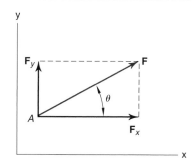

Resolution of a force into *x* and *y* components.

Equation (12.5) is used when the scalar components of a given vector are desired in the specified x and y directions. Equation (12.6) is used when the x and y components are known and the magnitude and direction of the resultant force are desired.

Example Problem 12.1   In Figure 12.7 suppose that the force vector **F** has a magnitude of 6.50(10²) lb and acts through point A at a slope of 2 vertical to 5 horizontal. Determine the x and y components of **F**.

Solution   We can compute the angle θ in Figure 12.7 from the tangent function.

$$\tan \theta = \frac{2}{5} = 0.40$$

$$\theta = 21.80°$$

$$F_x = |F_x| = |F| \cos 21.80° = 650(0.928\ 5) = 604 \text{ lb}$$

$$F_y = |F_y| = |F| \sin 21.80° = 650(0.371\ 4) = 241 \text{ lb}$$

We can write the actual forces F_x and F_y by showing a direction for each.

$$F_x = 604 \text{ lb} \rightarrow \qquad = 604\mathbf{i} \text{ lb}$$

$$F_y = 241 \text{ lb} \uparrow \qquad = 241\mathbf{j} \text{ lb}$$

An arrow indicates the direction in which the force acts. In this case the horizontal component of the force **F** acts in the positive x direction, and the vertical component acts in the positive y direction.

The force **F** can be written as

$$\mathbf{F} = 6.50(10^2) \text{ lb}$$

Or

$$\mathbf{F} = 6.50(10^2) \text{ lb} \qquad 21.8°$$

The latter expression for the force vector **F** is written in polar coordinate notation.

In Cartesian vector notation the force **F** would be

$$\mathbf{F} = 604\mathbf{i} + 241\mathbf{j} \text{ lb}$$

As the number of force vectors within a two-dimensional system increases, the complexity of the problem dictates that an orderly procedure be followed. Extending the application of Eqs. (12.5) and (12.6) to a system composed of several vectors, we have

$$\rightarrow \Sigma F_x = R_x \qquad\qquad (12.7)$$

$$\uparrow \Sigma F_y = R_y$$

$$|\mathbf{R}| = \sqrt{R_x^2 + R_y^2} \qquad\qquad (12.8)$$

$$\theta = \tan^{-1}\frac{R_y}{R_x}$$

Figure 12.8

285
Resolution of Forces

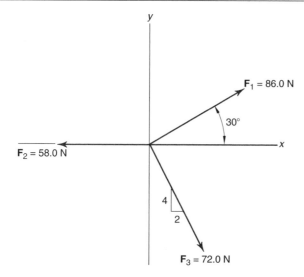

where **R** represents the resultant force. The arrows with the summation signs in Eq. (12.7) indicate the conventional positive directions for the scalar components.

Example Problem 12.2 Given the two-dimensional, concurrent, coplanar force system illustrated in Figure 12.8, determine the resultant, **R**, of this system.

Solution
1. It is convenient to make a table of each force and its components assuming conventional positive x and y directions.

Force	Magnitude	x-Component (→)	y-Component (↑)
F_1	86.0	86.0 cos 30.0° = 74.48 N	86.0 sin 30.0° = 43.0 N
F_2	58.0	58.0 cos 180° = −58.0 N	58.0 sin 180° = 0
F_3	72.0	$72.0\left(\dfrac{2}{\sqrt{20}}\right)$ = 32.2 N	$7.20\left(\dfrac{-4}{\sqrt{20}}\right)$ = −64.4 N

2. Applying Eq. (12.7)
$$\rightarrow \Sigma F_x = R_x = 74.48 - 58.0 + 32.2 = 48.68 \text{ N}$$
$$\uparrow \Sigma F_y = R_y = 43.0 - 64.4 = -21.4 \text{ N}$$

3. Applying Eq. (12.8)
$$|\mathbf{R}| = \sqrt{(48.68)^2 + (-21.4)^2}$$
$$= 53.2 \text{ N}$$
$$\theta_R = \tan^{-1}\frac{-21.4}{48.68}$$
$$= -23.7°$$

Figure 12.9

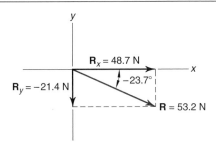

Alternatively, **R** can be expressed as

$$\mathbf{R} = (48.68\mathbf{i} - 21.4\mathbf{j})\ \text{N}$$

The resultant force vector, **R**, is shown in Figure 12.9.

Frequently it is necessary to resolve a force into nonrectangular components. A convenient procedure for accomplishing this is to first resolve the force into rectangular components, then equate the x-component value and y-component value respectively to the sum of the x and y components of the desired nonrectangular components. Example 12.3 will illustrate this procedure.

Example Problem 12.3 Find the magnitude of the nonrectangular component forces **F** and **G** of the 1250 N resultant force **R** in Figure 12.10.

Solution Equation 12.5 is used for the components of the 1250 N resultant force and **F** and **G**. The values of the cos and sin of the force directions are found from the given slopes.

1. The x and y components of the 1250 N force are:
   $R_x = (4/5)(1250) = 1000\ \text{N}$
   $R_y = (3/5)(1250) = 750\ \text{N}$
2. The x and y components of **F** and **G** are:
   $F_x = (3/5)|\mathbf{F}|$
   $F_y = (4/5)|\mathbf{F}|$
   $G_x = (12/13)|\mathbf{G}|$
   $G_y = (5/13)|\mathbf{G}|$
3. The sum of the x and y components of **F** and **G** are equal, respectively, to the x and y components of the 1250 N resultant.
   $F_x + G_x = (3/5)|\mathbf{F}| + (12/13)|\mathbf{G}| = R_x = 1000\ \text{N}$
   $F_y + G_y = (4/5)|\mathbf{F}| + (5/13)|\mathbf{G}| = R_y = 750\ \text{N}$
4. Simultaneous solution yields:
   $|\mathbf{F}| = 607\ \text{N}$
   $|\mathbf{G}| = 689\ \text{N}$

Figure 12.10

287

*Moments
and Couples*

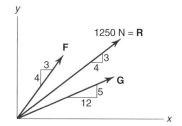

12.7 Moments and Couples

We now extend the effect of forces on rigid bodies at rest to moments and couples. You have likely approached an exit in a public building and pushed on the panic bar of the door only to find that you had chosen the wrong side of the door, the side closest to the hinges. No big problem. You simply moved your hands to the side opposite the hinges and easily pushed open the door. By doing this, you demonstrated the principle of the turning moment. The tendency of a force to cause rotation about an axis or a point is called the *moment*, **M**, of the force, **F**, with respect to that axis or point. The magnitude of the moment $|\mathbf{M}|$ is the product of the magnitude of the force $|\mathbf{F}|$ and the perpendicular distance from the line of action of the force to the axis or point. With respect to the door just mentioned, the same force may have been exerted in both attempts to open the door. In the second case, however, the moment was greater owing to the fact that you increased the distance from the force to the hinges, the axis about which the door turns.

Figure 12.11 illustrates how a moment is evaluated for a specific problem. The magnitude of the moment of the force **F** about a point B is Fd_B. The distance d_B is the perpendicular distance, called the *moment arm*, from the point of application to the line of action of **F**. Note that B may be considered a point or the point view of an axis about which the moment is calculated. The magnitude of the moment created by **F** about point A is Fd_A. The force **F** will tend to

Figure 12.11

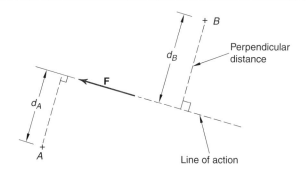

The force F and points *A* and *B* are in the same plane.

Figure 12.12

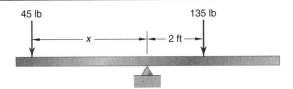

The forces are in the same vertical plane.

create a clockwise rotation about point *B* and a counterclockwise rotation about point *A*. Engineering convention is to assign a positive sign to counterclockwise moments and a negative sign to clockwise moments. Common units for moments are ft lb, in lb and Nm.

Figure 12.12 illustrates a physical interpretation of a moment. If we neglect the mass of the beam, we can determine the distance *x* from the fulcrum that will keep the beam level under the two applied forces. To maintain balance the moments of the two forces must be equivalent and opposite in the tendency to rotate the beam. Thus the 45 lb force must be applied 6 feet to the left of the fulcrum to create a moment of 270 ft lb counterclockwise and balance the 270 ft lb clockwise moment of the 135 lb force.

A couple is similar to a moment. Figure 12.13 shows a hand wheel used to close and open a large valve. The two forces shown form a *couple* because they are parallel, equal in magnitude, and opposite in direction. The perpendicular distance between the two forces is called the *arm* of the couple. The equivalent moment of a couple is equal to the product of one of the forces and the distance between the forces. In Figure 12.13 the couple has a magnitude of (10 lb) × (20 in), or 200 in lb.

We have seen that the components of a force can be determined from Eq. 12.5, and conversely, if the components are known we can find the resultant force from Eq. 12.6. This concept is extended to more than two concurrent forces in a two-dimensional system with the use of Eqs. 12.7 and 12.8. You will note that our analysis up to this point applies only to concurrent force systems where the lines of action of all the forces intersect at a common point.

Figure 12.13

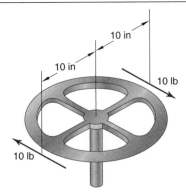

Illustration of a couple.

Now with the concept of moments, we can expand our analysis to two-dimensional nonconcurrent force systems by applying the principle of moments. **The principle of moments may be stated as: The magnitude of the moment of the resultant of a two-dimensional nonconcurrent force system with respect to any point or axis is equal to the sum of the moments of the forces of the system with respect to the same point or axis.** The applicable equations are:

$$\rightarrow \Sigma F_x = R_x \tag{12.9}$$

$$\uparrow \Sigma F_y = R_y$$

$$\hookleftarrow \Sigma M = \sqrt{R_x^2 + R_y^2} \ (d)$$

The value d is the perpendicular distance from the chosen point or axis to the line of action of the resultant **R**. Remember our accepted sign convention is positive to the right for x-direction forces, positive upward for y-direction forces, and positive counterclockwise for moments.

Example Problem 12.4   For the nonconcurrent force system shown in Figure 12.14, find the resultant and locate it with respect to the point A.

Solution   There are no forces having an x-component so only the y-component and moment equations are needed.

$$\uparrow \Sigma F_y = R_y$$

$$-60 - 50 + 150 = R_y$$

$$\mathbf{R}_y = 40 \text{ lbf} \uparrow$$

$$\hookleftarrow \Sigma M = \sqrt{R_x^2 + R_y^2} \ (d)$$

$$(60)(6) + (50)(2) + (150)(0) + (100) = (40)(d)$$

$$d = 14'$$

Thus the force system given may be replaced by a single 40 lbf upward force 14′ to the right of A with the same external effect on the beam.

Figure 12.14

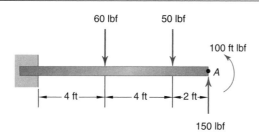

12.8 Free-Body Diagrams

The first step in solving a problem in statics is to draw a sketch of the body, or a portion of the body, and all the forces acting on that body. Such a sketch is called a *free-body diagram* (FBD). As the name implies, the body is cut free from all others; only forces that act upon it are considered. In drawing the free-body diagram, we remove the body from supports and connectors, so we must have an understanding of the types of reactions that may occur at these supports.

Examples of a number of frequently used free-body notations are illustrated in Figure 12.15. It is important that you become familiar with these so that each FBD you construct will be complete and correct.

Figure 12.15

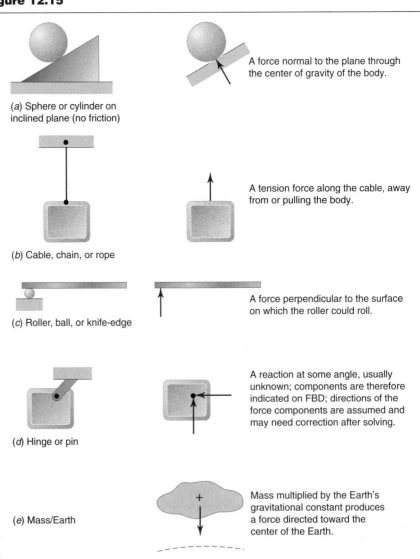

(*a*) Sphere or cylinder on inclined plane (no friction)

A force normal to the plane through the center of gravity of the body.

(*b*) Cable, chain, or rope

A tension force along the cable, away from or pulling the body.

(*c*) Roller, ball, or knife-edge

A force perpendicular to the surface on which the roller could roll.

(*d*) Hinge or pin

A reaction at some angle, usually unknown; components are therefore indicated on FBD; directions of the force components are assumed and may need correction after solving.

(*e*) Mass/Earth

Mass multiplied by the Earth's gravitational constant produces a force directed toward the center of the Earth.

Note that the force generated by the mass of an object is found by multiplying the local acceleration of gravity by the mass (units must be consistent). If the body is uniform, that is having the same material characteristics throughout, then the force can be placed at the geometric center and directed toward the center of the Earth. For example, a uniform beam 10 m long having a mass distribution of 65 kg/m would create a downward force of (65 kg/m)(10 m) $(9.807 \text{ m/s}^2) = 6400 \text{ kg} \cdot \text{m/s}^2 = 6.4 \text{ kN}$.

Example Problem 12.5   Construct a free-body diagram (FBD) for object A, shown in Fig 12.16a. Surfaces in contact are smooth. Object A is a homogeneous cylinder weighing 4.00×10^2 lbm.

Note: Weight is normally given in lbm but FBDs are visual representations of force, so weight (lbm) needs to be converted to force. Assume $g_L = 32.174\ 0$ ft/s^2.

Solution   A correct FBD will enable us to solve for unknown forces and reactions on an object. The steps to follow are:

(*a*) Isolate the desired object from its surroundings.
(*b*) Replace items cut free with appropriate forces.
(*c*) Add known forces, including weight.
(*d*) Establish a coordinate (*xy*) frame of reference.
(*e*) Add geometric data.

The result is shown in Figure 12.16b. A force vector of 400 lbf (lbf equals lbm when $g_L = 32.174\ 0$ ft/s^2) shown acting through the center of gravity. The cable is replaced with a tension force, **T**. The reaction of the smooth inclined plane on object A is shown by a normal force to the plane, **N**, acting through the center of gravity. The coordinate system and geometric characteristics complete the FBD.

Figure 12.16

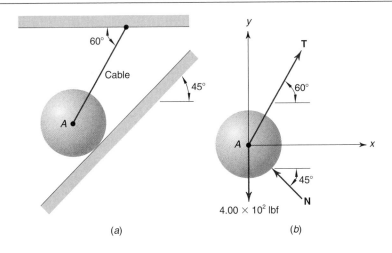

(a)   (b)

Figure 12.17

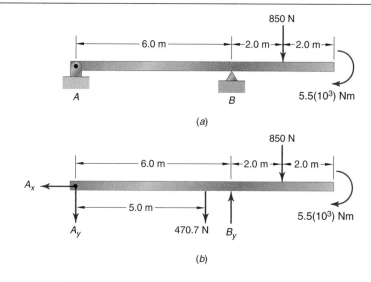

(a)

(b)

Example Problem 12.6 Construct a free-body diagram (FBD) for the beam shown in Figure 12.17a. There is a 5.50 kNm clockwise moment applied at the right end of the beam and the beam has a mass of 48 kg.

Solution The FBD is shown in Figure 12.17b. The pin reaction on the beam at A is shown as x and y component forces and the knife-edge support at B is a single vertical force, the same as a roller or ball. The directions chosen for the components at A are not important for the FBD. When calculations of the components are made, if the sign comes out negative, it means the force is in the opposite direction indicated on the FBD. The weight of the beam is computed from the mass as $(48 \text{ kg})(9.807 \text{ m/s}^2) = 470.7 \text{ N}$. The $5.5(10^3)$ Nm clockwise moment is applied at the end of the beam. All necessary dimensions are added.

12.9 Equilibrium

Newton's first law of motion states that if the resultant force acting on a particle is zero, then the particle will remain at rest or move with a constant velocity. This concept of an unchanging situation, or equilibrium, is essential to statics.

Combining Newton's first law and the concept of equilibrium, we can state that a body will be in equilibrium when the sum of all external forces and moments acting on the body is zero. This requires the body to be at rest or moving with a constant velocity.

In this chapter, we will consider only bodies at rest. To study a body and the forces acting upon it, one must first determine what the forces are. Some may be unknown in magnitude and/or direction, and quite often these unknown magnitudes and directions are information that is being sought. The conditions of equilibrium can be stated in equation form as follows:

$$\rightarrow \Sigma F_x = 0$$

$$\uparrow \Sigma F_y = 0 \qquad\qquad (12.10)$$

$$\circlearrowleft \Sigma M = 0$$

As before our sign convention is: F_x is positive to the right ($\rightarrow$) and F_y is positive upward ($\uparrow$), and counterclockwise moments are positive. When evaluating moments in a particular problem, the point from which the moment arms are measured is arbitrary. A point is usually selected that simplifies the computations.

The three independent Eqs. 12.10 enable solutions to equilibrium problems of no more than three unknowns. After construction of a correct FBD, check the unknowns and, if a solution is possible, write Eqs. 12.10 and solve in an efficient manner.

Example Problem 12.7   A beam, assumed weightless, is subject to the load shown in Figure 12.18a. Determine the reactions on the beam at supports A and B for the equilibrium condition.

Solution   This is an example of a beam problem, which commonly occurs in buildings and bridges. The loads on the beam are expressed with the prefix k signifying thousands of newtons; thus the loads shown are 1.2×10^4 N and 2.4×10^4 N, respectively. The force system is coplanar, and in this instance all forces are parallel. The solution is as follows:

1.   Construct the free-body diagram (FBD) (see Fig. 12.18b). The supports generate forces perpendicular to the length of the beam. Thus there are no forces in the x direction. The directions for the R_A and R_B reactions are

Figure 12.18

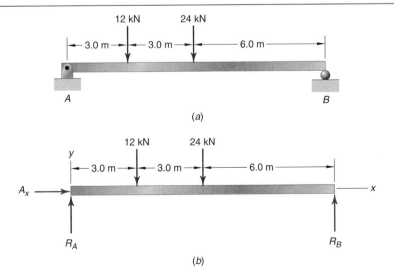

(a)

(b)

obvious in this example. However, the direction may be assumed either way, and the mathematics will produce the correct direction.

2. Write the equations of equilibrium (force units in kN)

$$\rightarrow \Sigma F_y = 0$$
$$R_A + R_B - 12 - 24 = 0$$
$$\curvearrowleft \Sigma M_A = 0$$
$$-12(3.0) - 24(6.0) + R_B(12.0) = 0$$

Note that we do not need to write $\Sigma F_x = 0$ because there are no x-direction forces. We have two equations and two unknowns, which we can solve for R_A and R_B.

From the moment equation:

$$12.0R_B = 24(6.0) + 12(3.0)$$
$$\mathbf{R_B} = 15 \text{ kN} \uparrow$$

Then from the sum of the vertical forces:

$$R_A + 15 - 12 - 24 = 0$$
$$\mathbf{R_A} = 21 \text{ kN} \uparrow$$

Since the numerical values for R_B and R_A came out positive, the reactions shown on the FBD were in the correct direction.

Example Problem 12.8   Solve Example Prob. 12.5 for the cable tension and reaction of the inclined surface on the cylinder.

Solution

1. The FBD is shown in Figure 12.16b. Observe that the force system on the cylinder is coplanar and concurrent. Since a moment equation will not produce any information for the equilibrium condition, we have two remaining equations to solve for the two unknowns **T** and **N**.

2. Write the equilibrium equations

$$\rightarrow \Sigma F_x = 0$$
$$T \cos 60.0° - N \cos 45.0° = 0$$
$$\uparrow \Sigma F_y = 0$$
$$T \sin 60.0° + N \sin 45.0° - 4.00 \times 10^2 = 0$$

Solving for T from the x-direction equation and substituting this result into the y-direction equation:

$$1.414 \, N (\sin 60.0°) + N \sin 45.0° - 4.00 \times 10^2 = 0$$
$$1.932 \, N = 4.00 \times 10^2$$

N = 207 lb

T = 293 lb

Example Problem 12.9 For the crane system shown in Figure 12.19a, determine the reactions on the crane at pin A and roller B. Neglect the weight of the crane.

Solution
1. The FBD is shown in Figure 12.19b. Note the two components for the pin reaction at A. The direction of each component was assumed.
2. There are three unknowns and three equations of equilibrium that can be written.

$$\rightarrow \Sigma F_x = 0$$
$$A_x + 94 \cos 60° = 0$$
$$A_x = -47 \text{ kN}$$
$$\mathbf{A}_x = 47 \text{ kN} \leftarrow$$

The negative sign indicates that the initial direction chosen for A_x was opposite the actual direction of the x-direction force on the crane.

Before we write the sum of moments equation with respect to point A, we will show a convenient procedure for evaluating the moment of the 94 kN

Figure 12.19

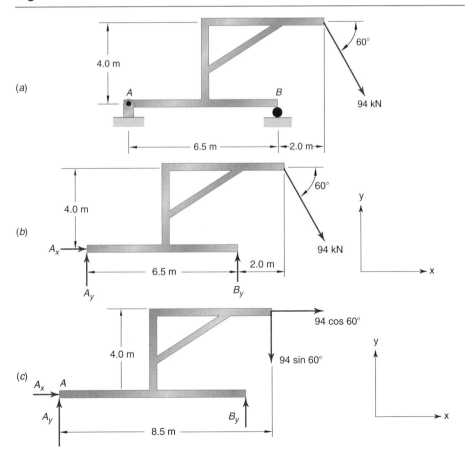

force. Finding the perpendicular distance from point A to the line of action of the 94 kN force requires trigonometry. It is less computation to first find the x and y components and apply these components at a convenient point on the line of action of the 94 kN force (principle of transmissibility). In this case the point where the force acts on the crane is the most convenient (see Fig. 12.19c). The moment arms of the horizontal and vertical components of the force with respect to the point A are 4.0 m and 8.5 m respectively.

$$\curvearrowright \Sigma M_A = 0$$
$$B_y (6.5) - (94 \cos 60°)(4.0) - (94 \sin 60°)(8.5) = 0$$
$$B_y (6.5) = 188 + 692$$
$$\mathbf{B}_y = 135 \text{ kN} \uparrow$$
$$\uparrow \Sigma F_y = 0$$
$$A_y + 135 - 94 \sin 60° = 0$$
$$A_y = -54 \text{ kN}$$
$$\mathbf{A}_y = 54 \text{ kN} \downarrow$$

Again, the initial direction for A_y was chosen opposite the actual direction of the y-direction force on the crane.

Example Problem 12.10   For the structure shown in Figure 12.20a, determine the pin reaction at G and the tension T in the cable. Assume the member, pinned at G, is weightless.

Solution
1. The FBD is shown in Figure 12.20b. Because of the orientation of the member pinned at G, there are some geometric calculations that must be made before writing the equilibrium equations.
2. Determine the geometry:

$$GH = 8.00 \text{ m from the right triangle } GHI$$
$$\alpha = 45 - \theta = 8.13°$$

3. Apply equations of equilibrium:

$$\curvearrowright \Sigma M_G = 0$$
$$-20.0(9.807)(12.0 \cos 8.13°) + 8.00H = 0$$
$$H = 291.3 \text{ N}$$
$$\mathbf{H} = 291.3 \text{ N}$$
$$\rightarrow \Sigma F_x = 0$$
$$G_x - H \cos 45° = 0$$
$$G_x = 2.06 = 10^2 \text{ N}$$
$$\mathbf{G}_x = 2.06 \times 10^2 \text{ N} \rightarrow$$
$$\rightarrow \Sigma F_y = 0$$
$$G_y + H \sin 45.0° - 20.0(9.807) = 0$$
$$G_y = -9.840 \text{ N}$$

Figure 12.20

297
Equilibrium

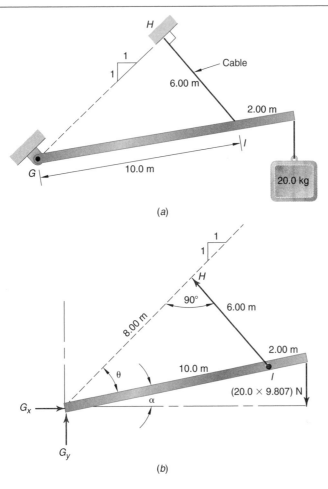

(a)

(b)

So that

$$\mathbf{G}_y = -9.840 \text{ N} \uparrow$$

or

$$\mathbf{G}_y = 9.840 \text{ N} \downarrow$$

4. Combine vector components into resultant G using Eq. (7.7):

$$G = 206.235 \text{ N}$$

$$\mathbf{G} = 206 \text{ N} \longrightarrow 2.73°$$

Problems

For problems 12.1 through 12.4, find the x and y components of the force **F**. The angle θ is measured positive counterclockwise from the positive x-axis. Include a sketch of the force **F** and its components.

	F	θ
12.1	$1.85(10)^2$ lbf	$32.5°$
12.2	$3.7(10)^3$ N	$105°$
12.3	$5.1(10)^2$ lbf	$-220°$
12.4	$4.6(10)^1$ kN	$68°$

For problems 12.5 through 12.7, find the resultant of the two concurrent forces **F**, which makes an angle θ with respect to the positive x-axis, and **G**, which makes an angle ϕ with respect to the positive x-axis. Show a sketch of **F** and **G** and the resultant.

	F	θ	**G**	φ
12.5	$8.6(10)^2$ N	$35°$	$5.7(10)^2$ N	$120°$
12.6	$4.6(10)^2$ lbf	$280°$	$3.9(10)^2$ lbf	$140°$
12.7	$3.2(10)^1$ kN	$-120°$	$7.2(10)^1$ kN	$45°$

For problems 12.8 through 12.10, find the resultant of the concurrent forces **R** and **S** for which the direction is specified by slope, expressed as rise and run values. Show a sketch of **R** and **S** and the resultant.

	R	**Rise**	**Run**	**S**	**Rise**	**Run**
12.8	$8.6(10)^3$ N	3.0	4.0	$6.2(10)^3$ N	-3.0	4.0
12.9	$1.3(10)^2$ lbf	12	-5.0	$1.3(10)^2$ lbf	12	5.0
12.10	$3.2(10)^4$ N	1.0	-3.0	$5.3(10)^4$ N	9.0	-16

12.11 Two forces $\mathbf{F}_1$ and $\mathbf{F}_2$ are applied as shown in Figure 12.21. The resultant **R** has a magnitude of 500 lbf and acts along the positive x-axis. Determine the magnitudes of $\mathbf{F}_1$ and $\mathbf{F}_2$.

12.12 Two forces $\mathbf{F}_1$ and $\mathbf{F}_2$ are applied as shown in Figure 12.22. The resultant **R** has a magnitude of 4 200 N and makes an angle of 28° with the positive y-axis. Determine the magnitudes of $\mathbf{F}_1$ and $\mathbf{F}_2$.

Figure 12.21

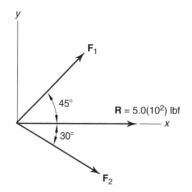

Figure 12.22

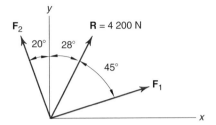

Figure 12.23

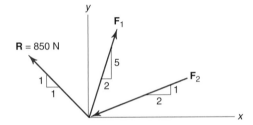

Figure 12.24

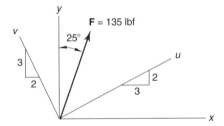

12.13 Two forces F_1 and F_2 are applied as shown in Figure 12.23. The resultant **R** has a magnitude of 850 N and acts in the direction shown in the figure. Determine the magnitudes of F_1 and F_2.

12.14 Force **F** is applied as shown in Figure 12.24.
    (*a*) Determine the x and y components of the force.
    (*b*) Determine the u and v components of the force.

For problems 12.15 through 12.20, the force **F** goes through the origin of the xy-coordinate system and makes an angle θ with the horizontal, as shown in Figure 12.25. Points A and B are at the coordinates indicated on the figure in units of feet. Calculate the moment of **F** about points A and B, assigning positive values to counterclockwise moments.

Figure 12.25

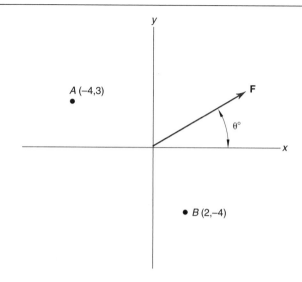

F	**θ**
12.15 $4.4(10)^3$ lbf	0°
12.16 $21.3(10)^2$ lbf	180°
12.17 $2.4(10)^3$ lbf	270°
12.18 $4.2(10)^2$ lbf	45°
12.19 620 lbf	-120°
12.20 7 450 lbf	330°

12.21 Determine the resultant of the coplanar force system shown in Figure 12.26 and locate it on a sketch of the system with respect to point *A*.

12.22 The resultant **R** of four forces is shown in Figure 12.27. One of the forces is not shown. Find the unknown force and locate it on a sketch of the system with respect to point *B*.

12.23 Find the resultant of the three-force system shown in Figure 12.28 and locate it with respect to point *A*.

12.24 Construct a free-body diagram of the system shown in Figure 12.29. The beam has a weight of 110 lbm. Determine the resultant reactions at the supports *A* and *B*. $g_L = 32.174\ 0$ ft/s^2.

Figure 12.26

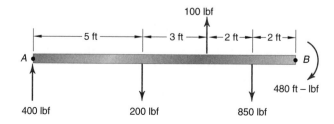

Figure 12.27

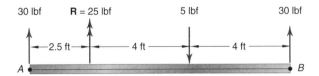

30 lbf   **R = 25 lbf**   5 lbf   30 lbf

←—2.5 ft—→|←——— 4 ft ———→|←——— 4 ft ———→

A   B

Figure 12.28

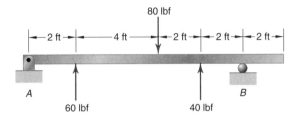

80 lbf

|←—2 ft—→|←——— 4 ft ———→|←—2 ft—→|←—2 ft—→|←—2 ft—→|

A

60 lbf   40 lbf

B

Figure 12.29

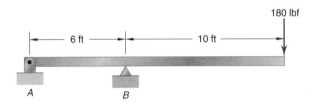

180 lbf

|←——— 6 ft ———→|←——— 10 ft ———→|

A   B

Figure 12.30

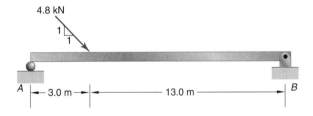

4.8 kN

1
1

A |←— 3.0 m —→|←——— 13.0 m ———→| B

12.25 Construct a free-body diagram of the system shown in Fig 12.30. The beam is uniform and has a mass distribution of 45 kg/m. Determine the reactions at the supports A and B.

12.26 Construct an FBD of the system shown in Fig 12.31. The beam is uniform and has a mass of 550 kg. Determine the moment **M** that will relieve roller A of any load. At this condition determine the resultant pin reaction at B. $g_L = 9.807 \ \text{m/s}^2$.

Figure 12.31

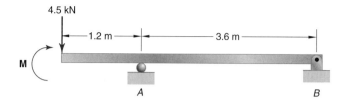

12.27 The beam in Figure 12.32 is weightless. A movable weight of 180 lbm is to be attached as shown. Determine the location x of the weight so that the vertical reactions at A and B are equal. $g_L = 32.174\ 0$ ft/s².

12.28 Assume the beam in Prob. 12.23 is weightless. (*a*) Find the reactions at A and B for the three forces acting on the beam and (*b*) Find the reactions using the resultant force found in Prob. 12.23. Compare the results and explain.

12.29 For the weightless rigid bar in Figure 12.33 construct a free-body diagram and find the resultant reactions at supports A and B.

12.30 For the weightless rigid structure in Figure 12.34 construct a free-body diagram and find the resultant reactions at supports A and B.

Figure 12.32

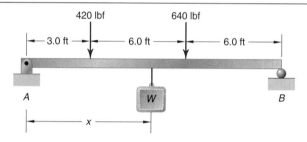

Figure 12.33

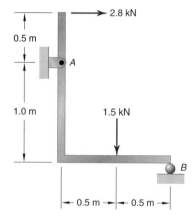

Figure 12.34

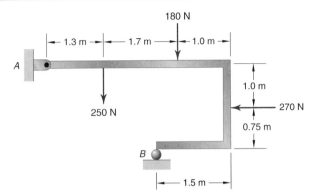

Figure 12.35

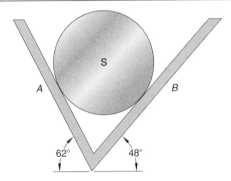

Figure 12.36

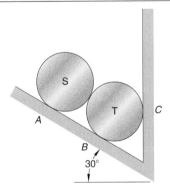

12.31 Construct a free-body diagram of the sphere S in Figure 12.35 resting against two smooth surfaces. Find the reactions of the smooth surfaces on the sphere at A and B. The sphere has a mass of 450 lbm. $g_L = 32.174\ 0$ ft/s².

12.32 Two cylinders S and T of unit length rest against smooth supports as shown in Figure 12.36. Each cylinder has a mass of 85 kg. Find the forces that the cylinders exert on the supports at A, B, and C.

Figure 12.37

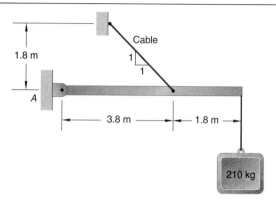

Figure 12.38

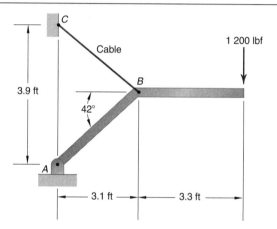

12.33 The uniform beam in Figure 12.37 has a mass of 180 kg. Determine the resultant pin reaction at A and the tension in the cable.

12.34 For the weightless bar in Figure 12.38, determine the resultant pin reaction and the tension in the cable.

12.35 The uniform beam AB in Figure 12.39 has a mass of 900 lbm. $g_L = 32.174\,0$ ft/s^2.
    (*a*) For a force **F** of 1200 lbf, construct a table listing the reactions at A and B if x varies from 0 to 32 ft in increments of 4 ft.
    (*b*) Determine the value of **F** that makes the reaction at A 75% of the reaction at B when $x = 13$ ft.

Figure 12.39

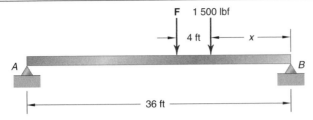

F   1 500 lbf

4 ft   x

A

B

36 ft

Figure 12.40

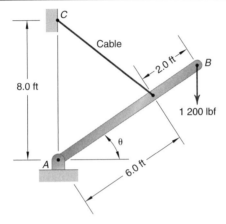

C

Cable

2.0 ft

B

8.0 ft

1 200 lbf

θ

A

6.0 ft

12.36 The Paul JK Company designs custom lifting devices. The winch shown in
Figure 12.40 was designed to lift weights of 1 200 lb. The weight of the boom AB
may be neglected for this analysis. The cable is reeled in and out at C. Develop
a table of magnitudes of cable tension **T** and the resultant pin reaction at A for θ
varying from 0° to 75° in increments of 5°.

CHAPTER 13

Mechanics: Strength of Materials

Chapter Objectives

When you complete your study of this chapter, you will able to:

- Define stress, yield strength, ultimate strength, modulus of elasticity, design stress, and safety factor
- Describe a stress-strain diagram
- Compute the required size of a structural member given the limiting stress and/or limiting deformation
- Describe the interrelationship of Statics and Strength of Materials in the design of structures and mechanisms

13.1 Introduction

Strength of materials, or *mechanics of materials,* is the branch of mechanics dealing with the deformation of a body due to the internal distribution of a system of forces applied to the body. We will investigate the effects of external forces on the internal structure of an object as well as the deformations created by the forces. In the design of structures and mechanisms that do not undergo accelerations, principles of both statics and strength of materials are required.

13.2 Stress

In statics, concepts are limited to rigid bodies. It is obvious that the assumption of perfect rigidity is not always valid. Forces will tend to deform or change the shape of any body, and an extremely large force may cause observable deformation. In most applications, slight deformations are experienced, but the body returns to its original form after the force is removed. One function of the engineer is to design a structure or mechanism within limits that allow it to resist permanent change in size and shape so that it can carry or withstand the force (load) and still recover.

In statics, forces are represented as having a magnitude in a particular direction. Structural members are usually characterized by their mass per unit length or their size in principal dimension, such as width or diameter. Consider the effect on a wire that is 5.00 mm in diameter and 2.50 m in length. It is suspended from a well-constructed support and has an object with a mass of 30.0 kg attached to its lower end (see Fig. 13.1). The force exerted by the mass

Figure 13.1

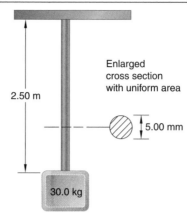

2.50 m

Enlarged
cross section
with uniform area

5.00 mm

30.0 kg

The circular wire is under a tensile stress due to the suspended mass.

is (30.0)(9.807) = 294 N. The wire has a cross-sectional area of π ($5^2/4$), or 19.6 mm^2. If it is assumed that every square millimeter equally shares the force, then each square millimeter supports 294/19.6, or 15.0 N/mm^2. This force per unit area (15.0 N/mm^2) is called the *stress* in the object (wire) due to the applied load. Note that this stress can also be expressed as (15.0 $\times$ 10^6 N/m^2 or 15.0 MPa). The force in the wire is literally trying to separate the atoms of the material by overcoming the bonds that hold the material together.

The relationship above is normally expressed as

$$\sigma = \frac{F}{A} \tag{13.1}$$

where σ = stress, Pa
  F = force, N
  A = cross-sectional area, m^2

Since stress is force per unit area, it is obvious that the stress in the wire (ignoring its own weight) is unaffected by the length of the wire. Such a stress is called *tensile stress* (tending to pull the atoms apart). If the wire had been a rod of 5.00 mm diameter resting on a firm surface and if the mass had been applied at its top, the force would have produced the identical stress, but it would be termed *compressive*.

The simple or direct stress in either tension or compression results from an applied force (load) that is in line with the axis of the member (axial loading). Also the cross-sectional area in both examples was constant for the entire 2.50 m length. If you have an axial load but the cross section varies, the stresses in separate cross sections are different because the areas are different.

A third type of stress is called *shear*. While tension and compression attempt to separate or push atoms together, shear tries to slide layers of atoms in the material across each other. (Imagine removing the top half of a stack of sheets of plywood without lifting.) Consider the pin in Figure 13.2a as it resists the force of 1.00 (10^5) N.

Figure 13.2

309
Strain

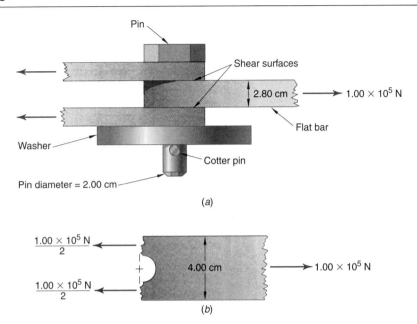

(a)

(b)

Using τ to represent a shear stress, the average shear stress in the pin is

$$\tau = \frac{F}{A} = \frac{1.00 \times 10^5}{\pi(2.00 \times 10^{-2})^2(2)/4}$$

$$= 159 \, \text{MPa}$$

Note: The (2) in the denominator indicates that two cross sections of the pin resist the force; hence, the area is twice the cross-sectional area of the pin. This is called double shear. The two pin shear surfaces are indicated in Figure 13.2a.

To complete the computation, the average tensile stress in the bar at the critical section through the pin hole, as shown in Figure 13.2b, is

$$\sigma = \frac{F}{A} = \frac{1.00 \times 10^5}{(2.00 \times 10^{-2})(2.80 \times 10^{-2})}$$

$$= 179 \, \text{MPa}$$

The actual stress in both cases is somewhat greater than the average because stresses tend to concentrate at the edges of the pin and hole. For this reason, engineers apply a factor of safety in the design process (see Sec. 13.5).

13.3   Strain

In the design of structures and mechanisms, it is important to consider not only the external forces but also the strength of each individual part or member. It is critical that each separate element be strong enough, yet not contain an

Figure 13.3

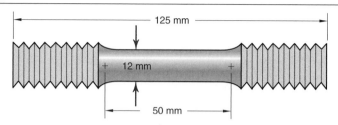

excessive amount of material. Thus in the solution of many problems a knowledge of the properties of materials is essential.

One important test that provides designers with certain material properties is called the *tensile test*. Figure 13.3 is a schematic of a tensile-test specimen. When this specimen is loaded in an axial tensile-test machine, the force applied and the corresponding increase in material length can be measured. This increase in length is called *elongation*. Next, in order to permit comparisons with standard values, the elongation is converted to a unit basis called *strain*.

Strain (ε) is defined as a dimensionless ratio of the change in length (elongation) to the original length:

$$\varepsilon = \frac{\Delta l}{l} = \frac{\delta}{l} \tag{13.2}$$

where ε = strain, mm/mm
δ = deformation, mm
l = length, mm

A *stress-strain diagram* is a plot of the results of a tensile test (see Fig. 13.4). This plot is the basis for design considerations of structures and mechanisms and should be studied carefully. Each notable point on the diagram has an important use in design. The shape of this diagram will vary somewhat for different materials, but in general there will first be a straight-line portion *OA*. Point *A* is the proportional limit—the maximum stress for which stress is proportional to strain.

At any stress up to point *A'*, called the *elastic limit,* the material will return to its original size once the load has been removed. At stresses higher than *A'*, permanent deformation (set) will occur. For most materials, points *A* and *A'* are very close together.

If the load is increased beyond the elastic limit to point *B*, and then returned to zero, the stress-strain curve will follow the dotted line, leaving a permanent deformation in the material called a permanent set. The stress at *B* in Figure 13.4 that causes a permanent set of 0.05 to 0.3% (depending on material) is termed the *yield strength*.

Point *C*, called the *ultimate strength,* is the maximum stress that the material can withstand. Between points *B* and *C*, a small increase in stress causes a significant increase in strain. At approximately point *C*, the specimen will begin to neck down sharply; that is, the cross-sectional area will decrease rapidly, and fracture will occur at point *D*.

Figure 13.4

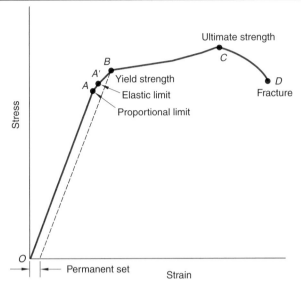

Stress-strain diagram.

Figure 13.5

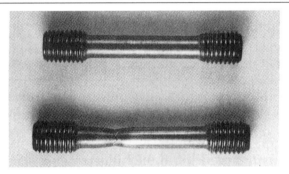

The lower portion of the photo shows the actual "necking-down" effect that occurs during a tensile test just prior to failure of the steel specimen.

Figure 13.5 shows a typical steel specimen prior to test and the specimen after it was pulled apart.

13.4 Modulus of Elasticity

Approximately 300 years ago, Robert Hooke recognized the linear relationship between stress and strain. For stresses below the proportional limit, Hooke's law can be written

$$\varepsilon = K\sigma \qquad (13.3)$$

Table 13.1 Modulus of elasticity for selected materials

	E, psi	**E, GPa**
Cold-rolled steel	30×10^6	210
Cast iron	16×10^6	110
Copper	16×10^6	110
Aluminum	10×10^6	70
Stainless steel	27×10^6	190
Nickel	30×10^6	210

where K is a proportionality constant. The modulus of elasticity E (the reciprocal of K) rather than K is commonly used, yielding

$$\sigma = E\varepsilon \tag{13.4}$$

Values of E for selected materials are given in Table 13.1.

13.5 Design Stress

Obviously, most products or structures that engineers design are not intended to fail or become permanently deformed. The task facing the engineer is to choose the proper type and size of material that will perform correctly under the conditions likely to be imposed. Since the safety of the user and the liability of the producer (including the engineer) are dependent on valid assumptions, the engineer typically selects a design stress that is less than the yield strength. The ratio of the yield strength to the design stress is called the *safety factor or factor of safety* (abbreviated F.S.). For example, if the yield strength is 210 MPa and the design stress is 70 MPa, the safety factor, based on yield strength, is 3. Care must be exercised in reporting and interpreting safety factors because they are expressed in terms of both yield strength and ultimate (tensile) strength. Table 13.2 lists typical values used in structural design. It should be noted that the United States still lists most of its standards in the U.S. Customary or Engineering systems, so conversions in this area will be necessary for some time to come.

Table 13.2 Ultimate and yield strength

	Ultimate Strength		**Yield Strength**	
	psi	**MPa**	**psi**	**MPa**
Cast iron	45×10^3	310	30×10^3	210
Wrought iron	50×10^3	345	30×10^3	210
Structural steel	60×10^3	415	35×10^3	240
Stainless steel	90×10^3	620	30×10^3	210
Aluminum	18×10^3	125	12×10^3	85
Copper, hard drawn	66×10^3	455	60×10^3	415

Example Problem 13.1   A round bar is 40.0 cm long and must withstand a force of 20.0 kN. What diameter must it have if the stress is not to exceed 140.0 MPa?

Solution

$$\sigma = \frac{F}{A}$$

$$A = \frac{F}{\sigma} = \frac{20.0 \times 10^3\,\text{N} \cdot \text{m}^2}{140.0 \times 10^6\,\text{N}} \, \frac{10^6\,\text{mm}^2}{1\,\text{m}^2} = 143\,\text{mm}^2$$

But $A = \dfrac{\pi d^2}{4}$

So $143 = \dfrac{\pi d^2}{4}$

$d = 13.5$ mm

Example Problem 13.2   Assume that the bar in Example Prob. 13.1 is made from cold-rolled steel and is permitted to elongate 0.125 mm. Determine the required diameter.

Solution

1.   We must use the quantities that define the relationship between stress and strain. From Eqs. (13.1), (13.3), and (13.4)

$$E = \frac{\sigma}{\varepsilon} = \frac{F/A}{\Delta l/l}$$

Therefore

$$\Delta l = \frac{Fl}{AE}$$

Δl is usually written as δ for deflection or elongation.

$$\delta = \frac{Fl}{AE}$$

and the necessary area for the given material and loading becomes

$$A = \frac{Fl}{\delta E}$$

2.   Substituting the numerical quantities, E is 21×10^4 MPa from Table 13.1

$$A = \frac{20 \times 10^3\,\text{N}}{125 \times 10^{-6}\,\text{m}} \, \frac{0.400\,\text{m} \cdot \text{m}^2}{21 \times 10^{10}\,\text{N}} \, \frac{10^6\,\text{mm}^2}{1\,\text{m}^2}$$

$$= 304.8\,\text{mm}^2 = \frac{\pi d^2}{4}$$

$d = 19.7$ mm

Figure 13.6

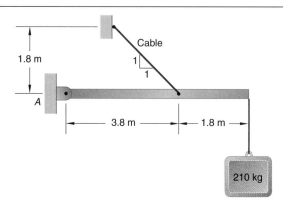

Note: The following two examples require a statics analysis before the specified components can be designed.

Example Problem 13.3   For the weightless beam shown in Figure 13.6, (a) Find the pin reaction at A and the tension in the cable, (b) Find the minimum cable diameter required if it has a design tensile stress of 95 MPa, and (c) Find the minimum pin diameter required if it is in single shear, is made of structural steel, and has a factor of safety of 4.0 based on yield strength.

Solution to (a)
The free-body diagram is shown in Figure 13.7. Applying the statics equations of equilibrium we can find the tension T and the magnitude of the pin reaction at A.

$$\rightarrow \Sigma F_x = 0$$
$$A_x = T(1/\sqrt{2}) = 0.7071T$$
$$\uparrow \Sigma F_y = 0$$
$$A_y + T(0.7071) - (210)(9.807) = 0$$
$$\circlearrowleft \Sigma M_A = 0$$
$$T(0.7071)(3.8) - (210)(9.807)(5.6) = 0$$
$$T = 4291.5 \text{ N}$$

Thus
$$A_x = 3034.5 \text{ N}$$
$$A_y = 975.5 \text{ N}$$
$$A = (A_x{}^2 + A_y{}^2)^{0.5}$$
$$= 3188 \text{ N}$$

Figure 13.7

315
Design Stress

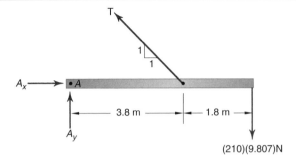

Solution to (b)

From Eq. 13.1

$$\text{Area} = \frac{T}{\sigma}$$

$$= \frac{4291.5 \text{ Nm}^2}{94 \times 10^6 \text{ N}}$$

$$= 45.65(10)^{-6} \text{ m}^2$$

$$\text{Area} = \frac{\pi d^2}{4}$$

$$d = \sqrt{\frac{4(\text{Area})}{\pi}}$$

$$= 7.62 \text{ mm}$$

From a cost standpoint, it is practical to choose the next largest commercially available diameter.

Solution to (c)

For structural steel the yield strength is 240 MPa. With a required factor of safety of 4.0, the design shear stress is 60 MPa.
Following the identical procedure in part (b)

$$\text{Area} = \frac{A}{\tau}$$

$$= \frac{3188 \text{ N} \cdot \text{m}^2}{60 \times 10^6 \text{ N}}$$

$$= 53.133(10^{-6}) \text{ m}^2$$

$$\text{Area} = \frac{\pi d^2}{4}$$

$$d = \sqrt{\frac{4(\text{Area})}{\pi}}$$

$$= 8.23 \text{ mm}$$

Again, it is practical to choose the next largest commercially available pin size.

Figure 13.8

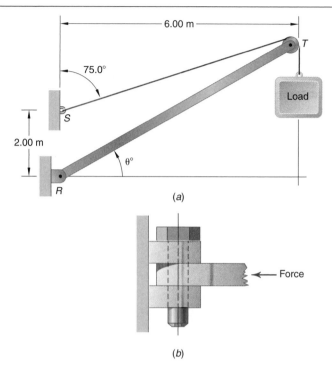

(a)

(b)

Example Problem 13.4 Given the configuration in Figure 13.8a, calculate the load that can be supported under the following design conditions:

(a) The pin at point R, enlarged in Figure 13.8b, is 10.0 mm diameter. What load can be supported by the pin if the ultimate shear strength of the pin is 195 MPa and a safety factor of 2.0 based on ultimate strength is required?

(b) Using the load condition from (a), size cable ST if it has a design stress of 207.5 MPa.

Solution to (a)

1. Inspection of the free-body diagram in Figure 13.9 reveals four unknown forces and three equations of equilibrium. We will write the cable tension Q, and the pin reactions, R_x and R_y, in terms of the load, L, and then use the pin design conditions to find the largest permissible L.

2. Determine the geometry of H and θ:

$$\tan 15° = \frac{H}{6.00}$$

$$H = 1.608 \text{ m}$$

$$\tan \theta = \frac{2 + H}{6}$$

$$\theta = 31.02°$$

3. Apply the equations of equilibrium:

$$\curvearrowleft \Sigma M_S = 0$$

$$2.00\,R_x - 6.00\,L = 0$$

$$R_x = 3.00\,L$$

$$\curvearrowleft \Sigma M_T = 0$$

$$(2.00 + 1.608)R_x - 6.00\,R_y = 0$$

$$(3.608)(3.00L) - 6.00\,R_y = 0$$

$$R_y = 1.804\,L$$

$$R = \sqrt{R_x^2 + R_y^2}$$

$$= \sqrt{(3L)^2 + (1.804L)^2}$$

$$= 3.501\,L$$

4. Determine pin limits from design conditions. Note that the pin is in double shear.

$$\text{Design stress} = \frac{\text{Ultimate strength}}{\text{Safety factor}}$$

$$= \frac{195\,\text{MPa}}{2.0} = 97.5\,\text{MPa}$$

$$\tau = \frac{R}{2A} \qquad \begin{array}{l}\text{Recall that double shear involves}\\ \text{twice the cross-sectional area}\end{array}$$

$$R = 2\tau A = 2\tau \left(\frac{\pi d^2}{4}\right)$$

$$= \frac{2\,(97.5 \times 10^6)\,\text{N}\,\pi\,(10\,\text{mm})^2}{4}\,\frac{1\,\text{m}^2}{10^6\,\text{mm}^2}$$

$$= 1.53 \times 10^4\,\text{N}$$

$$L = \frac{R}{3.501} = \frac{15\,300}{3.501} = 4.37\,\text{kN}$$

Figure 13.9

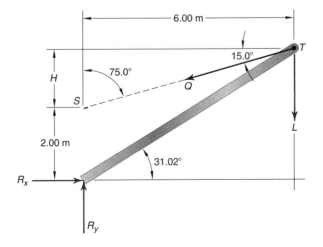

Solution to (b)

1. Determine cable load Q from part (a)

$$\curvearrowleft \Sigma M_R = 0$$

$$-6.00\,L + 2.00Q \cos 15° = 0$$

$$Q = 3.106\,L$$

$$= 13.57$$

$$= 13.6 \text{ kN}$$

2. Calculate cable size. The design stress is 207.5 MPa.

$$\sigma = \frac{F}{A}$$

$$A = \frac{F}{\sigma} = \frac{13570\text{N} \cdot \text{m}^2}{207.5 \times 10^6\text{N}}$$

$$= 6.54 \times 10^{-5} \text{ m}^2$$

$$\frac{\pi d^2}{4} = 6.54 \times 10^{-5} \text{ m}^2$$

$$d^2 = 8.32 \times 10^{-5} \text{ m}^2$$

$$d = 9.12 \times 10^{-3} \text{ m}$$

$$= 9.12 \text{ mm}$$

A practical choice would be the next larger size that is commercially available.

Problems

Note: Some problems require statics computations before application of strength of materials principles.

13.1 A round reinforcing bar constructed from cold-rolled steel and 320 mm long is required to support a tensile load of 65 kN. Determine the minimum cross-sectional area if the stress cannot exceed 240 MPa or elongate more than 0.42 mm.

13.2 Determine the minimum size for a pin if it is in double shear and the factor of safety is 5.0 based on ultimate strength. The pin is made of structural steel and supports a load of 85 kN.

13.3 In Figure 13.1, what is the maximum force in N that can be supported by the wire if the design stress is 95.5 MPa and allowable deformation is 1.3 mm? The wire is made from stainless steel.

13.4 The round, weightless bar in Figure 13.10 is $\frac{3}{4}$ in. in diameter. $E = 30(10^6)$ psi. Find the tensile stress in the bar and the deformation caused by the applied load.

13.5 Under certain loading conditions for the beam/cable configuration in Figure 13.7, static analysis determined $A_x = 43\,000$ N, $A_y = 5\,100$ N, and $T = 62\,000$ N. Find (a) the minimum allowable pin diameter if the pin is stainless steel, in single shear, and has a design stress of 75 MPa, and (b) the minimum cable diameter if the cable is structural steel with a factor of safety of 2.0 based on yield strength.

Figure 13.10

Figure 13.11

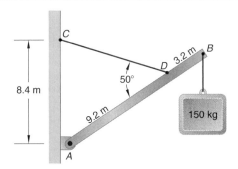

13.6 The boom *AB* in Figure 13.11 has a mass distribution of 8.2 kg/m.
  (a) Find the tension in the cable *CD*.
  (b) Find the minimum required diameter of cable if the design stress is 84 MPa.
  (c) If the cable is stainless steel, what is the factor of safety?
  (d) If the cable were to be replaced by an aluminum cable, what diameter must be used to maintain a factor of safety of 3.0?
13.7 A hollow aluminum circular cylinder, 32 cm tall with 5.0 cm outside diameter and a wall thickness of 0.35 cm, is used to support a compressive load *L* applied along the axis of the cylinder. Determine the maximum allowable value of *L* if the compressive stress is limited to 65 MPa and the deformation is limited to 0.044 cm.
13.8 Figure 13.12 represents a tower crane used by Tomas J. Contracting to handle materials at construction sites. The weight, *W*, is hoisted to the desired height and then moved along the boom AE from $6 \leq X \leq 86$. The counterweight, *CW*, weighs 8 500 lbm, and the crane motor and cab weigh 2 400 lbm acting at *B*. The crane structure weighs 5 500 lbm and can be considered to be acting at *G*. Pins at *C* and *D* secure the crane base. $g_L = 32.174\ 0$ ft/s^2.
  (a) If the weight to be lifted, *W*, is 3 200 lbm, plot the reactions at *C* and *D* versus the distance *x*.
  (b) Find the points where the loads at *C* and *D* are maximum and size the pins required. Each pin is in double shear and is made from stainless steel with a factor of safety of 5.0 based on ultimate strength.
  (c) If the load, *W*, is increased to 4 200 lbm, determine if this load can be lifted safely with the pin sizes calculated and, if so, the range of *x* for which it is possible.

Figure 13.12

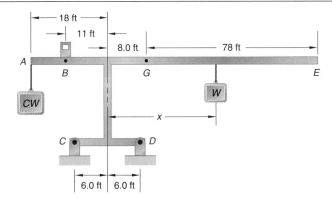

CHAPTER 14

Material Balance

Chapter Objectives

When you complete your study of this chapter, you will able to:

- Define the principle of the conservation of material (mass)
- Define systems to solve material (mass) balance problems
- Write the specific equations for overall and constituent balances
- Solve for unknowns using an independent set of mass balance equations

14.1 Introduction

In engineering analyses, it is extremely important to observe and account for all changes in physical quantities, such as mass, momentum, energy, charge, chemical composition, and other quantities. In this chapter the focus will be on accounting for material (mass) in engineering analyses of problems found in process industries.

We depend a great deal on industries that produce food, household cleaning products, energy for heating and cooling homes, fertilizers, and many other products and services. These process industries are continually involved with the distribution, routing, blending, mixing, sorting, and separation of materials. Figure 14.1 shows an oil refinery where crude oil is separated into many different petroleum products for use in various applications. The primary separation process is fractional distillation, in which hot crude oil is fed to the bottom of a distillation column. The weight and boiling point of the crude oil components (heavy and hot at the bottom and light and cool at the top of the column) allow bitumen, oils, diesel, kerosene, naptha, and gasoline to be withdrawn at appropriate points along the column.

Analysis of the process occurring in a typical oil refinery involves the application of *material balance,* a technique based on the *principle of conservation of mass.*

14.2 Conservation of Mass

The *principle of conservation of mass* states that in any process mass is neither created nor destroyed.

Antoine Lavoisier (1743–1794), regarded as the father of chemistry, conducted experiments demonstrating that matter, although undergoing

Figure 14.1

An oil refinery produces many petroleum products for consumers.

transformation from one form to another, was neither created nor destroyed during the experiments. The results were expressed as an empirical law called conservation of mass, material balance, or mass balance. Like many empirical laws, this one has an exception: nuclear reactions, in which mass is transformed into energy. To solve problems involving nuclear processes, both a material balance and an energy balance must be conducted jointly.

Before we apply the conservation-of-mass principle to a material balance problem, we will introduce additional concepts and terminology. Figure 14.2 illustrates a number of these terms.

Figure 14.2

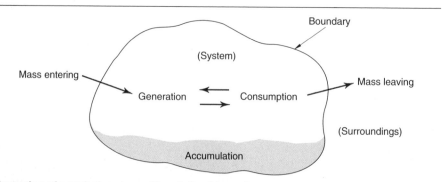

Illustration of a typical system with a defined boundary.

A *system* can be any designated portion of the universe with a definable boundary. Whenever mass crosses the boundary either into or out of the system, it must be accounted for. In certain situations, the amount of mass entering the system is greater than the amount leaving. In the absence of chemical reactions, this results in an increase of mass within the system called *accumulation*. If the mass leaving the system is greater than that entering, the accumulation is negative.

When chemical reactions occur within the system boundary, consumption of reactants through the formation or generation of reaction products is possible. A simple example would be the dehydrogeneration of ethane in a reactor: $C_2H_6 \rightarrow C_2H_4 + H_2$. One constituent is consumed and two others are generated. Thus, if chemical reactions occur, it is necessary to account for the consumption of some elements or compounds and the generation of others. It is important to understand that, even considering chemical reactions, mass is conserved. In the preceding example $(C_2H_6 \rightarrow C_2H_4 + H_2)$, the number of atoms of carbon and hydrogen remains constant.

In summary, a system is an arbitrary specification that must conform to the following:

1. Once specified, the system cannot change during the analysis.
2. The system boundary can be fixed or flexible and must be a closed surface.

Consider the following examples to more clearly visualize the definition of a system. First, the simple distillation system is shown in Figure 14.3. This distillation process involves three items of equipment, a heater, column and condenser with associated piping and connections. The feed, which may include several components is introduced to the column where it is heated. The component(s) with the lower boiling temperature are vaporized, drawn from the top of the column into a condenser where they are cooled to liquid form. Some of the concentrated product (distillate) is drawn off and part of it (reflux) is sent back to the column to be reboiled. The portion of the feed that is not vaporized is drawn off at the bottom of the column as waste. The dashed outline represents the system boundary. Therefore we must account for the feed, product and bottoms in a material balance analysis. The vapor and reflux piping do not need to be accounted for since they remain within the system.

Second, consider the dewatering process in Figure 14.4. It consists of a centrifuge that removes some of the water and a dryer that removes more water down to a very small percentage of the solid material. Three systems are available for analysis: the overall system in Figure 14.4(a), the centrifuge in Figure 14.4(b), and the dryer in Figure 14.4(c). Note here that each system shows inputs and outputs but no generation, consumption, or accumulation.

When all considerations are included, the conservation-of-mass principle applied to a system or to system constituents can be expressed, from Figure 14.2, as

$$\text{input} + \text{generation} - \text{output} - \text{consumption}$$
$$= \text{accumulation} \tag{14.1}$$

Actually Eq. (14.1) is more general than just a material (mass) balance. It is a convenient way to express a general conservation principle to account for the physical quantities mentioned at the beginning of this chapter. For a specified

Figure 14.3

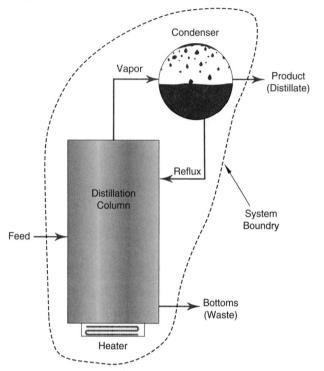

System designation for a distillation column.

Figure 14.4

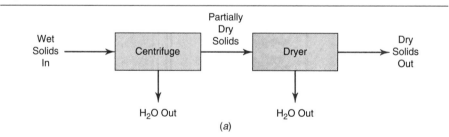

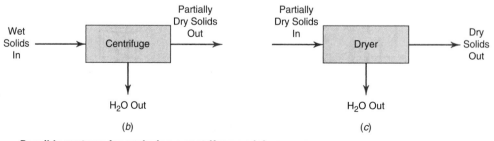

Possible systems for analyzing a centrifuge and dryer process.

system, such as a mass of air passing through a jet engine, an energy balance accounts for heat in and out and for work done on or by the system, heat and work being forms of energy. Heat input would come from burning of fuel, heat output would be the hot exhaust gases, work in would be compression of free stream air, and work out would be the thrust generated by the accelerated gases. Energy balances are studied in thermodynamics.

14.3 Processes

For our introductory look at material balance, we will now make an assumption that simplifies our investigations but will not reduce our understanding of the conservation of mass.

Many engineering problems involve chemical reactions, but if we assume no such reactions then there will be no generation and consumption of mass, and Eq. (14.1) can be reduced to

$$\text{input} - \text{output} = \text{accumulation} \tag{14.2}$$

Two types of processes typically analyzed are the batch process and the rate process. In a *batch process,* materials are put or placed into the system before the process begins and are removed after the process is complete. Cooking is a familiar example. Generally, you follow a recipe that calls for specific ingredients to be placed into a system that produces a processed food. Figure 14.5 illustrates a batch process for a concrete mixer.

Figure 14.5

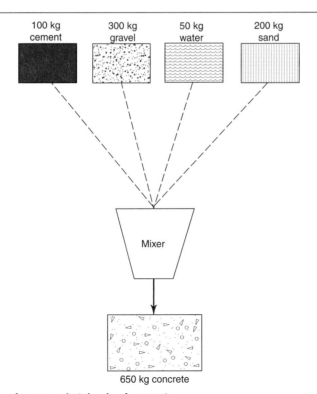

The constituents of a proper batch mix of concrete.

Figure 14.6

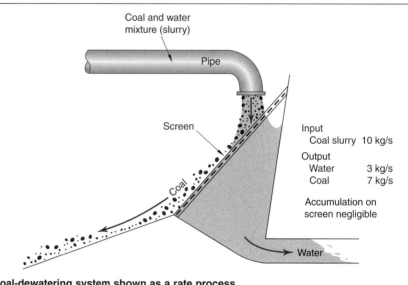

Coal and water
mixture (slurry)

Pipe

Screen

Coal

Input
    Coal slurry 10 kg/s
Output
    Water   3 kg/s
    Coal   7 kg/s

Accumulation on
screen negligible

Water

A coal-dewatering system shown as a rate process.

If we assume for a batch process that we take out at the end of the process all of the mass we placed into the system at the beginning, then the accumulation term is zero and Eq. (14.2) can be written as

$$\text{total input} = \text{total output} \tag{14.3}$$

A *rate-flow process* involves the continuous time rate of flow of inputs and outputs. The process is performed continuously as mass flows through the system. See Figure 14.6.

Rate processes may be classified as either *uniform* or *nonuniform, steady* or *unsteady*. A process is uniform if the flow rate is the same throughout the process, which means the input rate must equal the output rate. It is steady if rates do not vary with time. Solution of material balance problems involving nonuniform and/or unsteady flows may require the use of differential equations and will not be considered here.

A class of nonuniform, steady processes involves the filling or draining of tanks. For this situation we have a fixed rate of input and a different fixed rate of output, thus producing an accumulation. Equation (14.2) for such processes becomes

$$\text{rate of input} - \text{rate of output} = \text{rate of accumulation} \tag{14.4}$$

For continuous flow, if we assume a uniform, steady rate process, the accumulation term is also zero, so Eq. (14.2) written on a rate basis reduces to

$$\text{rate of input} = \text{rate of output} \tag{14.5}$$

Figure 14.6 is an example of a uniform, steady rate process.

Although Eqs. (14.3), (14.4), and (14.5) seem so overly simple as to be of little practical use, application to a given problem may be complicated by the need to account for several inputs and outputs as well as for many constituents in each input or output. The simplicity of the equations is in fact the advantage of a material balance approach, because order is brought to seemingly disordered data.

14.4 A Systematic Approach

Material balance computations require the manipulation of a substantial amount of information. Therefore, it is essential that a systematic procedure be developed and followed. If a systematic approach is used, material balance equations can be written and solved correctly in a straightforward manner. The following procedure is recommended:

1. Identify the system(s) involved.
2. Determine whether the process is a batch or rate process and whether a chemical reaction is involved. If no reaction occurs, all compounds will maintain their chemical makeup during the process. If a reaction is to occur, elements must be involved and must be balanced. In a process involving chemical reactions, additional equations based on chemical composition may be required in order to solve for the unknown quantities.
3. Construct a schematic diagram showing the feeds (inputs) and products (outputs).
4. Label known material quantities or rates of flow.
5. Identify each unknown input and output with a symbol.
6. Apply Eq. (14.3), (14.4), or (14.5) for each constituent as well as for the overall process. Note that not all equations written will be algebraically independent. This will become apparent in the example problems.
7. Solve the equations (selecting an independent set) for the desired unknowns and express the result in an understandable form.

Example Problem 14.1 Anna's Purification Co. produces drinking water from saltwater by partially freezing the saltwater to create salt-free ice and a brine solution. If saltwater is 3.50% salt by mass and the brine solution is found to be an 8.00% concentrate by mass, determine how many kilograms of saltwater must be processed to form 2.00 kg of ice.

Figure 14.7

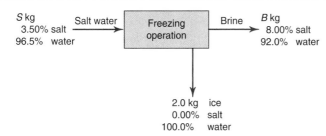

Schematic diagram of the saltwater freezing operation.

Solution

1. The system in this example problem involves a freezing process.
2. The freezing process is a batch process because a fixed amount of product (ice) is required. There are no chemical reactions.
3. A diagram of the process is shown in Figure 14.7.
4. Saltwater is the input to the system, with brine and ice taken out at the end.
5. Appropriate symbols are used to identify unknown quantities.
6. The material balance equation for each constituent as well as for the overall process is written. It is important to understand that the material balance equation (Eq. 14.3) is applicable for each constituent as well as for the overall process. In this example, three equations are written, but only two are independent. That is, the overall balance equation is the sum of the salt and water balance equations. Thus we have a good method of checking the accuracy of the equations we have written.

Equation	Input = Output
Overall balance	$S = B + 2.00$
Salt balance	$0.035S = 0.08B$
Water balance	$0.965S = 0.920B + 2.00$

7. The equations are solved by substitution for S from the overall balance equation into the salt balance equation.

$$0.035(B + 2.00) = 0.08B$$
$$0.045B = 0.070$$
$$B = 1.56 \text{ kg}$$

Since
$$S = B + 2.00, \text{ then}$$
$$S = 3.56 \text{ kg}$$

The water balance equation was not used to solve for B and S but can serve as a check of the results (that is, it should balance). Substituting the computed values for B and S into the water balance equation:

$$(0.965)(3.56) = [(0.920)(1.56)] + 2.00$$
$$3.435\ 4 = 3.435\ 2$$

which does balance within the round-off error.

Example Problem 14.2 A process to remove water from solid material consists of a centrifuge and a dryer. If 35.0 t/h of a mixture containing 35.0% solids is centrifuged to form a sludge consisting of 65.0% solids and then the sludge is dried to 5.00% moisture in a dryer, how much total water is removed in 24-hour period? *Note:* This problem is a specific application of the system definition shown in Figure 14.4.

Solution There are three possible systems involved in this problem: the centrifuge, the dryer, and the combination (see Fig. 14.8). The operation in this

Figure 14.8

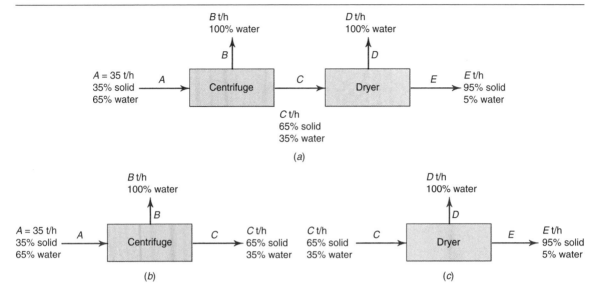

Systems defining the flow process inputs and outputs for centrifuge and dryer.

system is a continuous flow process. There are no chemical reactions and the process is steady and uniform, thus Eq. (14.5) applies.

$$\text{rate of input} = \text{rate of output}$$

The following equations are written for the process illustrated in Figure 14.8. The overall process is illustrated in Figure 14.8a, with subsystem diagrams for the centrifuge and the dryer shown in Figure 14.8b and Figure 14.8c, respectively. The overall balance equation for a selected system is the sum of the constituent balance equations for that system. This means that the set of equations written for a selected system are not all independent when the overall balance equation and the constituent balance equations are included.

For the entire system (Fig. 14.8a):

1. Solid balance $0.35(35) = 0.95E$
2. Water balance $0.65(35) = B + D + 0.05E$
3. Overall balance $35 = B + D + E$

For the centrifuge (Fig. 14.8b):

4. Solid balance $0.35(35) = 0.65C$
5. Water balance $0.65(35) = B + 0.35C$
6. Overall balance $35 = B + C$

For the dryer (Fig. 14.8c):

7. Solid balance $0.65C = 0.95E$
8. Water balance $0.35C = D + 0.05E$
9. Overall balance $C = D + E$

Solve for rate of mass out of centrifuge (C) from Eq. (4)

$$C = \frac{0.35(35)}{0.65}$$
$$= 18.85 \text{ t/h}$$

Solve for rate of water out of centrifuge (B) from Eq. (6):

$$B = 35 - C$$
$$= 35 - 18.85$$
$$= 16.15 \text{ t/h}$$

Solve for rate of mass out of dryer (E) from Eq. (7):

$$E = \frac{0.65C}{.095}$$
$$= \frac{0.65(18.85)}{0.95}$$
$$= 12.90 \text{ t/h}$$

Solve for rate of water out of dryer (D) from Eq. (9):

$$D = C - E$$
$$= 18.85 - 12.90$$
$$= 5.95 \text{ t/h}$$

Check the results obtained in the three balance equations for the entire system, Eqs. (1), (2), and (3).

Eq. (1) $(0.35)(35) = (0.95)(12.90)$
$12.25 = 12.25$ (checks)

Eq. (2) $(0.65)(35) = 16.15 + 5.95 + (0.05)(12.90)$
$22.75 = 22.75$ (checks)

Eq. (3) $35 = 16.15 + 5.95 + 12.90$
$35 = 35$ (checks)

All the results check, so then calculate total water removed in 24 h:

Total water $= (B + D)\, 24$
$= (16.15 + 5.95)\, 24$
$= (22.10)\, 24$
$= 5.30 \times 10^2 \text{ t}$

A general problem that would involve typical material balance consideration is a standard evaporation, crystallization, recycle process. Normally this type of system involves continuous flow of some solution through an evaporator. Water is removed, leaving the output stream more concentrated. This stream is fed into a crystallizer where it is cooled, causing crystals to form. These crystals are then filtered out, with the remaining solution recycled to join the feed stream back into the evaporator. This system is illustrated in the following example problem.

Example Problem 14.3 A solution of potassium chromate (K_2CrO_4) is to be used to produce K_2CrO_4 crystals. The feed to an evaporator is 2.50×10^3 kilograms per hour of 40.0% solution by mass. The stream leaving the evaporator is 50.0% K_2CrO_4. This stream is then fed into a crystallizer and is passed through a filter. The resulting filter cake is entirely crystals. The remaining solution is 45.0% K_2CrO_4, and is recycled. Calculate the total input to the evaporator, the feed rate to the crystallizer, the water removed from the evaporator, and the amount of pure K_2CrO_4 produced each hour.

Solution See Figure 14.9. There are different ways the system boundaries can be selected for this problem—that is, around the entire system, around the evaporator, around the crystallizer-filter, and so on. There are no chemical reactions that occur in the process, and the process is steady and uniform, thus

$$\text{Rate of input} = \text{rate of output}$$

Figure 14.9

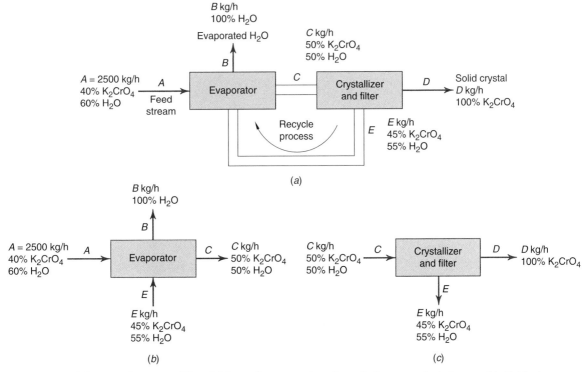

(a) Diagram of the overall system; (b) and (c) are diagrams where boundaries are selected around individual process components.

The following balance equations can be written, not all of which are independent.

For the entire system (Fig. 14.9a):

1. K_2CrO_4 balance $0.40\,(2\,500) = D$
2. H_2O balance $0.60\,(2\,500) = B$
3. Overall balance $2\,500 = B + D$

For the evaporator (Fig. 14.9b):

4. K_2CrO_4 balance $(0.40)(2\,500) + 0.45E = 0.50C$
5. H_2O balance $(0.60)(2\,500) + 0.55E = B + 0.50C$
6. Overall balance $2\,500 + E = B + C$

For the crystallizer-filter (Fig. 14.9c):

7. K_2CrO_4 balance $0.50C = D + 0.45E$
8. H_2O balance $0.50C = 0.55E$
9. Overall balance $C = D + E$

We will first develop the solution by selecting appropriate equations from the preceding set and solving them by hand. Then we will place a set of four equations with four unknowns in a math solver (Mathcad) to obtain (and verify in this case) the solution we have found by hand.

Solve for K_2CrO_4 out of crystallizer-filter from Eq. (1):

$$D = 0.4\,(2\,500)$$
$$= 1.00 \times 10^3 \text{ kg/h}$$

Solve for H_2O out of evaporator from Eq. (2):

$$B = 0.6\,(2\,500)$$
$$= 1.50 \times 10^3 \text{ kg/h}$$

Solve for recycle rate from crystallizer-filter from Eqs. (9) and (7):

(9) $C = D + E$
$$= 1\,000 + E$$

(7) $0.50C = D + 0.45E$
$$0.50(1\,000 + E) = 1\,000 + 0.45E$$
$$E = 10\,000 \text{ kg/h}$$

Calculate total input for evaporator:

$$\text{Total input} = E + A$$
$$= 10\,000 + 2\,500$$
$$= 1.25 \times 10^4 \text{ kg/h}$$

Calculate feed rate for crystallizer-filter from Eq. (6):

$$C = 2\,500 + E - B$$
$$= 2\,500 + 10\,000 - 1\,500$$
$$= 1.10 \times 10^4 \text{ kg/h}$$

55.0% sulfuric acid (H_2SO_4), and 21.0% water (H_2O). This acid is to be concentrated (strengthened in acid content) by adding sulfuric acid with 92.0% H_2SO_4 and nitric acid containing 89.0% HNO_3. The final product is to contain 31.0% HNO_3 and 58.0% H_2SO_4. Compute the mass of the initial acid solution and the mass of the concentrated acids that must be combined to produce $2.00(10^3)$ lbm of the desired mixture.

14.16 An industry is required to clean up 4.0 t of a by-product containing both toxic and inert material. The toxic content is 15%; the rest is inert. The by-product is treated with 45 t of solvent resulting in an amount of dirty solvent containing 0.35% toxic material and a discard containing 1.2% toxic material, some solvent, and all the inert materials. Determine the amount of dirty solvent produced, the percentage of solvent in the discard, and the percentage of toxic substance removed in the discard.

14.17 The drain valve on a 25 m³ water tank allows 0.60 kg/s to drain. The fill valve allows 1.2 kg/s to flow into the tank. If the tank is initially one-half full, how long will it take to fill the tank if both valves are open? If the tank is full and the fill valve is shut off, how long will it take the tank to empty?

14.18 A distillation column (see Fig. 14.3) is used to distill ethanol (chemical formula C_2H_5OH). If the feed to the column is 15 000 lb of 11% ethanol (the rest water) and the distillate amount is one-tenth of the feed and contains 62% ethanol, determine the amount of ethanol that is in the bottoms.

14.19 A batch of 250 lb of sulfuric acid (90% concentrate) is needed. Unlimited supplies of 98% and 40% concentrates of sulfuric acid are available. How much of each is needed to produce the required 90% concentrate?

14.20 Dilute sulfuric acid (19% acid, the rest water) is required for activating car batteries. A tank of weak acid (12.5% concentrate, the rest water) is available. If 450 lb of 78% concentrate acid is added to the tank to get the required 19% acid, how much of the 19% acid is now available?

14.21 A special mud formed by mixing dry clay and water is used to assist drilling through rock. The optimum amount of clay in the mud is determined from a chart that relates the percentage of clay to the hardness of the rock. If a supply of mud containing 42% clay and 58% water is available, how much of this mud and how much dry clay need to be mixed to produce one ton of a mix containing 48% clay?

14.22 A company has a large vat of brine containing 5.0% salt (95% water) and wishes to prepare 45 kg batches of brine having salt content varying from 6.0% to $3.0(10^1)$%. Prepare a graph showing the required amount of original brine and dry pure salt that must be mixed to obtain a brine with a specified salt content in the range. Consider the desired salt content of the 45 kg batch as the independent variable.

14.23 Andrew Foods, Inc., produces a sweet syrup made from corn syrup and beet syrup mixed with pure sugar. For each batch, 680 lbm of beet syrup (13% sugar, the remainder water) and 650 lbm of corn syrup (8.0% sugar, the remainder water) are mixed with pure sugar and 350 lbm of water is boiled off. Determine the amount of pure sugar that needs to be added to produce an 18% sugar content in the final product.

14.24 The feed to a distillation column contains 37% benzene, the remainder toluene. The overhead (product) contains 55% benzene and the bottoms (waste) contains 5.0% benzene. Calculate the percentage of the total feed that leaves as product.

14.25 In a mechanical coal-washing process, a portion of the inorganic sulfur and ash material in the mined coal can be removed. A feed stream to the process consists of 73% coal, 11% ash, 11% sulfur, and 5.0% water (to suppress dust). The clean coal side of the process consists of 88% coal, 6.0% sulfur, 3.0% ash, and 3.0%

water. The refuse side consists of 32% sulfur, with the remainder being ash, coal, and water. If $8.0(10^1)$ metric tons per hour of clean coal is required, determine the feed rate required for the mined coal and the composition of the refuse material.

14.26 Because of environmental concerns, your plant must install an acetone recovery system. Your task is to calculate the size of the various components of the system, which includes an absorption tank into which is fed $1.25(10^3)$ kg of water per hour and $7.00(10^3)$ kg of air containing 1.63% acetone. The water absorbs the acetone and the purified air is expelled. The water and acetone solution go to a distillation process where the solution is vaporized and then fed to a condenser. The resulting product is 98.9% acetone and 1.10% water. The bottoms (waste) of the distillation process contain 4.23% acetone and 95.77% water. To assist in the determination of the volume of a holding tank, calculate how much product is generated in kilograms per hour.

14.27 A process requires two distillation columns. The first column is fed a solution containing 21% component A, 31% component B, the rest component C. The product amount is one-fourth of the feed amount and contains 61% A, 5% B, the rest C. The bottoms amount is fed to a second column which then produces a product of 15% A, 79% B, the rest C. The waste from the second column contains 0.50% A, the rest B and C. If 2200 lb of solution is fed to the first column, determine the unknown compositions and amounts of the feed to and the product and waste from the second column.

14.28 Ether is used in the process of extracting cod-liver oil from livers. Livers enter the extractor at a rate of 1 500 lbm/h and consist of 32% oil and 68% inert material. The solvent is introduced to the extractor at a rate of 3 100 lbm/h and consists of 2.80% oil and 97.2% ether. The cod-liver oil extract leaves the process at a rate of 2 600 lbm/h and consists of 21% oil and 79% ether. Calculate the flow rate and composition of the leaving processed livers.

14.29 Two liquids, A and B, are mixed together. A is 3.0% solids, the remainder water. B is 8.0% solids, the remainder water. An amount S of pure, dry solids is mixed with A and B to form $2.0(10^2)$ kg of a mixture with 5.0% solids. Develop a table of combinations of amounts of A, B, and S that will satisfy the final mixture requirements. What is the maximum amount of B that can be used to meet the requirements without having to remove pure, dry solids from the mixture? Hint: Write a short computer program or use a spreadsheet and start with B = 0 and increase in increments of 2 kg.

14.30 A 2 200 lbm, finely ground mixture of salt and sand is to be separated by adding 1 500 lbm of water to dissolve the salt and then filtering the salt solution from the insoluble sand. The resulting salt solution is then separated by the distillation of 1 350 lbm of water, leaving a final solution that is 92% salt and 8.0% water. Determine the percentages of salt and sand in the original mixture.

CHAPTER 15

Energy Sources and Alternatives

Chapter Objectives

When you complete your study of this chapter, you will able to:

- Understand the source and use of fossil fuels in the United States, the world, and especially in developing countries
- Distinguish among the positive and negative facets of various alternative energy sources
- Use information gathered on the Internet to summarize the world dependence on petroleum, coal and natural gas today and 20 years into the future
- Evaluate alternatives to the dependence on fossil fuels in the transportation, home heating, and electricity generation sectors
- Use information from this chapter as a basis for expansion of your knowledge and ability to think, design, and live "GREEN"

15.1   Introduction

Energy is one of the world's most important commodities. If we were to look for one event that characterized the transformation of society during the past 200 years it would be the Industrial Revolution, which provided a direct substitution of machine power for muscle power. This transformation has been sustained by the rapid depletion of natural resources, namely, fossil fuels—primarily oil, coal and natural gas. Energy from these fossil fuels is converted into forms that can be stored, transported, and used at the appropriate time and place. To some extent the development of any society can be determined by the amount of energy usage. There is a strong correlation between productivity of a nation and its capability to generate energy. Figure 15.1 graphically illustrates the dramatic increase in the consumption of fossil fuels by the United States over the past 100 years. Notice in particular the demand for petroleum and natural gas.

Unfortunately, the heavy reliance that world economies and in particular the United States have placed on the use of fossil fuels presents a most unique challenge for science and engineering in the near future. As supplies dwindle, fossil fuels must be replaced by alternate sources such as nuclear, hydro, geothermal, solar, and wind as well as considerable conservation measures.

Figure 15.1

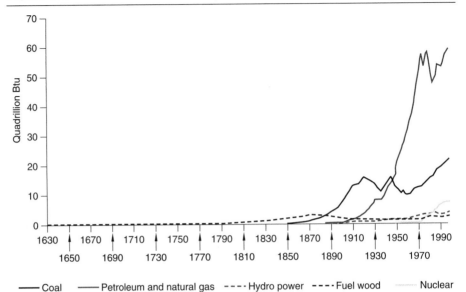

— Coal   — Petroleum and natural gas   ---- Hydro power   ---- Fuel wood   Nuclear

U.S. consumption of energy by resource. (Congressman Roscoe Bartlett, *"Our Dependence on Foreign Oil" Presentation to Congress, April 20, 2005.*)

We will begin this chapter with a short history of fossilized fuel formation and recovery. Then we will discuss the major areas of energy consumption in the United States and review some alternative sources of power and energy that will most certainly be a part of your lives and careers within the next 20 years.

15.2 Fossil Fuels

Over hundreds of millions of years, energy-rich substances were formed from buried plants and microorganisms. Eventually, when conditions were favorable—if there was sufficient temperature and pressure—this organic material chemically transformed into hydrocarbons. Depending on the time period, the underground formations, and the type of organic materials, these formations ultimately resulted in deposits of petroleum, coal, or natural gas.

15.2.1 Petroleum Formation and Recovery

Petroleum is formed from microscopic plants and bacteria that lived in the ancient oceans and saltwater seas. When these microorganisms died and settled to the seafloor, an organic-rich mud was formed. This mud was covered with heavy layers of sediment; over time, the resulting heat and compression chemically transformed the organic matter into petroleum and natural gas.

In the quest to find oil, geologists search for regions with three primary conditions necessary for petroleum formation: organic-rich source rock, relatively high temperatures that would generate petroleum from organic matter, and petroleum-trapping rock formations such as salt domes.

15.2.2 Coal Formation and Recovery

Coal is a solid fossil fuel formed from trees, ferns, and mosses that grew in ancient swamps, in bogs, and along coastal shorelines. The high pressure and temperature associated with the burial of these plants under heavy layers of sediment caused the original organic matter to become increasingly carbon-rich. The successive stages of coal formation range from peat, which is partially carbonized matter, to anthracite, which is hard coal with the highest carbon content and the lowest moisture content.

The majority of the world's coal beds have been located and are included in the world demonstrated reserve base (DRB). Not all coal in the DRB is recoverable, but the efforts of scientists and engineers in the last century have improved coal-mining methods and increased the recoverable percentage.

15.2.3 Natural Gas Formation and Recovery

Natural gas is formed from plankton (mainly algae and protozoans) that died and settled to the floor of the ancient oceans. Again, the organic matter was buried and compressed under layers of sediment for millions of years. Natural gas is primarily composed of methane and other light hydrocarbons.

Natural gas is much lighter (less dense) and forms a layer over the petroleum or coal deposits with which it is often found. Natural gas deposits are removed by wells drilled deep into the ground. Historically, natural gas was considered a waste byproduct of petroleum and coal mining, but demand has grown for this product because it can be piped directly to commercial plants and residences and because it is a cleaner-burning fuel than either petroleum or coal.

15.3   Finite Supply of Fossil Fuels

Since fossil fuels such as coal, oil, and natural gas took millions of years to form, energy derived from fossil fuels is a truly finite resource. Once fossilized fuels are consumed, they are gone forever and alternative energy sources must be identified and employed to supply our growing demand.

15.3.1 U.S. Oil Reserves, Consumption, and Production

We use the term "oil production" in this text as it refers to petroleum, but that is something of a misnomer. Whereas the objective of oil "producers" is to locate deposits, drill wells, refine, and distribute oil and its byproducts, the producers do not and cannot create or manufacture oil. Oil production by Exxon, British Petroleum, and others simply denotes the ability to make available a resource that is becoming more difficult to provide as supplies decrease and demands increase.

The world's first oil well was drilled in 1859 in Titusville, Pennsylvania. The main byproduct at that time was kerosene, which began to replace whale oil for use in lamps. Shortly thereafter, in 1861, a German entrepreneur invented the first gasoline-burning engine, and the demand for oil as a substitute for coal began to grow (see Fig. 15.1). In the mid-1950s petroleum was in high supply, and gasoline cost about $0.25 per gallon. Everyone thought that petroleum was available in limitless quantities and the supply would last for centuries.

One individual, however, did not agree. A geophysicist by the name of M. King Hubbert, working for the Shell Oil Company in the 1950s, predicted that at the rate oil was being extracted from wells in the United States, production would peak in the early 1970s and thereafter forever decline. Hubbert was correct—U.S. oil production peaked in 1970 at approximately nine million barrels per day and has declined to a 2008 level of about five million barrels per day. Hubbert's prediction, which came to be known as Hubbert's peak, proved to be very accurate. Figure 15.2 illustrates the peak of oil production in the United States, the rapid increase in consumption, and, most striking, the dramatic increase in net imports needed to offset lower production and increased consumption. As you review Figure 15.2 realize that the production line on the graph includes both crude oil and natural gas plant liquids.

Until the 1950s the United States produced all the petroleum it needed. Beginning in 1995, the United States imported more petroleum than it produced. In 2008 we consumed 19.5 million barrels of oil per day, of which over 11 million (57%) was from net imports. The United States continues to rely heavily on foreign sources. Worldwide consumption in 2008 was 85.4 million barrels per day, so the United States, which makes up 5% of the world population, consumes 23% of the world's total oil production.

Figure 15.2

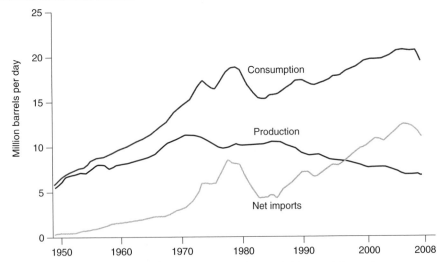

Consumption, Production, and Import Trends (1949–2008) *(Energy Information Administration, Annual Energy Review (June 2008).*

Table 15.1 Annual Consumption of Petroleum Products in 2008

	(Thousand barrels per day)	Percentage
Motor Gasoline	8,989	46.10
Diesel Fuel and Heating Oil	3,945	20.23
Liquefied Petroleum Gases (LPG)	1,954	10.02
Kero-Type Jet Fuel	1,539	7.89
Still Gas	670	3.44
Residual Fuel Oil	622	3.19
Petroluem Coke	464	2.38
Asphalt and Road Oil	417	2.14
Feedstock Oils	304	1.56
Naptha for Feedstocks	248	1.27
Lubricants	131	0.67
Misc. Products	67	0.34
Special Napthas	44	0.23
Aviation Gas	15	0.08
Kerosene	14	0.07
Waxes	9	0.05
Total	19,498	100.00
Consumption of 19,498,000 barrels of oil per day		

Source: Energy Information Administration, U.S. Department of Energy.

Since we are listing production and consumption in both barrels per day and quadrillion (10^{15}) Btu, let's define a barrel. A barrel of oil contains 42 U.S. gallons of crude oil, equivalent to 5 800 000 Btu. Due to a reduction in the density during the refining process, a barrel of 42 U.S. gallons of crude oil yields nearly 45 gallons of petroleum products.

Table 15.1 illustrates the actual deposition of the 19.5 million barrels of petroleum products used each day in the US. Motor gasoline accounts for 46% of the total which is used primarily for cars and light trucks.

15.3.2 World Oil Reserves, Consumption, and Production

Scientists and geologists began to extend Hubbert's principles to predict when world petroleum production would peak and then forever decline. Recall that U.S. oil production peaked in 1970 at approximately nine million barrels per day. Most scientists agree that the peak will occur within the next 10 to 20 years. Remember, the peak of oil production ("peak oil") occurs when approximately half of the oil has been extracted. After the peak has been reached it will take some number of years to extract the remaining oil. The exact time period from peak oil until the last drop depends primarily on rate of extraction.

Within a few years, half of the world's oil reserves will have been depleted. The remaining reserves, estimated at one trillion barrels are being consumed at a rate of 30 billion barrels annually. That suggests it will be 30 to 35 years until the complete depletion of petroleum occurs, assuming that usage stays constant. A more likely scenario is that world demand will increase with a decrease in supply. Additional pressure on oil supply will come not only from the United States, where demand is projected to grow by 2% annually, but even more so from the rapidly expanding economies in countries like China and India, where demand is currently expanding between 5 and 10% annually.

15.3.3 U.S. Coal Reserves, Consumption, and Production

The United States has vast reserves of coal. The demonstrated reserve base (DRB) was estimated in 2008 to be 490 billion short tons. However, due to property rights, land use constraints, and environmental restrictions, only about half of the DRB is considered recoverable, in other words 200–300 billion short tons. In recent years approximately 1.0 billion short tons of coal has been consumed in the United States. Figure 15.3 illustrates changes in the percentage of coal used by major consumers over a nearly 60-year period. Currently, the majority of the coal consumed in the United States (93%), is for the generation of electricity.

However, as oil and natural gas supplies dwindle, additional coal will likely be consumed in a variety of additional applications. For instance, North Dakota currently has a coal gasification plant that uses 6.3 million tons of lignite coal per year to produce 54 billion cubic feet of synthetic natural gas annually.

15.3.4 World Coal Reserves and Consumption

Worldwide, coal is the most abundant of the fossil fuels and its reserves are the most widely distributed. The world's total recoverable reserves are approximately one trillion short tons. The United States has 28% of the global coal reserves, Russia 19%, China 14%, and India 7%. At current rates of consumption (five billion metric tons per year) worldwide reserves could, in theory, last

Figure 15.3

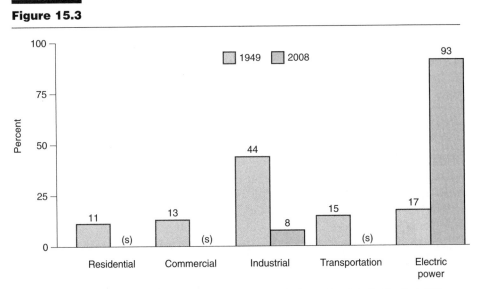

U.S. coal consumption (s = small percentage). *(Energy Information Administration, U.S. Department of Energy.)*

Figure 15.4

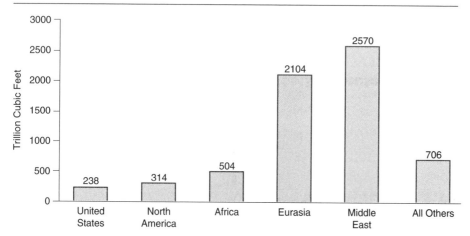

Natural Gas Reserves, Trillions of cubic ft. *(Energy Information Administration, U.S. Department of Energy)*

for another 200 years; however, as might be expected, developing countries are increasing their annual consumption of coal.

15.3.5 U.S. Natural Gas Reserves and Consumption

The vast majority of natural gas reserves exist outside the United States (see Fig. 15.4). Note that only 4% or 238 trillion cubic feet of the world natural gas reserves are located the US. However, the United States accounts for 24% of the world annual consumption.

The United States and Russia alone account for 40% of the total world consumption.

15.3.6 World Natural Gas Reserves and Consumption

The world reserve of natural gas is estimated to be 6 430 trillion cubic feet with an annual worldwide use of approximately 96 trillion cubic feet, suggesting depletion in about 65 years. But depletion is likely to come sooner than that: In 20 years the world annual consumption is expected to reach 150 trillion cubic feet, a projected increase of nearly 60%, primarily due to increased demand from developing countries.

15.3.7 Conclusion

As authors of this text we have no way of precisely predicting the actual undiscovered reserves or the demand for petroleum, coal and natural gas, but we do know that it took millions of years to create these fossils fuels and that they are rapidly being depleted. What we can do is make perfectly clear that if the United States continues to use large amounts of energy, then a need for the immediate development of alternative forms most certainly exists.

15.4 Major Areas of Energy Consumption in the United States

The four primary areas of energy consumption in the United States are transportation, industrial, residential/commercial, and electric power. Figure 15.5 provides an overview of the quantities of energy from supply sources and the amounts consumed by each sector. Notice in particular the percentage of renewable energy. We see that 85% of the nearly 100 quadrillion Btu of energy consumed by the United States in 2008 was provided by fossil fuels.

15.4.1 Transportation

As mentioned earlier, the United States consumes 19.5 million barrels (820 million gallons) of petroleum products each day, almost half of it in the form of gasoline. Consumption in 2008 was 138 billion gallons, which is an average of 380 million gallons per day. There are approximately 250 million vehicles in the U.S. that consume gasoline and they each travel over 12 000 miles per year. The majority of gasoline is used in cars and light trucks

While getting the oil out of the ground and refining it is complicated, moving it from the point of production to the refinery and on to the final consumer is just as complex. The refining process usually involves (1) distillation, or separation of the hydrocarbons that make up crude oil so that the heavier

Figure 15.5

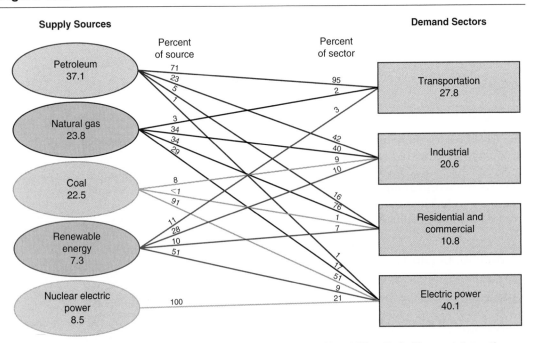

U.S. Primary Energy Consumption by Source and Sector in 2008, (Quadrillion Btu). *(Energy Information Administration, U.S. Department of Energy)*

Figure 15.6

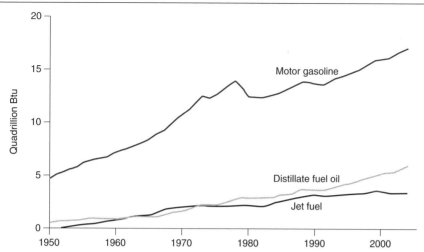

Transportation sector demand. *(Energy Information Administration, U.S. Department of Energy.)*

products such as asphalt are separated from the lighter products like kerosene; (2) conversion, or cracking of the molecules to allow the refiner to squeeze a higher percentage of light products such as gasoline from each barrel of oil; and (3) treatment or enhancement of the quality of the product, which could entail removing sulfur from such fuels as kerosene, gasoline, and heating oil. The addition of blending components to gasoline is also a part of this process.

After crude oil is refined into gasoline and other petroleum products, the products must be distributed to consumers. The majority of gasoline is delivered first by pipeline—today, there are more than 160 000 miles of pipeline in the United States—to storage terminals near consuming areas, and then loaded into trucks for delivery to individual gas stations.

Transportation (automobile, truck, train, aircraft, and military) demands are nearly 100% exclusively dependent on crude oil derivatives. Price levels and economic growth trends both influence the demand for petroleum products (see Fig. 15.6). High prices tend to provide incentives for individuals and industry to adopt short-term conservation measures such as reducing discretionary driving as well as long-term measures like design changes that increase fuel efficiency.

Technology is beginning to develop alternatives to the gasoline and diesel vehicles that are the primary sources of transportation in the U.S. today. Research on a wide variety of alternative fuel vehicles includes hybrid, biodiesel, flex-fuel, ethanol, nature gas, propane, hydrogen, electric, and fuel cell.

For example, the hybrid-electric vehicles were some of the first alternatives to the standard gasoline engines, by coupling an electric motor with a gasoline engine to improve fuel economy.

Unlike the all-electric vehicles the hybrid does not need to be plugged into an external source of electricity to be charged.

15.4.2 Generation of Electricity

Approximately 40 quadrillion (40 x 10^{15}) Btu of energy are consumed annually in the United States to generate electricity. Roughly one-third of that amount is converted into usable electricity provided by utilities to the end user. The other two-thirds is lost as waste heat and other inefficiencies.

American consumers expect electricity to be available whenever they plug in an appliance or flip a switch. Satisfying these instantaneous demands requires an uninterrupted flow of electricity. In order to meet this requirement, utilities operate several types of electric generating facilities, powered by a wide range of fuel sources. Notice in Figure 15.7 that nearly 70% of U.S. electrical power generation totaling 4 175 billon kilowatt-hours results from burning fossil fuels.

Dependence on fossil fuels can be replaced with increased use of alternatives like nuclear and renewable resources. It is much more reasonable to envision a changeover in the source fuel for electrical generation sector than it is for changeover in source fuel in the transportation sector.

This use of energy to produce electricity can be achieved by a variety of fuels and generating techniques, including

- *Steam power plants.* In a boiler, water is heated to a high temperature forming high-pressure steam. The steam is sent through a turbine that turns an electric generator.
- *Gas turbine power plant.* The fossil fuel is burned to create a hot gas, which goes through the turbine.
- *Internal-combustion equipment.* This method uses petroleum products to drive the internal-combustion equipment that turns the generator.

Figure 15.7

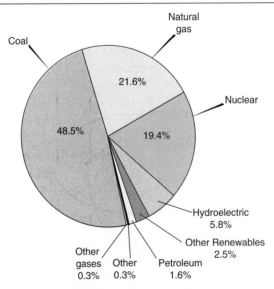

U.S. Electric Power Industry Net Generation 2007. *(EIA, Form EIA-923, "Power Plant Operations Report" and predecessor form(s) including Form EIA-906, "Power Plant Report," and Form EIA-920, "Combined Heat and Power Plant Report.")*

- *Nuclear reactors.* A nuclear reactor is a controlled fission process that provides energy (heat) to water, which in turn delivers high-temperature, high-pressure steam to the turbine, which drives the electric generator.

Wind turbines, hydroelectric turbines, and geothermal power plants can produce limited amounts of electricity as well. These together with municipal waste heat combustion, solar energy, and wood provide the renewable sources of electrical generation.

15.4.3 Home Heating

The most popular home heating fuel is gas. Fifty-seven percent of U.S. households are heated with natural gas or liquefied petroleum gas (LPG). The second most frequently used energy source is electricity (31.3%); the remaining homes use heating oil, kerosene, or wood.

The United States has two sources of heating oil: domestic refineries and imports from foreign countries. Refineries produce heating oil as a part of the "distillate fuel oil" product family, which includes heating oils and diesel fuel. Distillate products are shipped throughout the United States by pipelines, barges, tankers, trucks, and rail cars. Most imports of distillate come from Canada, the Virgin Islands, and Venezuela.

Recall that natural gas is withdrawn from the Earth's interior. It is primarily methane (90%) with propane, butane, and ethane. LPG is also a member of the family of light hydrocarbons. It consists of a mixture of propane and butane and it can be obtained from either natural gas or petroleum refinement.

The relative high cost of gas both natural and LPG has resulted in more new and existing homes using electric. Heat pumps can be used in southern climates for both heating and cooling. In the more northern areas geothermal heat pumps are becoming increasingly popular. Instead of using air as the heat transfer medium, they use water piped underground.

Conservation methods may include adjusting thermostats and improved insulation in newly constructed and existing homes. Again, it is reasonable to consider alternative sources for the U.S. dependence on natural gas as worldwide supplies dwindle.

15.4.4 Energy Projections over the Next 25 Years

The Energy Information Administration (EIA), while preparing projections for its annual energy review, evaluated a number of trends and issues that could affect tomorrow's energy demands. Future trends in energy supply and demand are influenced by factors that make predictions difficult. Factors include public policy decisions, energy prices, economic growth, and technology advances.

The information presented in this section is an abbreviated review of EIA projections to the year 2030. For the complete text of the overview, visit www. eia.doe.gov. Total primary energy consumption as illustrated in Fig.15.8 projects an average rate of increase of 1.2% per year. The United States consumed approximately 100 quadrillion Btu of energy in 2008; that figure is expected to increase to 135 quadrillion Btu annually by 2030.

Figure 15.8

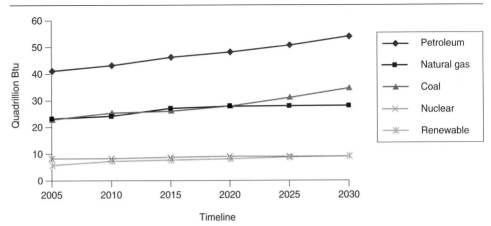

U.S. energy consumption projection 2005–2030. *(Energy Information Administration, Annual Energy Outlook 2006.)*

The EIA projects that electricity generation will increase from 3 800 billion kilowatt-hours in 2004 to 5 620 billion kilowatt-hours in 2030. To meet this projection, an increase in the consumption of coal is expected, especially since electrical generation from natural gas is expected to decline as a result of higher prices.

15.5 Alternate Energy Sources

This section of the text is intended to suggest energy sources other than petroleum, coal, and natural gas. As outlined before, either life as we know it will have to change or alternative fossil fuel sources must be developed. Most certainly energy sources of the future will include nuclear, but they will also include an emerging group of renewable energy sources such as biomass, hydropower, wind, geothermal, and solar. Figure 15.9 illustrates the percentage of each renewable energy source that contributes to the approximately 100 Quadrillion Btu's consumed each year in the U.S. Recall that Figure 15.5 illustrates the demand sectors for this energy.

The engineer must continue to design mechanisms that will perform work by conversion of energy. However, the source of energy used to produce work and its corresponding conversion efficiency of energy into work will play a much greater role in the design procedure.

More is being written today regarding the depletion of fossil fuels and the immediate need to develop alternative energy sources to solve the problem. Many opinions are given with regard to how we should proceed to seek new energy sources and how we should use existing sources. Environmental concerns about energy acquisition and consumption become stronger by the day. The engineer must be able to discriminate between facts and opinions. The following discussion provides an insight into some energy sources that will be utilized in the future.

Figure 15.9

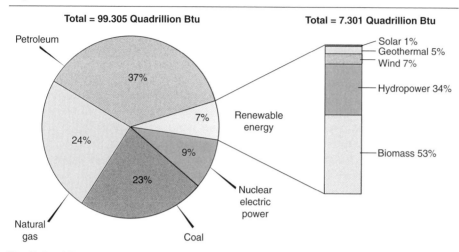

Total = 99.305 Quadrillion Btu   Total = 7.301 Quadrillion Btu

Petroleum
37%
24%
7%   Renewable
energy
9%
23%
Natural
gas   Coal
Nuclear
electric
power

Solar 1%
Geothermal 5%
Wind 7%
Hydropower 34%
Biomass 53%

The Role of Renewable Energy in the Nation's Energy Supply, 2008 *(Note: Sum of components may not equal 100% due to independent rounding. Source Energy Information Administration, Renewable Energy Consumption and Electricity Preliminary Statistics 2008, Table 1: U.S. Energy Consumption by Energy Source, 2004-2008 (July 2009)).*

15.5.1 Nuclear Power

As we reach the end of the age of fossil fuels, the world will have to reconsider increasing the use of nuclear power. Nuclear fission is a well-established technology and it may well be the only proven technology capable of providing an adequate amount of electric power as fossil fuels become either extremely expensive or nonexistent. Issues of safety and disposal of nuclear waste will continue to demand the talents and resources of many engineers.

Nuclear fission is the splitting of an atom of nuclear fuel, usually uranium 235 or plutonium 239, by bombarding it with neutrons. The splitting process releases a great quantity of energy. As the atom splits, it divides approximately in half, releasing two or three free neutrons. The mass of the product after the split is always less than the mass of the reactants. This mass deficiency represents the released energy determined by Einstein's equation $E = mc^2$.

The liberated neutrons are capable of successive fission, which is called a chain reaction. The minimum quantity of fissionable material needed to sustain a chain reaction is called its critical mass. Chain reactions can be controlled or uncontrolled. The atomic bomb is an example of a uncontrolled reaction. To make fission a useful source of energy, a nuclear reactor utilizes a controlled reaction.

As of 2008 there were 104 commercial nuclear reactors at 65 nuclear power plants in 31 states across our country. They produce approximately 20% of the total electrical energy generated in the United States. Palo Verde in Arizona is the largest nuclear plant in the United States (see Fig. 15.10). Its three-unit system provides 3 733 megawatts annually, enough power to meet the needs of four million customers. The reactors are used to heat water for use in steam turbines that drive the generators.

Nuclear reactions take place in the reactor's core. A moderator, normally graphite or heavy water, is used to slow the neutrons that were released from the initial fission process to a speed and temperature that will cause succeeding reactions. Steel rods containing boron or cadmium are inserted into the reactor to control the number of free neutrons. The reactor is also shielded with lead and concrete to prevent the escape of dangerous radiation.

Nearly all reactors in use today producing electric energy are fission reactors using U-235. However, U-235 constitutes only 0.7% of the natural uranium supply and thus it would quickly become a scarce resource if we were totally dependent on it as a source of energy. Other isotopes, such as U-238 and thorium-232, are relatively abundant in nature. A new series of reactors that can produce new fissionable material and energy at the same time are called breeder reactor. They have been designed to use isotopes that are generally a waste product in current reactors.

The radioactive waste material from nuclear power generation has created disposal problems. The half-life of radioactive materials can be 1 000 years or more, thus creating a perpetual need for secure disposal of the waste materials from a nuclear power–generation facility. Radiation leaks and other environmental concerns have led to the shutdown of some facilities. This has made nuclear power generation a politically sensitive issue. Nuclear engineers and scientists continue to work on the complex problems associated with this source of energy. As these problems are solved and the general public becomes aware of the potential afforded by nuclear energy, increased use of nuclear power is likely.

Nuclear fusion—the energy gained from fusing light nuclei into heavier ones—has produced both the hope that it will be the ultimate energy source of the future and frustration. The Sun is the best example of nuclear fusion as it converts hydrogen into helium. The only problem is that the reaction occurs at a temperature of millions of degrees.

Nuclear fusion would use a fuel supply that is nearly inexhaustible, and we know of no scientific principle that forbids it from working. However, it has proved to be remarkably elusive, presenting a multitude of technical problems that have yet to be solved. It has been said that nuclear fusion is the energy source of the future—but it may not be advisable to rely on this technology to solve our immediate energy situation.

15.5.2 Renewable Energy Sources

Biomass (Ethanol and Biodiesel)

Biomass is a term that includes all energy materials that emanate from biological sources and are available on a renewable or recurring basis. Biomass includes agricultural crops and trees, wood and wood wastes, plants, grasses, residues, fibers, animal wastes, and municipal wastes. The Environmental Protection Agency estimates that, on a Btu-output basis, 75% of municipal trash contains biomass.

At its most basic, ethanol is grain alcohol, produced from a variety of crops but mainly corn. Because it is produced from renewable resources, it could assist in reducing U.S. dependence on foreign sources of energy. Pure ethanol

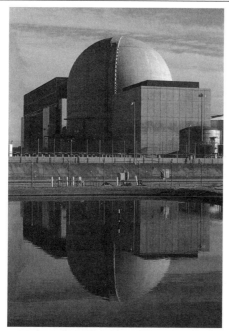
Figure 15.10

353

Alternate Energy
Sources

A nuclear power plant facility in Arizona.

is not generally used as a motor fuel; rather, it is combined with unleaded gasoline in two primary blends:

- E10, 10% ethanol and 90% unleaded gasoline, is approved for use in any vehicle sold in the United States since 1980. Today, approximately one-third of America's gasoline was blended with ethanol.
- E85, 85% ethanol and 15% unleaded gasoline, is an alternative fuel for use in flexible-fuel vehicles (FFVs). There are currently more than four million FFVs on American roads, and automakers are producing more of these vehicles each year. More and more gas stations are installing E85 pumps; when E85 is not available, FFVs can operate on straight gasoline or any ethanol blend up to 85%.

Current U.S. ethanol production capacity is four billion gallons (just over 95 million barrels) per year.

Ethanol is a clean-burning fuel that reduces carbon monoxide and hydrocarbon tailpipe exhaust as well as an oxygenate, which allows it to burn more cleanly and more completely than gasoline. It is considered a high-performance fuel; pure ethanol has an octane rating of 113, so adding 10% ethanol to gasoline raises the overall octane by 2 to 3 points. Ethanol also has a lower Btu value than gasoline, meaning that ethanol burns cooler and is gentler on the engine, leading to longer engine life.

Biodiesel is a clean-burning alternative fuel, produced from renewable resources—primarily soybean oil. One bushel of soybeans produces about 1.5 gallons of biodiesel. Biodiesel can be used in its pure form, called "neat biodiesel" or B100, or it can be blended at any level with petroleum diesel. The most common mix is B20 (20% biodiesel and 80% petrodiesel). Biodiesel fuels can be used in regular diesel vehicles without any modification to the engine. Biodiesel has exceptional lubricating qualities that contribute to longevity and cleanliness of diesel engines, and in its pure form is biodegradable and nontoxic.

Algae that is grown in farm-like conditions is another possible alternative fuel source. The extracted oil from algae plants is currently an important consideration for the Biofuel industry. Algae, like corn, soybeans, sugar cane, and other plants, use photosynthesis to convert solar energy into chemical energy. They store this energy in the form of oils, carbohydrates, and proteins. This plant oil can be converted to biodiesel. It turns out the more efficient a particular plant is at converting solar energy into chemical energy, the better it is from a biodiesel perspective, and algae are among the most efficient plants at photosynthesis on earth.

Biomass (Waste-to-Energy and Wood)

Waste-to-energy facilities produce clean, renewable energy through the combustion of municipal solid waste in specially designed power plants equipped with modern pollution control equipment. Today there are 90 waste-to-energy facilities located in 27 states that handle about 13% of U.S. trash, or about 95 000 tons per day. These facilities generate about 2 500 megawatts of electricity to meet the power needs of nearly two million homes. Turning garbage into energy makes sense, as there is a constant need for trash disposal as well as an equally constant demand for reliable electricity generation.

On an annual basis, waste-to-energy facilities remove and recycle more than 700 000 tons of ferrous metals and more than three million tons of glass, metal, plastics, batteries, ash, and yard waste. In addition, communities served by these facilities tend to be more conscious of recycling and therefore sort, reuse, and recycle at a higher rate than the national average rate of 30%.

Wood energy is primarily derived from the following sources. Roundwood is a term used by the industrial and electric utilities for timber poles, decking, floors, etc. Wood byproducts are also used in the residential sector for fuel, and wood waste is used in the industrial sector. (See Fig. 15.11; Biomass.)

Figure 15.11

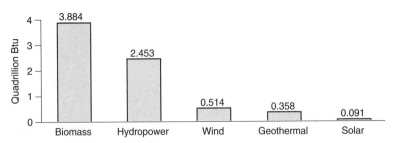

Renewable Energy Sources, Quadrillion Btu. (*Energy Information Administration, U.S. Department of Energy*)

Hydroelectric

Currently, about 6% of U.S. electric power is generated from hydroelectric dams on rivers. The largest hydroelectric facility in the United States is the Grand Coulee Dam, located in the state of Washington, with a generating capacity of 7 100 megawatts. We are, however, near the end of our ability to increase electrical generation by hydropower as dams have already been built nearly everywhere in the world where they are viable. In addition, environmental concerns for the protection of scenic rivers and wildlife will limit the construction of new hydroelectric facilities. (See Fig. 15.11; Hydropower.)

Tides

The difference between the elevation of the ocean at high tide and at low tide varies from 1 to 2 m in most places up to 15 to 20 m in some locations. The idea of using this energy source is not new. This concept is technically feasible—plants are located in France, Canada, and the former Soviet Union—but the economic feasibility is still in serious doubt. The toughest problems seem to be that vast storage volumes are required and suitable basins are rare.

Ocean Currents and Waves

We are also aware of the warming effect that the Gulf Stream provides for the British Isles and Western Europe. This and other ocean currents possess massive amounts of kinetic energy even though they move at very slow velocities. It has been proposed that a series of large (170 m in diameter) turbines be placed in the Gulf Stream off Florida. Ten such turbines could produce power equivalent to that produced by one typical coal-fired power plant. Detractors warn that such installations may reduce the stream velocity to the point that the Gulf Stream's warming of Western Europe would be lost.

No doubt you have watched the surf smash into the beach and have been in awe of the obvious power being displayed. Machines have been made that produce power from the wave action. But to be successful, the installations must be located where the magnitude of the wave action is high and somewhat uniform, and it is difficult and expensive to design such installations against major storms. The best sites in the United States are on the coasts of Washington and Oregon.

Wind Power

Heated equatorial air rises and drifts toward the poles. This phenomenon coupled with the Earth's rotation results in a patterned air flow. In the United States, we see this pattern as weather systems moving from west to east across the country. These weather systems possess enormous energy, but the energy is diffused, variable, and difficult to capture.

Most are familiar with the historical use of the wind to pump water and to grind grain. These methods are still viable, and today a great deal of research is underway in an attempt to capture energy from the wind in an economically feasible manner. Engineers have learned that the power output from a windmill is approximately proportional to the square of the blade diameter and to the cube of the wind velocity. This suggests that to generate high power output, the windmills must be large and must be located in areas where the average wind velocity is high. Thus, coastal regions and the Great Plains are promising locations. However, many people find wind farms unsightly and undesirable near their property.

About one-half of one percent of U.S. electric power is now generated by the wind. That amount will likely increase as efforts continue toward reducing the costs of wind-power installations and as tax breaks for power producers have made wind power more economically competitive. The supply/demand problem related to wind may be solved by, for example, using excess power generated when the wind is blowing to pump water to an elevated storage area and use the resulting potential energy to drive a water turbine for power generation when the wind is not blowing. (See Fig. 15.11; Wind.)

Geothermal Power

It is commonly accepted that the Earth's core is molten rock with a temperature of 10 000 to 12 000 °F. This source has the potential to provide a large portion of our energy needs. The Earth's crust varies in thickness from a few hundred feet to perhaps 20 mi. It is composed mostly of layers of rock—some solid, some porous, and some fractured. Engineers are now exploring the use of the Earth's heat (geothermal energy) to produce power from hot water, steam, and heated rock.

There are many areas throughout the world where large amounts of hot water are available from the Earth. The hot water varies in many ways—in quantity, temperature, salinity, and mineral content.

At only a few locations in the world is steam available in sufficient quantities to be used to produce electricity. The only area in the United States producing large quantities of steam is in northern California. Low-cost electricity has been produced there for nearly 50 years. This steam is captured by drilling from 500 to 10 000 ft deep, and it has a temperature of about 350°F but at low pressure. Besides the corrosive nature of the steam, it contains several gases, including ammonia and hydrogen sulfide, that have objectionable odors and are poisonous.

It is clear that using geothermal energy has potential, but many problems need to be solved and solutions will probably come only after large-scale pilot plants have been in operation long enough to obtain reliable efficiency data and operating costs. (See Fig. 15.11; Geothermal.)

Solar Power

The sun is an obvious source of energy that we have employed in different ways since the beginning of time. It supplies us with many, many times as much energy as we need. Our problems lie with collecting, converting, and storing the inexhaustible supply. Efforts are being directed primarily toward direct heating, charging batteries, and heat engines—such as turbines that can generate electricity. Others are experimenting with crop drying and metallurgical furnaces. You are no doubt aware of the increasing use of solar energy in building heating and in domestic hot-water supplies. This use will surely continue to increase, particularly in new installations (see Fig. 15.12). But we will have to improve on current technology and/or have tax relief and low-interest loans or other incentives to make solar installations economically feasible.

Figure 15.12

357
Problems

Solar panels being used in new construction.

There are two general methods by which energy from the sun is used as a renewable source.

Solar thermal devices use heat directly from the sun, concentrate that heat in some fashion, then use it for numerous applications.

Photovoltaic (PV cells) is a method of converting sunlight directly into electricity using various semiconducting materials. However, sunlight loses much of its intensity as it travels through our atmosphere. Solar cells have been shown to work well on space vehicles but at the present time they are extremely costly. We must increase the efficiency of PV cells and work to reduce their costs in order for this application to come into general use. (See Fig. 15.11; Solar.)

Problems

15.1 Conduct a study detailing current production and consumption of fossil fuels. Determine the estimated annual usage of the fuels on a national and international basis, the geographical locations of the sources, and estimated time until the sources are exhausted at current usage rates. Compare fossil fuel usage rates in the United States with those in Japan, England, Germany, China, India, Russia, and South Africa. Working in teams, develop written reports or oral presentations according to your instructor's direction.

    (*a*) coal

    (*b*) petroleum

    (*c*) natural gas

15.2 Prepare a report or oral presentation on the status of nuclear power in the United States. Include usage data, public opinion on the use of nuclear power, regulations on the industry, and current information on the storage/disposal of spent fuel. Work in teams.

15.3 Prepare a report and give an oral presentation on the status of one of the following renewable sources of power in the United States. Include usage data, geographic locations of these sources, and the ultimate form of energy that is consumed from these sources. Work in teams.

(*a*) Biomass to include: (*i*) Ethanol and biodiesel (*ii*) Waste-to-energy and wood
(*b*) Hydropower
(*c*) Wind power
(*d*) Geothermal
(*e*) Solar power

15.4 The transportation sector in the United States presently runs almost exclusively on petroleum-based derivatives (gasoline, diesel, jet fuel). Research the following alternatives.

■ *Hybrid*   ■ *Propane*
■ *Diesel & biodiesel*   ■ *Hydrogen*
■ *Flex-fuel*   ■ *Electric*
■ *Ethanol*   ■ *Fuel cell*
■ *Natural gas*

Prepare a one-page paper outlining what you believe may be viable alternatives to the use of fossil fuels. May be assigned as a team or an individual assignment.

15.5 Conservation is another logical method by which the U.S. can reduce its demand for foreign-based petroleum and domestic supplies of coal and natural gas. Prepare a report that will include at least five practical steps that you as an individual could take to reduce energy usage. As an example, consider the house where you grew up, the automobile you drive, the electricity that you consume, etc.

15.6 Explore the possibility of large solar farms in the southwest. Prepare a paper that examines both Thermal and Photovoltaic. How many acres are available? What would be the cost versus conventional electric power delivery? How does it compare with nuclear power generation?

15.7 A wind farm is a cluster of at least three turbines with generating capacities in the order of hundreds of megawatts. Investigate the capabilities of the wind farm nearest your college and determine the how much of your school's electrical demand could be met by this resource. Write a paper that considers technical issues like the appropriate spacing of multiple turbines as well as control, stability, and power balance in relation to the power grid.

15.8 Considering the five major sources of renewable energy—biomass, hydropower, wind, geothermal, and solar—prepare a paper outlining what you believe will be the major source of renewable energy in 2015.

15.9 As a college student, you have the opportunity to experiment with various "Green" initiatives. Look into ways to reduce, reuse, and conserve energy in your place of residence, your college classrooms, your recyclables, and your transportation choices. Consider how individual changes that you and other classmates might consider could impact energy consumption.

15.10 As a class, develop a "Green" initiative, design and engineer its impact, market the initiative to the college, university, or city, then implement. Remember, a very small adjustment to the rudder of a large ship can turn the vessel 360 degrees. It only takes time.

CHAPTER 16

Fundamental Energy Principles

Chapter Objectives

When you complete your study of this chapter, you will able to:

- Identify and define the different forms in which energy can be found: potential, kinetic, internal, chemical, and nuclear
- Demonstrate an understanding based on the Conservation of Energy principles as they relate to energy transfer and conversion
- Discuss the limitations of energy conversion into useful work
- Compare actual cyclic efficiencies to Carnot efficiency
- Distinguish between available and unavailable forms of energy
- Conceptualize the fact that a refrigeration cycle is a reversed heat engine

16.1   Introduction to Thermodynamics

Energy is a fundamental concept of thermodynamics and plays a significant role in engineering analysis. If engineers are to provide solutions to energy-related problems, we must understand energy's fundamental principles.

What is energy? It cannot be seen; it has no mass or defining characteristics; it is distinguished only by what it can produce. Energy can be stored within a system in various forms. Energy can also be transformed from one form to another and transferred between systems. In a broad sense, energy may be defined as an ability to produce an effect or change on matter.

Thermodynamics is one of the major areas of engineering science. It is usually introduced to engineering students in a one-semester course, with students in energy-related disciplines continuing with one or more advanced courses. This text will outline some concepts that will help solve basic problems involving the transformation of energy. Our discussion is limited to the first law for closed systems and a brief introduction to the second law, which governs efficiency and power. These terms are explained in the following sections.

16.2   Stored Energy

Let's consider some of the different forms in which energy can be found. Stored energy exists in distinct forms: potential, kinetic, internal, chemical, and nuclear. The ultimate usefulness of stored energy depends on how efficiently the energy can be converted into a form that produces a desirable result.

When an object or mass m is elevated to a height h in a gravitational field (See Fig. 16.1) a certain amount of work must be done to overcome the gravitational attraction. Energy is the capacity to do work. (Work is considered further in Sec. 16.3.) The object may be said to possess the additional energy that was required to elevate it to the new position. In other words, its potential energy has been increased. The quantity of work done to increase its energy is the amount of force multiplied by the distance the object moved in the direction in which the force acts. It is a result of a given mass going from one condition to another. *Gravitational potential energy* is thus stored-up energy due to a change in elevation. It is derived from force and height above a datum plane, not from the means by which the height was attained. Mass m stores up energy as it is elevated to height h and loses this advantage when it comes back to its starting point. When an object is raised, a force is needed to overcome the effect of gravity. The object's increase in potential energy depends upon the distance or change in elevation experienced by the object. Thus,

$$PE = \frac{mg_L h}{g_C} \tag{16.1}$$

The units for mass m can be kilograms or lbm, those for the local acceleration of gravity g_L can be either meters per second squared or feet per second squared, and those for height h can be either meters or feet with g_c being the appropriate constant of proportionality. The units for potential energy, PE, are newton-meters (joules) or ft · lbf.

It is unnecessary in most cases to evaluate the total energy of an object; however, it is customary to evaluate its energy changes. In the case of potential energy, this is accomplished by establishing a datum plane (see Fig. 16.1) and evaluating the energy possessed by objects in excess of that possessed at the datum plane. Any convenient location, such as sea level, may be chosen as the datum plane. The potential energy at any other elevation is then equal to the work required to elevate the object from the datum plane.

A second form of stored energy can be realized by virtue of an object's velocity (V). The energy possessed by the object at a given velocity is called *kinetic energy.* It is equal to the energy required to accelerate the object from rest

Figure 16.1

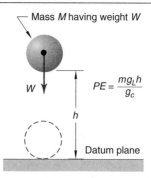

Mass *M* having weight *W*

W

$PE = \dfrac{mg_L h}{g_c}$

h

Datum plane

Potential energy.

Figure 16.2

361
Stored Energy

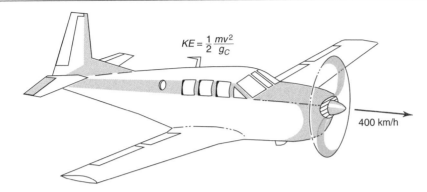

$$KE = \frac{1}{2} \frac{mv^2}{g_C}$$

400 km/h

Kinetic energy.

to its given velocity (considering the Earth's velocity to be the datum plane). In equation form,

$$KE = \frac{mV^2}{2g_c} \qquad (16.2)$$

where

m = mass, kg
V = velocity, m/s
$g_c = \dfrac{1.0 \text{kg} \cdot m}{N \cdot s^2}$

KE = kinetic energy, in newton-meters or joules. See Figure 16.2.

In many situations in nature, an exchange in the form of energy is common. Consider a ball thrown vertically into the air. It is given kinetic energy when thrown upward. When the ball reaches its maximum altitude, its velocity is zero, but it now possesses a higher potential energy because of the increase in altitude (height). When it begins to fall, the potential energy decreases and the kinetic energy increases until the ball is caught. If other effects such as air friction are neglected and the ball is caught at the same altitude from which it was thrown, there is no change in the total energy although the energy of the ball was transformed from kinetic to potential and then back to kinetic. This phenomenon is called *conservation of energy*, which is discussed in Sec. 16.4.

All matter is composed of molecules that, at finite temperatures, are in continuous motion. In addition, there are intermolecular attractions which vary as the distance between molecules changes. The energy possessed by the molecules as a composite whole is called *internal energy*, designated by the symbol U, which is largely dependent on temperature.

When a fuel is burned, energy is released. When food is consumed, it is converted into energy that sustains human efforts. The energy that is stored in a lump of coal or a loaf of bread is called *chemical energy*. Sunlight is transformed by a natural process called photosynthesis, that is, a process that forms chemical compounds with the aid of light. The stored energy in combustible fuels is generally measured in terms of heat of combustion, or heating value. For example, gasoline has a heating value of $47.7(10^6)$ J/kg, or $20.5(10^3)$ Btu/lbm.

Certain events change the atomic structure of matter. During the processes of *nuclear fission* (breaking the nucleus into two parts, which releases high amounts of energy) and *fusion* (combining lightweight nuclei into heavier ones, which also releases energy), mass is transformed into energy. The stored energy in atoms is called *nuclear energy,* which may need to play an increasing role in meeting energy needs of the future.

16.3 Energy in Transit

Energy is transferred from one form to another during many processes, such as the burning of fossil fuels to generate electricity or the converting of electric energy to heat by passing a current through a resistance. Like all transfer processes, there must be a driving force or potential difference in order to effect the transfer. In the absence of the driving force or potential difference, a state of equilibrium exists and no process can take place. The character of the driving force enables us to recognize the forms of energy in transit, that is, work or heat.

Energy is required for the movement of an object against some resistance. When there is an imbalance of forces and movement against some resistance, mechanical work is performed according to the relationship

$$W = (\text{force})(\text{distance})$$

$$= Fd \tag{16.3}$$

where the force is in the direction of movement. See Figure 16.3. If force has units of newtons and distance is expressed in meters, then work has units of joules.

Let's review the relationship between work and kinetic energy. When a body is accelerated by a resultant force, the work done on the body can be considered a transfer of energy, where it is stored as kinetic energy. For example, imagine a body with a given mass (m) moving with a velocity (V_1) and a resultant force (F) with no other external effects is applied to the body. This action alone results in the body moving with a velocity V_2. The work done on the body can be considered a transfer of energy in the form of kinetic energy.

Figure 16.3

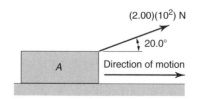

$(2.00)(10^2)$ N

$20.0°$

Direction of motion

A

An applied force doing work.

Electric energy is another form of work. This form of energy is transferred through a conducting medium when a difference in electric potential exists in the medium.

Other examples of energy transfer that are classified as work include magnetic, fluid compression, extension of a solid, and chemical reactions. In each case a driving function exists that causes energy to be transferred during a process.

Heat is energy that is transferred from one region to another by virtue of temperature difference. The unit of heat is the joule. The large numerical values occurring in energy-transfer computations has led to the frequent use of the megajoule (10^6 J), or MJ. The symbol used for heat is Q.

The relationship between the energy forms of heat and work during a process is given by the first law of thermodynamics, which is discussed in Section 16.4. In addition, every form of work carries with it a corresponding form of friction that may change some of the work into heat. When this happens, the process is irreversible, meaning that the energy put into the process cannot be totally recovered by reversing the process. Another way of stating this is that heat is a low-grade form of energy and cannot be converted completely to another form such as work. This concept is basic to the second law of thermodynamics (discussed in Section 16.5). The first and second laws of thermodynamics are fundamental to the study of processes involving energy transfer.

Several examples are provided below that illustrate both stored forms of energy and energy in transition. Particular attention should be given to the units and unit conversions in the examples.

Example Problem 16.1   A boulder with mass of 1.000 (10^3) kg rests on a ledge 200.0 m above sea level. ($g_L = 9.807$ m/s²) What type of energy does the object possess and what is its magnitude?

Solution   The object possesses potential energy, so from Eq. (16.1),

$$PE = \frac{mg_Lh}{g_c}$$

$$= \frac{1\ 000\ \text{kg}\,(9.807\ \text{m})\,(200.0\ \text{m})\,\text{N}\cdot\text{s}^2}{1.0\text{kg}\cdot\text{m}\cdot\text{s}^2}$$

$$= 1.961(10^6)\ \text{N}\cdot\text{m}$$

$$= 1.961(10^6)\ \text{J}$$

$$= 1.961\ \text{MJ}$$

Example Problem 16.2   A 2.00 (10^4) lbm semitrailer is traveling at sea level at a speed of 50.0 mph. Determine the energy form and magnitude possessed by the truck. Express the magnitude in Engineering System units.

Solution   Convert mph to ft/s:

$$50.0\,\text{mi/h} = \frac{50.0\,\text{mi}}{\text{h}} \times \frac{5280\,\text{ft}}{\text{mi}} \times \frac{\text{h}}{3600\,\text{s}} = 73.33\,\text{ft/s}$$

The truck possesses kinetic energy; therefore, from Eq. (16.2),

$$KE = \frac{mV^2}{2g_c}$$

$$= \frac{20\,000\,\text{lbm}}{2} \times \frac{(73.33)^2\text{ft}^2}{\text{s}^2} \times \frac{\text{lbf} \cdot \text{s}^2}{32.174\,0\,\text{lbm} \cdot \text{ft}} = 1\,671\,315\,\text{ft} \cdot \text{lbf}$$

$$= 1.67 \times 10^6\,\text{ft} \cdot \text{lbf}$$

If metric units are desired, this can be converted to SI as follows:

$$1.0\,\text{ft} \cdot \text{lbf} = 1.355\,8\,\text{joules, so}$$

$$KE = 2.27(10^6)\,\text{N} \cdot \text{m}$$

$$= 2.27\,\text{MJ}$$

Thus 2.27 MJ is the amount of energy that must be absorbed by the truck brakes in order to bring the vehicle to zero velocity in an emergency stop. If the stop is gradual, some of the energy can be absorbed by the engine and by road friction; in an emergency nearly all of the energy will be absorbed by the brakes and road friction.

Example Problem 16.3   A mass of water is heated from 10.0 to 20.0°C by the addition of 5.00(10³) Btu of energy. What is the final form of the energy? Express the final form of the energy in megajoules.

Solution   The final form of the heat energy added appears as increased internal energy (U) of the water. If state 1 of the water is prior to heating and state 2 is after heat has been added, then $U_2 - U_1 = 5\,000$ Btu.

Converting to SI, we get

$$U_2 - U_1 = 5.00\,(10^3)\,\text{Btu} = (5.00 \times 10^3\,\text{Btu})(1055.1\,\text{J/Btu})$$

$$= 5.28\,(10^6)\,\text{J}$$

$$= 5.28\,\text{MJ}$$

Example Problem 16.4   A force of 2.00×10^2 N acting at an angle of 20.0° with the horizontal is required to move block A along the horizontal surface (see Fig. 16.3). How much work is done if the block is moved 1.00×10^2 m?

Solution   Work is computed as the product of the force in the direction of motion and the distance moved, as in Eq. (16.3):

$$W = Fd$$

$$= (2.00 \times 10^2 \text{ N})(\cos 20.0°)(1.00 \times 10^2 \text{ m})$$

$$= 18\ 800 \text{ N} \cdot \text{m}$$

$$= 18\ 800 \text{ J}$$

$$= 18.8 \text{ kJ}$$

What happened to the energy released by the work done on Block A? There is no increase in potential energy, since height was not changed. There is no velocity change, so the change in kinetic-energy is zero. The energy of the work in this example is dissipated as heat in the form of friction between the block and surface; therefore, the temperature of the block and the surface in the immediate vicinity of the block increased.

Would you believe that the reverse process is possible? That is, could the molecules in the block and surface that are moving faster than the surrounding molecules due to an increased temperature randomly move the object back to its original position as they return to their initial temperature? The first law of thermodynamics does not place any restrictions on the reverse process, other than conservation of energy. That is why there is a second law of thermodynamics, which does not allow certain processes.

16.4   First Law of Thermodynamics: The Conservation of Energy

Conservation of energy in nonnuclear processes means simply that energy can never be created or destroyed, only transformed. In effect, energy is converted from one form to another without loss. Careful measurements have shown that during energy transformations there is a definite relationship in the quantitative amounts of energy transformed. This relationship—the first law of thermodynamics—is a restatement of the principle of the conservation of energy.

When applying the first law to substances undergoing energy changes, it is necessary to define a system and write a mathematical expression for the law. In (Fig. 16.4a), a generalized closed system is illustrated. In a *closed system*, no material (mass) may cross the boundaries, but the boundaries may change shape. Energy, however, may cross the boundaries and/or the entire system may be moved intact to another position.

Some applications involve analysis of an *open system*, in which mass crosses the defined boundaries and a portion of the energy transformation is carried in or out of the system with the mass (Fig. 16.4b) An example is an air compressor, which takes atmospheric air, compresses it, and delivers it to a storage tank. (Because of the complexity of analysis and explanation, problems involving open systems will not be considered in this chapter.)

For the generalized closed system, the first law is written as

Energy in + Energy stored at condition 1 = Energy out + Energy stored at condition 2

where conditions 1 and 2 refer to initial and final states of the system.

In thermodynamics all changes in the total energy of a closed system are considered to be made up of three contributions. Two of these changes are kinetic energy and gravitational potential energy. Both of these changes are associated with the motion and position of the system as a whole. All other energy changes are combined together into internal energy of the system.

Energy in any of its forms may cross the boundaries of the system. If the assumption is made that no nuclear, chemical, or electrical energy is involved, then the change in total energy can be written as

$$\Delta E = \Delta KE + \Delta PE + \Delta U$$

If the closed system as a whole is stationary, that is, not changing velocity or changing its elevation, then ΔKE and ΔPE are zero, and the first law can be stated as:

Heat in + Work done on system + Internal energy at condition 1 = Heat out + Work done by the system + Internal energy at condition 2

or, combining heat, work, and internal energy quantities at conditions 1 and 2:

$$_1Q_2 = U_2 - U_1 + {_1W_2} \tag{16.4}$$

where

$$_1Q_2 = \text{heat } added \text{ to system}$$

$$_1W_2 = \text{net mechanical work } done\ by \text{ system}$$

$$U_2 - U_1 = \text{change in internal energy from state 1 to state 2}$$
$$\text{with } \Delta PE \text{ and } \Delta KE \text{ zero.}$$

Equation 16.4 is the first law for a closed system when the potential and kinetic energy terms for the system are zero.

Example Problem 16.5   The internal energy of a system decreases by 108 J while 175 J of work is done by the system on the surroundings. Determine if heat is added to or removed from the system.

Solution
$$_1Q_2 = U_2 - U_1 + {_1W_2}$$

$$= -108 + 175$$

$$= +67 \text{ J (heat is added to system)}$$

Example Problem 16.6   Analyze the energy transformations that can take place when the piston in Figure 16.5 moves in either direction (work) and/or heat is transferred.

Figure 16.4

367
*First Law
of Thermodynamics:
The Conservation
of Energy*

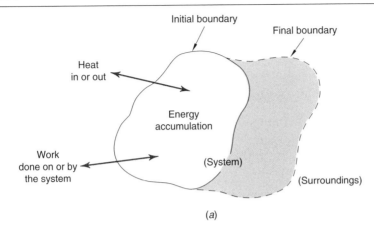

(a)

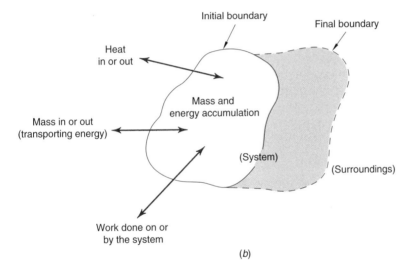

(b)

(a) Closed and (b) open thermodynamic systems.

Solution   The air within the piston is considered to be the system. If the position shown is condition 1, there must be a force applied to the piston toward the left to hold its position. The magnitude of the force is

$$F = \text{(pressure of air)(area of piston)}$$

$$= PA \text{ (in newtons)}$$

At this position, with no change in volume of the air, heat could be added or removed, which would increase or decrease, respectively, the internal energy of the system. In terms of the first law:

$$_1Q_2 = U_2 - U_1$$

Note the absence of the work term. Work cannot take place without action of a force through a distance. The force (PA) is present, but no movement takes place; thus no work is transferred.

Figure 16.5

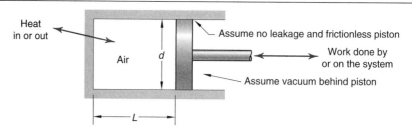

Piston-cylinder combination.

It is a different situation if the force to the left is increased slightly, moving the piston left. Work is done because movement occurs as a result of a force. The air temperature may change depending on how much heat is added or removed from the air. The first law for this situation can be written as

$$_1Q_2 = U_2 - U_1 + {_1W_2}$$

The work term, $_1W_2,$ as the system moves from state 1 to state 2 will be equal to $(PA)L$ (at constant pressure) as the piston moves a distance L to the left. Since external work is required to move the piston, the work term is negative. In other words, it was work done by the surroundings on the system.

For the case of the piston moving to the right, the first law is still

$$_1Q_2 = U_2 - U_1 + {_1W_2}$$

but $_1W_2$ is positive, because work is done by the system on the surroundings.

If no heat is allowed to transfer across the boundaries, then the first law becomes

$$_1W_2 + U_2 - U_1 = 0$$

This process, where no heat crosses the boundary, is called an *adiabatic process.*

Many engineering applications assume that air and other gases behave ideally, that is, the pressure, volume, and temperature obey an equation of state derived from the laws of Boyle and Charles. One characteristic of a perfect gas is that internal energy is a function of temperature only. If the ideal-gas assumption is made, then a constant-temperature (*isothermal*) process implies that there is no change in internal energy, and therefore the first law for an isothermal process involving an ideal-gas becomes

$$_1Q_2 = {_1W_2}$$

Example Problem 16.7 Let the piston in Figure 16.5 be moved so that the volume of air is reduced to one-half its original value. During this process, 116 MJ of heat is removed from the system. If the process is isothermal and we assume an ideal gas, how much work is done on the system?

Solution

$$_1Q_2 = {_1}W_2$$

$$-116 \text{ MJ} = {_1}W_2$$

The negative sign indicates that work has been done on the air by the surroundings.

Example Problem 16.8 Determine the pressure of the air within the cylinder (Fig. 16.5) if the force (F) on the piston is 1.00×10^2 N and its diameter (d) is 0.100 m.

Solution

$$P = F/A$$

$$= 12\ 732.4 \text{ N/m}^2$$

$$= 1.27\ (10^4)\text{N/m}^2$$

$$= 1.27\ (10^4) \text{ Pa}$$

$$= 12.7 \text{ kPa}$$

16.5 Second Law of Thermodynamics

The second law is not an actual proven law; rather, it is an axiom whose verification is in the fact that all experimental evidence about it is always true. Let's begin by considering the similarities between the energy contained in a river and the energy contained in high-temperature, high-pressure steam. Both contain a certain quantity of energy but both have limits on how much useful work can be obtained from them.

The water in the river contains energy in the form of flow rate and elevation. The Sun heats the Earth's surface, causing evaporation that deposits water in the form of rain and snow at high elevations. The water migrates from mountaintop to sea level. In the process, its energy is transformed from potential to kinetic to internal, and eventually the water arrives at the sea with exactly the same amount of energy with which it started. Energy was transformed but conserved.

The questions to be explored: Did the water do any useful work along its path to the sea? What is the maximum amount of work that it could possibly do? If the water is stationary when it starts at the mountaintop and stationary when it arrives at the sea, then its maximum work output is a function of the change in elevation; the constraints on work availability are the height of the mountain and the level of the sea. Did the water do any useful work along the way—for example, generate electricity? That would depend on the construction of mechanical devices to convert water energy into electric energy. If potential and/or kinetic energy is converted into useful work during the process, it is still only possible to capture the portion called available energy. If we assume no mechanical devices to convert water energy to electric energy, then the total energy was conserved and remains in the mass of water at sea level. However, it is now unavailable energy.

High-temperature, high-pressure steam has a parallel limitation. Work available from this fluid is limited by the maximum temperature and pressure that we can safely produce as a starting point and the temperature of the atmosphere into which it is released. These limitations are analogous to the water on the mountaintop and at sea level.

Heat cannot be completely converted to work, but work can be completely converted to heat. Heat can perform work only when it passes from a higher to a lower temperature. In other words, heat will not flow spontaneously from a colder to a hotter substance. This limitation on heat conversion forms the basis of the second law of thermodynamics.

Heat energy at a high temperature is capable of doing work, but the same amount of energy at low temperature is not capable of doing useful work. The total amount of energy is still the same, but its *entropy* has changed. If we think of entropy as a property, then high temperature has low entropy and low temperature has high entropy.

There are numerous statements of the second law. Rudolf Clausius, a German mathematical physicist, in 1865 was the first person to combine the fact that heat will always flow from high temperature to low temperature together with the law of conservation of energy. Clausius stated: "It is impossible for any device to operate in such a way that the sole effect would be an energy transfer of heat from a cooler to a hotter body." Another statement of the second law would read: "No device can completely and continuously transform all of the heat supplied to it into work." The heat that can be transformed into work is called *available energy;* the remaining portion is termed *unavailable energy.*

As indicated earlier, whenever work is performed, friction downgrades some available energy to unavailable energy and some available energy is released into the atmosphere, becoming unavailable energy. This energy released into the atmosphere is heat that is at too low a temperature to perform work under the conditions specified by the system and its surroundings.

For a given system and surroundings, the available and unavailable energies can be computed. The procedures for this computation are beyond the scope of this text.

16.6 Efficiency

It is important to understand what is meant by a cycle. When a system at a given initial state goes through a sequence of processes and then returns to its initial state, the system has executed a thermodynamic cycle.

$$\Delta E_{cycle} = Q_{cycle} - W_{cycle} \qquad (16.5)$$

During a cycle the system is returned to its initial state at the completion of the cycle, so there is no net change in energy; that is, $\Delta E = 0$. This expression can then be written as

$$W_{cycle} = Q_{cycle} \qquad (16.6)$$

or

$$\dot{W}_{cycle} = \dot{Q}_{cycle} \tag{16.7}$$

In Eq. 16.7, the overdot represents an expression for time rate of change or power.

Systems undergoing power cycles deliver or transfer a net amount of energy to their surroundings. The net work output of the system equals the net heat transfer to the cycle.

$$W_{cycle} = Q_{in} - Q_{out} \tag{16.8}$$

where Q_{in} or (Q_H) represents the heat transfer of energy into the system from a hot body and Q_{out} or (Q_L) represents heat transfer out of the system to a cold body.

The high side or energy supplied by heat transfer to a system undergoing a power cycle is normally obtained from the combustion of fossil fuels. The energy Q_{out} is generally discharged to the surrounding atmosphere or a local body of water. The extent of the energy conversion from heat to work is commonly called thermal efficiency:

$$\eta = \frac{W_{cycle}}{Q_{in}} = \frac{Q_{in} - Q_{out}}{Q_{in}} = 1 - \frac{Q_{out}}{Q_{in}} \tag{16.9}$$

This relationship provides a measure of efficiency for any heat engine.

Another application of the second law is the determination of the maximum efficiency of any device that converts heat into work. These heat engines cannot attain 100% efficiency because of the second law.

A French engineer named Nicolas Leonard Sadi Carnot proposed in 1824 an ideal engine cycle that had the highest attainable efficiency within thermodynamic laws. In reality, any engine following his proposed cycle, called the Carnot cycle, could not be constructed, but the theory provides a basis of comparison for practical engines. The Carnot efficiency can be shown to be

$$\text{Carnot efficiency} = 1 - \frac{T_L}{T_H} \tag{16.10}$$

where T_H is the absolute temperature at which the engine receives heat (high temperature) and T_L is the absolute temperature at which the engine rejects heat (low temperature) after performing work. Absolute temperatures are determined by adding 273° to a reading in °C or 460° to a reading in °F. Temperatures on the absolute scales are in degrees kelvin (K) for centigrade-size degrees, and in degrees Rankine (°R) for Fahrenheit-size degrees.

Example Problem 16.9 A steam engine is designed to accept steam at 300°C and exhaust this steam at 100°C. What is its maximum possible efficiency?

Solution

$$\text{Carnot efficiency} = 1 - \frac{T_L}{T_H}$$

$$= 1 - \frac{100 + 273}{300 + 273}$$

$$= 1 - \frac{373}{573}$$

$$= 35\%$$

We have learned that all the energy put into a system does not end up producing useful work. According to the second law, a certain amount of energy is unavailable for productive work. That is not the entire story; in fact, it gets worse. The available energy promised by the Carnot engine does not perform an equivalent amount of work because of losses incurred during the transfer of energy from one form to another. An automobile engine converts chemical energy in the form of gasoline to mechanical energy at the axle; however, some of the energy is lost through bearing friction, incomplete combustion, cooling water, and other thermodynamic and mechanical losses.

An engineer designing a device to convert heat into work must also be concerned with the overall efficiency of a proposed system. Equation 16.9 can be written to include the overall efficiency of an entire plant as

$$\text{Overall efficiency}\,(\eta) = \frac{\text{useful output}}{\text{total input}} = \frac{W_{\text{cycle}}}{Q_{\text{in}}} \qquad (16.11)$$

From Eq. (16.10), it is clear that for a given process the maximum (Carnot) efficiency can be increased by lowering the exhaust temperature T_L and/or increasing the input temperature T_H. In theory this is true, but practical design considerations must include available materials for construction of the heat engine. The engineer thus attempts to obtain the highest possible efficiency using existing technology. Examples of overall efficiency are 17 to 23% for automobile engines (gasoline), 26 to 38% for diesel engines, and 20 to 33% for turbojet aircraft engines. Thus, in the case of the gasoline automobile engine, for every 80 L (21 gal) tank of gasoline, only 20 L (5.3 gal) ends up moving the automobile.

Care must be exercised in the calculation and use of efficiencies. To illustrate this point, consider Figure 16.6, which depicts a steam power plant operating from the burning of fossil fuel for steam generation to driving a turbine attached to an electric generator. Efficiencies of each stage or combinations of stages in the power plant may be calculated by comparing energies available before and after the particular operations. For example, combustion efficiency can be calculated as 0.80/1.00, or 80%. The turbine efficiency is 0.33/0.50, or 66%. The overall power plant efficiency up to the electric generator is 0.33/1.00, or 33%. By no means is the 33% a measure of the efficiency of generation of electricity for use in a residential home. There will be losses in the generator and line losses in the transmission of the electricity from the power plant to the home. The overall efficiency from fuel into the boiler at the power plant to the electric oven in the kitchen may run as low as 25 to 30%. It is interesting to note

Figure 16.6

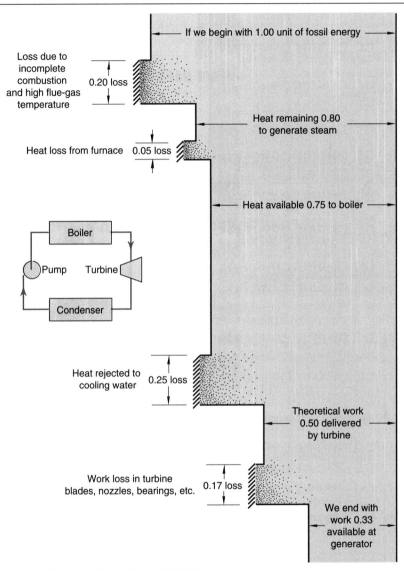

Energy losses in a typical steam power plant.

that the cycle is complete when the oven converts electricity back into heat, which is where the entire process began.

16.7 Power

Power is the rate at which energy is transferred, generated, or used. In many applications it may be more convenient to work with power quantities rather than energy quantities. The SI unit of power is the watt (one joule per second), but many problems will have units of horsepower (hp) or foot-pound-force per second (ft · lbf/s) as units, so conversions will be necessary.

Example Problem 16.10   A steam power plant produces 3 500 net hp at the shaft. The plant uses coal as fuel (12 000 Btu/lbm). Using the overall efficiency presented in Figure 16.6 (33%), determine how many metric tons of coal must be burned in a 24-h period to run the turbine.

Solution From Eq. (16.11),

$$\text{Total input} = \frac{\text{useful output}}{\text{overall efficiency}}$$

$$= \frac{3\,500\ \text{hp}}{0.33} \times \frac{2\,545\ \text{Btu}}{1.0\ \text{hp} \cdot \text{h}}$$

$$= 2.7(10^7)\ \text{Btu/h}$$

The amount of coal needed for one day of operation is therefore

$$\text{Coal required} = \frac{2.7\,(10)^7\,\text{Btu}}{1\,\text{h}} \times \frac{1\,\text{lbm}}{12\,000\,\text{Btu}} \times \frac{1\,\text{kg}}{2.205\,\text{lbm}} \times \frac{1\,\text{t}}{10^3\text{kg}} \times \frac{24\,\text{h}}{1\,\text{day}}$$

$$= 24\ \text{t/day}$$

Example Problem 16.11   Estimate the area in acres required for an array of solar cells to collect and provide enough electrical energy for a community of 4 500 people in a central Arizona town. Assume a solar cell conversion efficiency of 8%.

Solution   This example requires some assumptions in order to obtain a meaningful result.

1. Each home consumes about 13 000 kWh of electricity on average each year.
2. An average of three people live in each home.
3. The sun shines an average of 8 h/d.
4. The typical solar heat transfer rate is 1.0 kW/m². Equation (16.11) may be used.

$$\text{Efficiency} = \frac{\text{useful output}}{\text{total input}}$$

$$\text{Input} = \frac{\text{output}}{\text{efficiency}}$$

Output:

    # of homes in town = 4 500/3 = 1 500

    # of kWh needed for town/year = 13 000 × 1 500 = 19 500 000 kWh/year

    # of kWh needed for town/day = 19 500 000/365 = 53 425 kWh/day

Input:

Input = (area)(solar heat-transfer rate) = area × 1.0 kW/m²

Input/8 h day = area × 1.0 kW/m² × 8 h = 8 kWh/m² × area

Input = area (8.0 kW/m²)

Therefore, the area can be computed for an 8 h day as follows

$$\text{Area} = \frac{\text{m}^2}{8.0\,\text{k Wh}} \times \frac{53\,425\ \text{kWh}}{0.08}$$

$$= 83\ 476\ \text{m}^2 = 898\ 500\ \text{ft}^2 = 20.6\ \text{acres}$$

16.8 Refrigeration Cycles

For many years a measure of efficiency commonly used in the refrigeration and air-conditioning fields has been the *energy efficiency ratio,* abbreviated EER. In essence, a refrigerating machine is a reversed-heat engine; that is, heat is moved from a low-temperature region to a high-temperature region, requiring a work input to the reversed-heat engine. The expression for efficiency of a reversible-heat engine is

$$\text{Refrigeration efficiency} = \frac{\text{refrigerating effect}}{\text{work input}} \qquad (16.12)$$

Refrigeration efficiency is also called *coefficient of performance* (CP). The numerical value for CP may be greater than 1.

More recently, the U.S. Department of Energy has developed a testing method that rates performance of a unit over a wide range of operating conditions. The new rating system is called the seasonal energy efficiency rating (SEER).

Industrial-size refrigeration units are measured in tons of cooling capacity. The ton unit originated with early refrigerating machines and was defined as the amount of refrigeration produced by melting 1 ton of ice in a 24-h period. If the latent heat of ice is taken into account, then a ton of refrigeration is equivalent to 12 000 Btu/h. Home-size units are generally rated in British thermal units per hour of cooling capacity.

The EER takes into account the normal designations for refrigerating effect and work input and expresses these as power rather than energy. Thus, EER is the ratio of a refrigerating unit's capacity to its power requirements:

$$\text{EER} = \frac{\text{refrigerating effect, Btu/h}}{\text{power input}} \qquad (16.13)$$

Example Problem 16.12 Compare the costs of running two 36 000 Btu/h air-conditioning units (3-ton) an average of 8 h/d for 1 month if electricity sells for $0.10/kWh. One unit has an EER of 8 and a second has an EER of 10.0.

Solution

$$EER = \frac{\text{refrigerating effect}}{\text{power input}}$$

$$8 = \frac{36\,000\,\text{Btu/h}}{\text{power input}} \quad 10 = \frac{36\,000\,\text{Btu/h}}{\text{power input}}$$

Power input = 4 500 Btu/h   Power input = 3 600 Btu/h

1 watt = 3.414 4 Btu/h

Power input = 1 054.36 W   Power input = 1 317.95 W

$$\frac{\text{Dollars}}{\text{month}} = \frac{1.317\,95\,\text{kW}}{1} \times \frac{8\,\text{h}}{1\,\text{day}} \times \frac{30\,\text{day}}{1\,\text{month}} \times \frac{0.10\,\text{dollars}}{1\,\text{kWh}}$$

= $25.30 for a 10 EER unit

= $31.63 for an 8 EER unit

PROBLEMS

Potential and Kinetic Energy

16.1 A new flexible-fuel vehicle (FFV) is traveling at 65 mph and has a kinetic energy of 625 kJ. Determine its mass in kilograms.

16.2 An SUV weighs 2 345 kg and has a kinetic energy of 815 kJ. Determine its velocity in miles per hour (mph).

16.3 A Chevrolet Silverado with a mass of 5 520 lbm is being driven on a mountain road 6 250 ft above sea level ($g_L = 31.\,0$ ft/s^2) at 65 mph. Determine the total energy possessed by the vehicle using sea level as the datum for potential energy. Express the answer in joules.

16.4 A Ford Ranger with a mass of 2 410 kg is traveling at a velocity of 65 mph down a mountain road in Colorado. The brakes fail and the driver elects to use a "runaway" truck tramp to stop the pickup. What distance in feet will it travel if the ramp has a 22% incline? Neglect air and road friction ($g_L = 32$ ft/s^2).

16.5 A bobsled with two passengers, total mass of 225 lbm, races down a 35% slope. Neglecting air and snow friction, what is the speed (mph) after the sled has traveled 125 ft ($g_L = 32.2$ ft/s^2)?

16.6 A new military aircraft, the Superhornet, with a weight of 5 000 lbm is traveling 1 350 mph at an altitude of 28 550 ft ($g_L = 30.5$ ft/s^2). What is the total energy in joules? How many 50-ton tanks at sea level have an equivalent amount of energy, if they have an average ground speed of 35 mph?

16.7 A cliff diver, mass 165 lbm, dives from a 135 ft cliff into the sea. Neglecting air friction:

(a) What is the diver's total energy in joules at the following heights: 100 ft, 50 ft, surface of the water?

(b) What is the diver's velocity in mph at the first contact with the water?

16.8 During the World Series a ball is hit straight up from the batter's box and reaches a height of 275 ft. Neglecting air friction, what speed (mph) will the baseball attain if it is caught by the catcher at the same height it was hit?

16.9 A father pulls his children in a wagon for 25 minutes covering 1.25 miles on a level sidewalk. He pulls with an average force of 35 lbf at an angle of 45 degrees to the sidewalk.

 (*a*) How much work is done (in ft · lbf)?

 (*b*) What is the average horsepower required?

16.10 How much work in N · m is done when raising a mass of 478.5 kg to an altitude of 85.0 m above the surface of the Moon? Use g_L on surface of the Moon as one-sixth that of Earth at sea level.

16.11 If the mass in Prob. 16.10 fell back to the surface of the Moon, with what velocity, in ft/s, would it strike the surface?

Pressure and Force

16.12 Air in a piston–cylinder configuration is heated until an external force of 1 025 lbf is required to hold the piston stationary. If the piston diameter is 2.85 inches, what is the pressure exerted by the air normal to the piston face (in lbf/ft²)? (See Fig. 16.5.)

16.13 The piston-cylinder arrangement in Prob. 16.12 was designed and rated with an internal pressure of 7.50 MPa. Prior to shipping, the cylinder is tested by increasing the internal pressure to 3.0 times the design pressure (safety factor of 3.0). What maximum force in newtons must be applied to the piston to hold it stationary during the test?

Closed Systems

16.14 If 12 500 ft · lbf of work is done by a closed system while 85.0 Btu of heat is removed, what is the change in internal energy (in joules) of the system?

16.15 During a process, 21.6 MJ of heat is added to a closed system. If the internal energy is increased by 59.6 MJ, how much work in Btu was done? Is the work done on or by the system?

16.16 In an adiabatic process, 102 kJ of work is done on 25 kg of water. Express the change in internal energy (Btu) per pound of mass of fluid. *Hint:* $u2 - u1 = U2 - U1/m$ (kJ/kg). Lowercase denotes energy per unit mass.

Cycles

16.17 A V-6 internal combustion engine burns fuel at the rate of 4.0 gallons per hour. The fuel has a heating value of 42.7 kJ/mL. Calculate the useful output to the power train (in kilowatts) if the overall efficiency is 35%.

16.18 Compute the Carnot efficiency of a heat engine operating between 465 and 65°F.

16.19 An inventor claims to have developed a power cycle capable of delivering a net work output of 420 kJ for an energy input by heat transfer of 1 000 kJ. The system receives its heat transfer from a fossil fuel at 235°C and rejects heat at 55°C. Is this a reasonable claim? Why or why not?

16.20 Heat is delivered to a Carnot engine at the rate of 285 kJ/s and heat is rejected at a rate of 8.35 (10³) Btu/min to a 32°C low-temperature reservoir. What is the temperature (in °C) of the high-temperature source? *Hint:* $Q_{in} - Q_{out}$ = Work (useful output).

16.21 A gasoline engine is used to drive an electric generator. To operate the engine 9.3 L/h of gasoline with a heating value of 45 MJ/L is burned. The engine delivers 55 kW to an electric generator that produces 110 amps at 240 volts DC. Determine:

 (*a*) engine efficiency

(b) generator efficiency

(c) overall efficiency

16.22 An electric motor with an efficiency of 78% drives a water pump. What input power is required of the motor (in kilowatts) to pump 22 250 gals of water from a lake to a storage tank 65 ft above the lake surface in 4.0 hours?

16.23 A pump with an efficiency of 65% has an output capacity of 25×10^3 ft · lbf/s and is used to deliver a grain-slurry mixture. An electric motor with an efficiency of 85% drives the pump. Determine:

(a) the input horsepower to the pump

(b) the power, in kilowatts, required for the motor.

16.24 An 85% efficient electric motor has a useful output of 0.75 hp and it drives a 68% efficient water pump.

(a) How much power, in watts, does the motor require?

(b) How much water, in gallons, could this combination lift 7.5 ft in one hour?

16.25 A heating value of 90 TJ has been measured from the fission of 1.0 kg of U-235.

(a) How many gallons of water would need to be stored on a hillside 225 ft above sea level to provide an energy equivalence?

(b) If the heating value of 1.0 kg of coal is 25 MJ, how many short tons of coal would need to be consumed to provide the heating equivalence of 1.0 kg U-235?

16.26 Estimate the number of households that could receive electric energy from 35 kg of U-235. Assume the average household requires 17 000 kWh per year, that the conversion of U-235 to electricity is 68% efficient, and that the energy equivalent of U-235 is 10^8 MJ/kg.

16.27 Using the data given in Problem 9.30, determine the surface area (in hectares) of solar collectors operating at an efficiency of 12% to supply electric energy to the same households. Assume an average of 8 hours of sunlight per day and a transfer rate of one kW per square meter.

16.28 A one-ton window air conditioner (12 000 Btu/h) is installed in a mobile home. If all the following appliances were operating for one hour, what fraction of the unit's capacity would be needed to remove the heat generated by four 150 W lamps, six 100 W lamps, two 75 W lamps, a 500 W refrigerator, an 900 W oven, and four people watching TV (120 W)? How many kilowatt-hours of electric energy are used during each hour for these conditions? What is the hourly cost to the mobile home owner if electricity costs $0.10/kWh? *Note:* The average heat output of a person at rest is 480 Btu/h.

16.29 Develop a spreadsheet and produce a table of values that represents the solar collector area, in acres, needed to supply electric energy for cities with populations of 5 000, 20 000, 50 000, 100 000, and 1 000 000. Provide results for conversion efficiencies from 5 to 15% in increments of 1%. Use the same assumptions as used in Example Problem 9.11. Interpolate the data to determine collector area for the city in which you attend college.

16.30 Develop a spreadsheet that will determine the average monthly cost of operating 215 individual air-conditioning units in a manufacturing facility. Each unit is rated at 24 000 Btu/h. Consider EER values from 7.0 to 11.0 in increments of 0.5 and prepare a table of monthly costs versus EER. The electrical rate structure is as follows:

$$\text{First 5 000 kWh} = 11.0 \text{¢/kWh}$$

$$\text{Next 15 000 kWh} = 8.5 \text{¢/kWh}$$

$$\text{Over 20 000 kWh} = 8.0 \text{¢/kWh}$$

Assume that the units run an average of eight hours per day on weekdays and 10 hours on Saturdays (closed Sunday). Use a 31-day month beginning on a Monday.

CHAPTER 17

Electrical Theory*

Chapter Objectives

When you complete your study of this chapter, you will able to:

- Compute the equivalent resistance of resistors in series and in parallel
- Apply Ohm's law to a resistive circuit
- Determine the power provided to a DC circuit and the power used by circuit components
- Use Kirchhoff's laws to solve resistive networks
- Utilize mesh currents to solve resistive networks

17.1 Introduction

Electricity is universally one of our most powerful and useful forms of energy. It affects our world in many useful ways. It affects our lives through communication systems, computer systems, control systems, and power systems. Certainly electrical engineers, but to some extent all engineers, must understand how to design, analyze, and maintain such systems.

Electrical and computer engineering are very large and diverse fields of study. This chapter provides an introduction to one small aspect called *circuit theory*. It is important to the study of engineering because many products or systems that are designed involve the application of the electrical theory. This area that is fundamental to computer and electrical engineers is also important to all engineering disciplines.

Circuit theory does many things for us. (1) It provides simple solutions to practical problems with sufficient accuracy to be useful. (2) It allows us to reduce the analysis of large systems to a series of smaller problems that we can conveniently handle. (3) It provides a means of synthesizing (i.e., building up) complex systems from basic components.

In this chapter we will review a number of the elementary concepts of electricity that were first learned in physics. We also will introduce and apply some fundamental circuit-analysis equations such as Ohm's law and Kirchhoff's laws. Applications, however, will be restricted to those involving steady-state direct current (DC).

---

*Users will find Appendix A and F useful reference material for this chapter.

17.2 Structure of Electricity

Matter consists of minute particles called molecules. Molecules are the smallest particles into which a substance can be divided and still retain all the characteristics of the original substance. Each of these particles will differ according to the type of matter to which it belongs. Thus a molecule of iron will be different from a molecule of copper.

Looking more closely at a molecule, we find that it can be divided into still smaller parts called atoms. Each atom has a central core, or nucleus, that contains both protons and neutrons. Moving in a somewhat circular motion around the nucleus are particles of extremely small mass called electrons. In fact the entire mass of the atom is practically the same as that of its nucleus since the proton is approximately 2.0×10^3 times more massive than the electron.

To understand how electricity works, bear in mind that electrons possess a negative electric charge and protons possess a positive electric charge. Their charges are opposite in sign but numerically the same magnitude. The neutron is considered neutral, being neither positive nor negative.

The typical atom in its entirety has no net electric charge because the positive charge of the nucleus is exactly balanced by the negative charge of the surrounding electron cloud. That is, each atom contains as many electrons orbiting the nucleus as there are protons inside the nucleus.

The actual number of protons depends on the element of which the atom is a part. Hydrogen (H) has the simplest structure, with one proton in its nucleus and one orbital electron. Helium (He) has two protons and two neutrons in the nucleus; and since the neutrons exhibit a neutral charge, there are two orbital electrons (see Fig. 17.1). More complex elements have many more protons, electrons, and neutrons. For example, gold (Au) has 79 protons and 118 neutrons, with 79 orbital electrons. As the elements become more complex, the orbiting electrons arrange themselves into regions, or "shells," around the nucleus.

The maximum number of electrons in any one shell is uniquely defined. The shell closest to the nucleus contains a maximum of two electrons, the next eight, and so on. There are a maximum of six shells, but the last two shells are

Figure 17.1

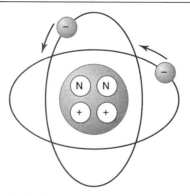

Schematic representations of a helium atom.

never completely filled. Atoms therefore can combine by sharing their outer orbital electrons and thereby fill certain voids and establish unique patterns of molecules.

Atoms are extremely minute. In fact, it is difficult to imagine the size of an atom, since a grain of table salt is estimated to contain 10^{18} atoms. However, it is possible to understand the size relationship between the nucleus and the orbital electrons. Assume for purposes of visualization that the diameter of the hydrogen nucleus is a 1.0 mm sphere. To scale accurately the electron and its orbit, the electron would revolve at an average distance of 25 m from the nucleus. Although the relative distance is significant, this single electron is prevented from leaving the atom by an electric force of attraction that exists because the proton has a positive charge and the electron has an opposite but equal negative charge.

How closely the millions upon millions of atoms and molecules are packed together determines the state (e.g., solid, liquid, or gas) of a given substance. In solids, the atoms are packed closely together, generally in a very orderly manner. The atoms are held in a specific lattice structure but vibrate around their nominal positions. Depending on the substance, some electrons may be free to move from one atom to another.

17.3 Static Electricity

History indicates that the Greeks were the first to have a word for electricity. They discovered that after rubbing certain items together, the materials would exert a force on one another. It was concluded that during the rubbing process, the bodies were "charged" with some unknown element, which the Greeks called electricity. For example, they concluded that by rubbing silk over glass, electricity was added to the substance. We realize today that during the process of rubbing, electrons are displaced from some of the surface atoms of the glass and are added to the surface atoms of the silk. The branch of science concerned with static, or stationary, charged bodies is called electrostatics.

The charge of an electron can be measured, but the value is extremely small. A large unit, the coulomb, has been selected to denote electric charge. Charles Augustin Coulomb (1736–1806) was the first individual to measure an electric force. In recognition of his work, the SI unit of charge is called the *coulomb*. A coulomb is defined in terms of the force exerted between unit charges. A charge of one coulomb will exert on an equal charge, placed one meter away in air, a force of about 8.988×10^9 N. The magnitude of this force is very large, equivalent to the weight of 15 million people. Because of the large size of this unit of measure, the charge on a proton is only $+1.6 \times 10^{-19}$ C.

17.4 Electric Current

Earlier we noted that electrons are prevented from leaving the atom by the attraction of the protons in the nucleus. It is entirely possible, however, for an electron to become temporarily separated from an atom. These free electrons drift around randomly in the space between atoms. During their random travel many of them will collide with other atoms; when they do collide with sufficient force, they dislodge electrons from those atoms. Since electrons are frequently colliding with other atoms, there can be a continuous movement of free electrons in a solid. If the electrons drift in a particular direction instead of moving randomly, there is movement of electricity through the solid. This continuous movement of electrons in a direction is called an electric current. If the electron drift is in only one direction, it is called direct current (DC). If the electrons periodically reverse direction of travel, then we have alternating current (AC).

The ease with which electrons can be dislodged by collision as well as the number of free electrons available varies with the substance. Materials in which the drift of electrons can be easily produced are said to be good conductors; those in which it is difficult to produce an electron drift are good insulators. For example, copper is a good conductor, whereas glass is a good insulator. Practically all metals are good conductors. Silver is a very good conductor but it is expensive. Copper and aluminum also are good conductors and are commonly used in electric wire.

17.5 Electric Potential

Both theory and experimentation suggest that like charges repel and unlike charges attract. Consequently, to bring like charges together, an external force is necessary, and therefore work must be done. The amount of work required to bring a positive charge near another positive charge from a large distance is used as a measure of the electric potential at that point. This amount of potential is measured in units of work per unit charge or joules per coulomb (J/C). By definition, one joule per coulomb is one volt, a unit of electric potential.

Devices such as electric batteries or generators are capable of producing a difference in electric potential between two points. Such devices are rated in terms of their ability to produce a potential difference, and this difference in potential is measured in volts (V). When these devices are connected to other components in a continuous circuit, electric current flows. You also could think of batteries as devices that change chemical energy to electric energy whereas generators convert mechanical energy to electric energy.

17.6 Simple Electric Circuits

When electric charge and current were initially being explored, scientists thought that current flow was from positive to negative. They had no knowledge of electron drift. By the time it was discovered that the electron flow was from negative to positive, the idea that current flow was from positive to negative had become so well established that it was decided not to change the convention. See Figure 17.2.

When battery or generator terminals are connected to a conducting material, the battery or generator creates a potential difference across the load measured as a voltage. It follows that the random movement of the negatively

Figure 17.2

383

Resistance

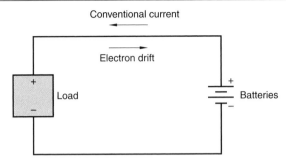

Conventional current

Electron drift

Load

+

−

+

−

Batteries

A simple DC electric circuit.

charged electrons will have a drift direction induced by this potential difference. The resulting effect will be the movement of electrons away from the negative terminal of the battery or generator. Electrons will travel in a continuous cycle around the circuit, reentering the battery or generator at its positive terminal.

The speed by which any individual electron moves is relatively slow, less than 1 mm/s. However, once a potential difference is connected into a circuit, the "flow" of electrons starts almost instantaneously at all points. Individual electrons at all locations begin their erratic movement around the circuit, colliding frequently with other atoms in the conductor. Because electron activity starts at all points practically simultaneously, electric current appears to travel about 3×10^8 m/s.

Electric current is really nothing more than the rate at which electrons pass through a given cross section of a conductor. The number of electrons that migrate through this cross section is gigantic in magnitude in that approximately 6.28×10^{18} electrons pass a point per second per ampere. Since this number is very large, it is not convenient to use the rate of electron flow as a unit of current measurement. Instead, current is measured in terms of the total electric charge (coulomb) that passes a certain point in a unit of time (second). This unit of current is called the ampere (A). That is, one ampere equals one coulomb per second (C/s).

17.7 Resistance

Another critical component of circuit theory is *resistance*. George Simon Ohm (1789–1854), a German scientist investigating the relation between electric current and potential difference, found that, for a metal, the current in the conductor was directly proportional to the potential difference across the conductor. This important relationship has become known as Ohm's law, which is stated as:

At constant temperature, the current I *in a conductor is directly proportional to the potential difference between its ends,* E.

The ratio E/I describes the "resistance" to electron flow and is denoted by the symbol R:

$$R = \frac{E}{I} \tag{17.1}$$

and is called Ohm's Law. It is more often presented as

$$E = R\,I$$

This is one of the simplest but most important relations used in the electric circuit theory. A conductor has a resistance R of one ohm (Ω) when the current I through the conductor is one ampere (A) and the potential difference E across it is one volt (V).

When a specific value of resistance is required in a circuit, a resistor is used. Resistors come in many sizes and tolerances. Resistance and tolerance values of resistors are marked on the body of the resistors with coded color bands.

The reciprocal of resistance is called *conductance* (G):

$$G = \frac{1}{R} \tag{17.2}$$

Conductance is measured in siemens (S).

Example Problem 17.1   The current in an electric aircraft instrument heater is measured as 2.5 A when connected to a battery with a potential of 60.0 V. Calculate the resistance of the heater.

Solution

$$R = \frac{E}{I}$$

$$= \frac{60.0}{2.5}$$

$$= 24\ \Omega$$

17.8   DC Circuit Concepts

A considerable amount of information about electric circuits can be presented in a compact form by means of circuit diagrams. Figure 17.3 illustrates three typical symbols that are used in such diagrams.

Figure 17.3

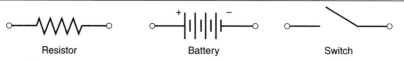

Resistor   Battery   Switch

Symbols.

Figure 17.4

385
DC Circuit Concepts

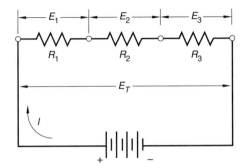

Series circuit.

Circuits may have resistors that are connected either end to end or parallel to each other. When resistors are attached end to end, they are said to be connected in series, and the same current flows through each. When several resistors are connected between the same two points they are connected in parallel.

For the series circuit illustrated in Figure 17.4, there will be a potential across each resistor, since a potential must exist between the ends of a conductor if current is to flow. This potential difference E is related to current and resistance by Ohm's law. For each unknown voltage (potential), we can write

$$E_1 = R_1 I$$

$$E_2 = R_2 I$$

$$E_3 = R_3 I$$

Since the total voltage drop E_T across the three resistors is the sum of the individual drops, and since the current is the same through each resistor, then

$$E_T = E_1 + E_2 + E_3$$

$$= IR_1 + IR_2 + IR_3$$

$$= I(R_1 + R_2 + R_3)$$

or

$$R_1 + R_2 + R_3 = \frac{E_T}{I}$$

Since E_T is the total potential difference across the circuit and I is the circuit current, then the total resistance must be

$$R_T = \frac{E_T}{I}$$

Therefore

$$R_T = R_1 + R_2 + R_3 \qquad (17.3)$$

This total resistance R_T is sometimes called the equivalent resistance R_E.

These steps demonstrate that when resistors are connected in series, their combined resistance is the sum of their individual values.

Figure 17.5

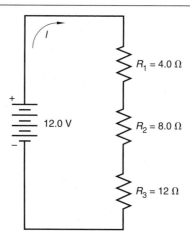

R₁ values shown in figure: $R_1 = 4.0\ \Omega$, $R_2 = 8.0\ \Omega$, $R_3 = 12\ \Omega$, 12.0 V

Example Problem 17.2 The circuit in Figure 17.5 has three resistors connected in series with a 12.0 V source. Determine the line current I and the voltage drop across each resistor (E_1, E_2, and E_3).

Solution   For resistors in series

$$R_T = R_1 + R_2 + R_3$$
$$R_T = 4.0 + 8.0 + 12$$
$$= 24\ \Omega$$

Ohm's law gives

$$E = RI$$
$$I = \frac{E_T}{R_T} = \frac{12.0}{24} = 0.50\ \text{A}$$

Then

$$E_1 = 0.50(4.0)$$
$$= 2.0\ \text{V}$$
$$E_2 = 0.50(8.0)$$
$$= 4.0\ \text{V}$$
$$E_3 = 0.50(12)$$
$$= 6.0\ \text{V}$$

Check: $E_T = 12.0 = 2.0 + 4.0 + 6.0$

When several resistors are connected between the same two points, they are said to be in parallel. Figure 17.6 illustrates three resistors in parallel.

Figure 17.6

387
DC Circuit Concepts

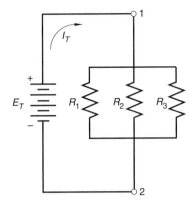

Parallel circuit.

The current between points 1 and 2 divides among the various pathways formed by the resistors. Since each resistor is connected between the same two points, the potential difference across each of the resistors is the same.

In analyzing the problem, we let I_T be the total current passing between the points 1 and 2 with I_1, I_2, and I_3 representing the branch currents through R_1, R_2, and R_3, respectively.

Using Ohm's law we can write

$$I_1 = \frac{E_T}{R_1}, I_2 = \frac{E_T}{R_2}, I_3 = \frac{E_T}{R_3}$$

or

$$I_1 + I_2 + I_3 = E_T\left(\frac{1}{R_1} + \frac{1}{R_2} + \frac{1}{R_3}\right)$$

But we know that

$$I_T = I_1 + I_2 + I_3$$

Applying Ohm's law to total circuit values reveals that

$$I_T = \frac{E_T}{R_T}$$

Therefore,

$$\frac{E_T}{R_T} = E_T\left(\frac{1}{R_1} + \frac{1}{R_2} + \frac{1}{R_3}\right)$$

or

$$\frac{1}{R_T} + \frac{1}{R_1} + \frac{1}{R_2} + \frac{1}{R_3} \qquad (17.4)$$

This equation indicates that when a group of resistors are connected in parallel, the reciprocal of their combined resistance is equal to the sum of the reciprocals of their separate resistances.

Figure 17.7

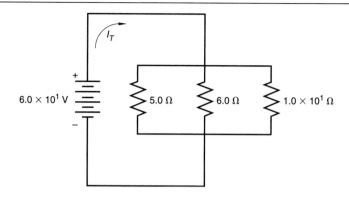

Example Problem 17.3   In Figure 17.7 three resistors are connected in parallel across a 6.0×10^1 V battery. What is the equivalent resistance of the three resistors and the line current?

Solution   For resistors in parallel:

$$\frac{1}{R_T} = \frac{1}{R_1} + \frac{1}{R_2} + \frac{1}{R_3}$$

$$= \frac{1}{5.0 \times 10^0} + \frac{1}{6.0 \times 10^0} + \frac{1}{1.0 \times 10^1}$$

$$= \frac{28}{6.0 \times 10^1}$$

$$R_T = \frac{6.0 \times 10^1}{28}$$

$$= 2.1 \ \Omega$$

From Ohm's law

$$E = RI$$

$$E_T = R_T I_T$$

So that

$$I_T = \frac{E_T}{R_T}$$

$$= (6.0 \times 10^1)\frac{2.8 \times 10^1}{6.0 \times 10^1}$$

$$= 28 \ \text{A}$$

Figure 17.8

389
DC Circuit Concepts

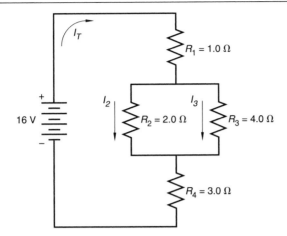

Electric circuits may involve combinations of resistors in parallel and in series. The next example problem demonstrates a solution of that nature.

Example Problem 17.4   Determine the line current I_T, the circuit equivalent resistance R_E, and the voltage drop E_4 across the resistor R_4 for the circuit in Figure 17.8. What resistance should be substituted for R_1 to reduce the line current by one-half?

Solution

Ohm's law: $E = RI$

Resistors in series: $R_T = R_1 + R_2 + \ldots + R_N$

Resistors in parallel: $\dfrac{1}{R_T} = \dfrac{1}{R_1} + \dfrac{1}{R_2} + \ldots + \dfrac{1}{R_N}$

For parallel resistors R_2 and R_3

$$\frac{1}{R_T} = \frac{1}{R_2} + \frac{1}{R_3}$$

$$= \frac{1}{2.0} + \frac{1}{4.0}$$

For series resistors

$$R_E = R_1 + R_T + R_4$$

$$= 1.0 + \frac{4.0}{3.0} + 3.0$$

$$= \frac{16}{3.0}$$

$$= 5.3\ \Omega$$

Then the line current is

$$E_E = R_E I_T$$

$$I_T = \frac{E_E}{R_E}$$

$$= 16\left(\frac{3.0}{16}\right)$$

$$= 3.0 \text{ A}$$

and the voltage drop across R_4 is

$$E_4 = R_4 I_T$$

$$= 3.0(3.0)$$

$$= 9.0 \text{ V}$$

If we reduce the line current by one-half, the new line current is $3.0/2 = 1.5$ A. The new equivalent resistance then must be

$$R_E = \frac{E_t}{I_T} = \frac{16}{1.5} = 10.667 \ \Omega$$

But

$$R_E = R_{new} + R_T + R_4$$

so the resistor replacing R_1 is

$$R_{new} = R_E - R_T - R_4 = 10.667 - \frac{4.0}{3.0} - 3.0 = 6.3 \ \Omega$$

17.9 DC Electric Power

Consider the following illustration of a simple DC circuit. A common household lantern, as illustrated in Figure 17.9, is represented by a section view (a) and a circuit diagram (b). As the switch is closed, current (I) flows through the bulb producing light and heat. Both light and heat are forms of energy, thus the chemical energy stored in the battery is converted first to electrical energy and then to light (useful energy) and heat (waste energy).

The rate at which energy is consumed by the bulb is a product of the voltage and current.

$$\text{Power } (P) = E(I) \tag{17.5}$$

Voltage (E) is in volts (V) or joules/coulomb (J/C) and current (I) is in amperes (A) or coulombs/second (C/s). The units of power, therefore, are joules/second or watts (W) (after James Watt, 1736–1819). Using Ohm's law, we can express power in the following alternative forms:

$$\text{Power } (P) = \frac{E^2}{R} \tag{17.6}$$

Figure 17.9a

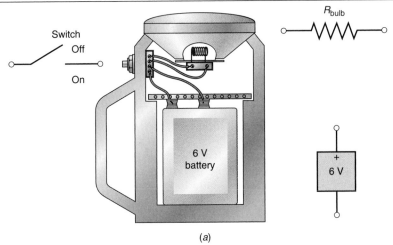

(a)

Household lantern.

Figure 17.9b

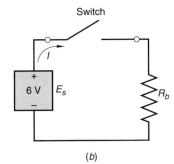

(b)

Schematic diagram.

$$\text{Power } (P) = I^2 R \tag{17.7}$$

If the rate of energy production or consumption (power) is constant, then the total energy produced or consumed can be computed from

$$\text{Energy} = (\text{Power})(\text{Time}) = EIt \tag{17.8}$$

and is expressed in units of joules (J).

Example Problem 17.5   The lantern in Figure 17.9 contains a lamp with a measured resistance of 2.0 Ω. When the switch is closed, what is the power consumed by the lamp?

Solution

$$P = \frac{E^2}{R}$$

$$= \frac{(6.0)^2}{2.0}$$

$$= 18 \text{ W}$$

Example Problem 17.6   A gallon of gasoline has the potential energy of approximately 131.8 MJ. Assume that a gasoline engine is driving a generator and the generator is supplying electricity to a 100 W lamp. The overall efficiency (from gasoline to electrical energy) of the engine-generator set is 20%.

  (*a*) How long will the lamp provide light from one gal of gasoline?
  (*b*) If the system operates at 120 volts, what is the lamp current?
  (*c*) How much electric charge passes through the lamp in 10 seconds?

Solution

  (*a*) Electrical energy produced from 1 gal of gasoline is:

$$131.8 \times 10^6 \text{ J} \times 0.2 = 26.36 \times 10^6 \text{ J}$$

From the equation:

Energy = (power)(time)

$$\text{time} = \frac{26.36 \times 10^6 \text{ J}}{100 \text{ J/s}} = 0.2636 \times 10^6 \text{ s} = 73 \text{ hr, } 13 \text{ min, } 20 \text{ s}$$

  (*b*) $I = \dfrac{P}{E} = \dfrac{100 \text{ W}}{120 \text{ V}} = 0.833 \text{ A}$

  (*c*) 0.833 A = 0.833 C/s or 8.33 C in 10 s

17.10   Terminal Voltage

Figure 17.10 is a basic circuit in which there can be different sources of electric potential. The storage battery and the electric generator are two familiar examples.

The potential does work of amount E in joules per coulomb on charges passing through the voltage source from the negative to the positive terminal. This results in a difference of potential E across the resistor R, which causes current to flow in the circuit. The energy furnished by the voltage source reappears as heat in the resistor.

Current can travel in either direction through a voltage source. When the current moves from the negative to the positive terminal, some other form of energy (such as mechanical or chemical energy) is converted into electric

Figure 17.10

393
Terminal Voltage

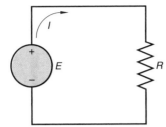

Basic circuit.

energy. If we were to impose a higher potential in the external circuit—for example, forcing current backward through the voltage source—the electric energy would be converted to some other form. When current is sent backward through a battery, electric energy is converted into chemical energy (which can be recovered in certain types of batteries; that is, charge the battery). When current is sent backward through a generator, the device becomes a motor.

A resistor, on the other hand, converts electric energy into heat no matter what the direction of current. Therefore, it is impossible to reverse the process and to regain electric energy from the heat. Electric potential always drops by the amount IR as current travels through a resistor. This drop occurs in the direction of the current.

In a generator or battery with current flowing negative to positive, the positive terminal will be E volts above the negative terminal minus the voltage drop due to internal resistance between terminals. There always will be some energy converted to heat inside a battery or generator no matter which direction the current flows.

When a battery or a generator is driving the circuit, the internal current passes from the negative to the positive terminal. Each coulomb of charge gains energy E from chemical or mechanical energy but loses IR in heat dissipation. The net gain in joules per coulomb can be determined by $E - IR$.

For a motor or for a battery being charged, the internal current passes from the positive to the negative terminal. Each coulomb loses energy E and IR. The combined loss can be determined by $E + IR$.

In the case of a motor, the quantity E is commonly called back-emf (the electromotive force, or potential), since it represents a voltage that is in a direction opposite the current flow.

Figure 17.11 shows a circuit wherein a battery is being charged by a generator. E_G, E_B, R_G, and R_B indicate the potentials and internal resistances, respectively, of the generator and battery. Each coulomb that flows around the circuit in the direction of the current I gains energy E_G from the generator and loses energy E_B in the form of chemical energy to the battery. Heat dissipation is realized as IR_G, IR_1, IR_B, and IR_2.

Figure 17.11

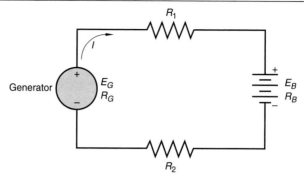

Generator charging a battery.

17.11 Kirchhoff's Laws

To analyze a DC circuit network that consists of more than simple elements in series or parallel, we will use Kirchhoff's network laws. These laws recognize the conservation of current at a node and potential drop in a closed loop and state these concepts in algebraic form. Thus a network solution requires that one write these algebraic equations and then solve the resulting set of simultaneous, linear, algebraic equations.

Kirchhoff's current law:

The algebraic sum of all of the currents coming into a node (junction) in a network must be zero.

Kirchhoff's voltage law:

The algebraic sum of the voltages (potential drops) around any closed loop in a network equals zero.

Example Problem 17.7 Given the circuit illustrated in Figure 17.12, determine the currents I_x, I_y, and I_z.

Figure 17.12

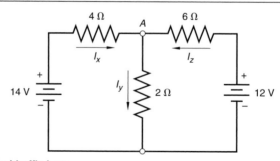

Application of Kirchhoff's laws.

Solution   From Kirchhoff's current law we can write (at point A)

$$I_y = I_x + I_z$$

Applying Kirchhoff's voltage law around the left loop in a clockwise fashion beginning at A, we find

$$-I_y(2) + 14 - I_x(4) = 0$$

Likewise, Kirchhoff's voltage law for the right loop applied counterclockwise from A gives

$$-I_y(2) + 12 - I_z(6) = 0$$

We then have three equations in I_x, I_y, and I_z which yield

$$I_x = 2 \text{ A}$$
$$I_y = 3 \text{ A}$$
$$I_z = 1 \text{ A}$$

To check the solution, write voltage drops in the left loop

$$-(3)(2) + 14 - (2)(4) = 0$$

and in the right loop.

$$-(3)(2) + 12 - (1)(6) = 0$$

As a further check, write voltage drops around the entire outside loop, moving clockwise from A

$$(6)(1) - 12 + 14 - (2)(4) = 0$$

Example Problem 17.8   A 220 V generator is driving a motor drawing 8.0 A and charging a 170 V battery. Determine the back-emf of the motor (E_m), the charging current of the battery (I_2), and the current through the generator (I_1). See Figure 17.13.

Figure 17.13

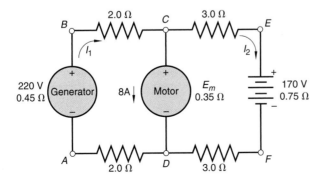

Application of Kirchhoff's laws.

Solution   Since the current through the motor is given as 8.0 A, we can see by applying Kirchhoff's current law to junction C that $I_1 = I_2 + 8.0$.

Kirchhoff's voltage law dictates that we select a beginning point and travel completely around a closed loop back to the starting point, thereby arriving at the same electric potential.

Applying Kirchhoff's voltage law to the circuit $ABCDA$ in Figure 17.13:

$$\left(E_{A-B}\right)_{emf} + \left(E_{A-B}\right)_{loss} + \left(E_{B-C}\right)_{loss} + \left(E_{C-D}\right)_{emf} + \left(E_{C-D}\right)_{loss} + \left(E_{D-A}\right)_{loss} = 0$$

By applying correct algebraic signs according to the established convention, we get the results given in Table 17.1.

Substituting the values from Table 17.1, we find that the equation becomes

$$220 - 0.45(I_1) - 2.0(I_1) - E_m - 0.35(8.0) - 2.0(I_1) = 0$$

Substituting $I_1 = I_2 + 8.0$ and simplifying, we get

$$-4.45I_2 - E_m + 181.6 = 0$$

This equation cannot be solved because there are two unknowns. However, a second equation can be written around a different loop of the circuit. For loop *DCEFD*

$$\left(E_{D-C}\right)_{emf} + \left(E_{D-C}\right)_{loss} + \left(E_{C-E}\right)_{loss} + \left(E_{E-F}\right)_{emf} + \left(E_{E-F}\right)_{loss} + \left(E_{F-D}\right)_{loss} = 0$$

From this we can develop Table 17.2 and therefore we have a second equation:

$$E_m + 0.35(8.0) - 3.0I_2 - 170 - 0.75I_2 - 3.0I_2 = 0$$

$$E_m - 6.75I_2 - 167.2 = 0$$

Solving these two equations simultaneously, we obtain the following results:

$$I_2 = 1.3 \text{ A}$$

$$E_m = 1.8 \times 10^2 \text{ V}$$

Then

$$I_1 = I_2 + 8.0 = 1.3 + 8.0 = 9.3 \text{ A}$$

These values can be checked by writing a third equation around the outside loop, *ABCEFDA*:

$$\left(E_{A-B}\right)_{emf} + \left(E_{A-B}\right)_{loss} + E_{B-C} + E_{C-E} + \left(E_{E-F}\right)_{emf} + \left(E_{E-F}\right)_{loss} + E_{F-D} + E_{D-A} = 0$$

Table 17.1   Voltage summation for loop *ABCDA*

Symbols	Quantities	Notes
$\left(E_{A-B}\right)_{emf}$	+220 V	Potential of generator
$\left(E_{A-B}\right)_{loss}$	$-0.45\,I_1$	Loss in generator
$\left(E_{B-C}\right)_{loss}$	$-2.0\,I_1$	Loss in line
$\left(E_{C-D}\right)_{emf}$	$-E_m$	Back-emf of motor
$\left(E_{C-D}\right)_{loss}$	$-0.35(8.0)$	Loss in motor
$\left(E_{D-A}\right)_{loss}$	$-2.0\,I_1$	Loss in line

Table 17.2 Voltage summation for loop *DCEFD*

Symbols	Quantities	Notes
$(E_{D-C})_{emf}$	$+E_m$	Back-emf of motor
$(E_{D-C})_{loss}$	$+0.35(8.0)$	IR rise in motor
$(E_{C-E})_{loss}$	$-3.0I_2$	Loss in line
$(E_{E-F})_{emf}$	-170 V	Drop across battery
$(E_{E-F})_{loss}$	$-0.75I_2$	Loss in battery
$(E_{F-D})_{loss}$	$-3.0I_2$	Loss in line

Using values from Tables 17.1 and 17.2, we obtain

$$+220 - 0.45(I_1) - 2.0\,(I_1) - 3.0\,(I_2) - 170 - 0.75\,(I_2) - 3.0\,(I_2) - 2.0\,(I_1) = 0$$

With $I_1 = 9.3$ A and $I_2 = 1.3$ A, we have

$$+220 - 0.45(9.3) - 2.0(9.3) - 3.0(1.3) - 170 - 0.75(1.3) - 3.0(1.3) - 2.0(9.3) = 0$$
(within roundoff error)

By the preceding procedure, a set of simultaneous equations may be found that will solve similar problems, provided the number of unknowns is not greater than the number of circuit paths or loops.

The following general procedure is outlined as a guide to systematically applying Kirchhoff's laws.

1. Sketch a circuit diagram and label all known voltages, currents, and resistances. Show $+$ and $-$ signs on potentials.
2. Indicate a current direction in each branch of the circuit. If the direction is not known, choose a direction. A negative current solution will indicate that the current is flowing in the opposite direction to the direction assumed.
3. Assign symbols to all unknown currents, voltages, and resistances.
4. Apply Kirchhoff's voltage law to circuit loops and Kirchhoff's current law at junctions to obtain as many independent equations as there are unknowns in the problem.
5. Solve the resulting set of equations.
6. Check results from Kirchhoff's voltage law written on a loop that was not used earlier.

Example Problem 17.9 A 115-V generator provides energy for a 95-Ω resistive load and a motor. See Figure 17.14. Determine the currents through the motor and resistive load and the back-emf of the motor and power consumed by the resistive load.

Solution Apply Kirchhoff's current law at point A.

$$20 - I_2 - I_3 = 0$$

Figure 17.14

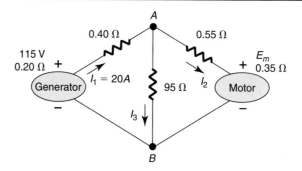

Applying Kirchhoff's voltage law to the left loop, moving clockwise from B, we have

$$115 - (0.20)(20) - (0.40)(20) - 95(I_3) = 0$$

Kirchhoff's voltage law for the right loop, moving clockwise from A, gives

$$-(0.55)(I_2) - E_m - (0.35)(I_2) + 95(I_3) = 0$$

The left loop equation can be solved for I_3:

$$I_3 = (115 - (0.20)(20) - (0.40)(20))/95 = 1.0842 \text{ A}$$

Then the current equation gives

$$I_2 = 20 - I_3 = 20 - 1.0842 = 18.9158 \text{ A}$$

E_m can be found from the right loop equation:

$$E_m = 0.95(I_3) - (0.55)(I_2) - (0.35)(I_2)$$
$$= (95)(1.0842) - (0.55)(18.9158) - (0.35)(18.9158)$$
$$= 85.9748 \text{ V}$$

Write Kirchhoff's voltage law for the outside loop as a check (clockwise from B):

$$115 - 0.20(20) - 0.40(20) - 0.55(18.9158) - 85.9748 - 0.35(18.9158)$$
$$= 9.8 \times 10^{-4} \text{ (should be zero, okay within roundoff error)}$$

Power consumed by the resistive load is

$$P = 95 \, I_3^2 = 95(1.0842)^2 = 111.6715 \text{ W}$$

The results considering significant figures then are

Current through the motor $= I_2 = 19$ A

Current through the resistive load $= I_3 = 1.1$ A

Back-emf of the motor $= E_m = 86$ V

Power consumed by the resistive load $= P = 1.1 \times 10^2$ W

Individual elements or components can be connected to form unique circuits. The interconnectivity of each element can be described in terms of nodes, branches, meshes, paths, and loops.

> *A node is a specific point or location within a circuit where two or more components are connected.*
> *A branch is a path that connects two nodes.*
> *A mesh is a loop that does not contain any other loops within itself.*

A mesh current is defined as a current that exists only in the perimeter of the mesh. Mesh currents are selected clockwise for each mesh. A mesh current is considered to travel all the way around the mesh.

Notice that Figure 17.15 is the same circuit diagram as Figure 17.12 without specific values. Referring to Figure 17.15, we can apply Kirchhoff's voltage law in the direction of the mesh currents around the two meshes expressing voltages across each component in terms of the mesh currents, I_a and I_b.

$$E_1 - I_a R_1 - (I_a - I_b) R_3 = 0$$

and

$$-E_2 - (I_b - I_a) R_3 - I_a R_2 = 0$$

Collecting and rearranging these two equations gives us

$$E_1 - I_a (R_1 + R_3) + I_b R_3 = 0$$

and

$$-E_2 + I_a R_3 - I_b (R_3 + R_2) = 0$$

By comparing Figures 17.12 and 17.15, the branch currents can be expressed in terms of the mesh currents:

$$I_1 = I_a \qquad (I_1 \text{ is the current through } R_1, \text{ to the right in Fig. 17.15})$$

$$I_2 = I_b \qquad (I_2 \text{ is the current through } R_2, \text{ to the left in Fig. 17.15})$$

$$I_3 = I_a - I_b \quad (I_3 \text{ is the current through } R_3, \text{ downward in Fig. 17.15})$$

Figure 17.15

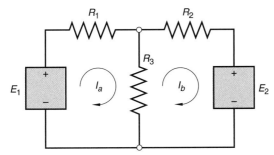

Application of mesh analysis.

Once we know the mesh currents, we know the branch currents, and once we know the branch currents, we can compute any voltage.

Example Problem 17.10   Substitute the specific values given in Figure 17.12 into the mesh current equations and verify the answers obtained in Example Problem 17.7.

Solution   Writing Kirchhoff's voltage law for the two loops gives

$$E_1 - I_a (R_1 + R_3) + I_b R_3 = 0$$
$$-E_2 + I_a R_3 - I_b (R_3 + R_2) = 0$$

Substituting values we see that

$$14 - 6I_a + 2I_b = 0$$
$$-12 + 2I_a - 8I_b = 0$$

Thus

$I_b = -1$ A (meaning that mesh current I_b is counterclockwise rather than clockwise as first assumed)

and

$I_a = 2$ A (meaning that mesh current I_a is clockwise as assumed)

Therefore

$I_1 = 2$ A (to the right in the diagram)

$I_2 = -(-1) = 1$A (to the left in the diagram)

$I_3 = 2 - (-1) = 3$A (downward in the diagram)

Example Problem 17.11   Solve Example Problem 17.9 using mesh currents.

Solution   Apply Kirchhoff's voltage law in the direction of the mesh currents I_a and I_b around the two meshes. See Figure 17.16.

$$115 - 0.20I_a - 0.40I_a - 95(I_a - I_b) = 0$$

and

$$-0.55I_b - E_m - 0.35I_b - 95(I_b - I_a) = 0$$

Collecting terms

$$115 - 95.6I_a + 95I_b = 0$$
$$-E_m - 95.9I_b + 95I_a = 0$$

But $I_a = I_1 = 20$A

Figure 17.16

401

Problems

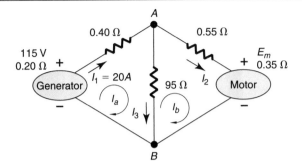

therefore

$$I_b = -\left[\frac{115 - 20(95.6)}{95}\right] = 18.9158 \cong 19 \text{ A}$$

Then

$$E_m = 95(20) - 18.9158(95.9) = 85.9748 \cong 86 \text{ V}$$

Current through the resistive load is

$$I_a - I_b = 20 - 18.9158 = 1.0842 \cong 1.1 \text{ A}$$

Power consumed by the resistive load is

$$P = 95(I_a - I_b)^2 = 95(20 - 18.9158)^2 = 111.6715 \cong 1.1 \times 10^2 \text{ W}$$

Problems

17.1 What is the average current through a conductor that carries 4 375 C during a 2.00-min time period?

17.2 How many coulombs are supplied by a battery in 36 h if it is supplying current at the rate of 1.5 A?

17.3 Assuming that the current flow through a conductor is due to the motion of free electrons, how many electrons pass through a fixed cross section normal to the conductor in 1 h if the current is 1 650 A?

17.4 Five 1.5-V batteries in series are required to operate an 8.0 W portable radio. What is the current flow? What is the equivalent circuit resistance?

17.5 A 120-V generator delivers 12 kW to an electric furnace. What current is the generator supplying?

17.6 A small vacuum cleaner designed to be plugged into the cigarette lighter of an auto has a 12.6-V DC motor. It draws 4.0 A in operation. What power must the auto battery deliver? What size resistor would consume the same power?

17.7 A portable electric drill produces 1.2 hp at full load. If 85% of the power provided by the 9.6-V battery pack is useful, what is the current flow? How much power goes into waste heat?

17.8 A DC power supply of 95 V is connected across three resistors in series. $R_1 = 12\ \Omega$, $R_2 = 15\ \Omega$, $R_3 = 25\ \Omega$.
  (a) Draw the circuit diagram.
  (b) Determine the equivalent resistance of the three resistors.
  (c) What is the current through the power supply?
  (d) What is the voltage drop across each resistor?

17.9 Three resistors—1.0, 5.0, and 15 MΩ—are connected in series to a 75-V ideal DC voltage source.
  (a) Draw the circuit diagram.
  (b) Determine the equivalent circuit resistance.
  (c) Calculate the line current.
  (d) Find the voltage drop across each resistor.
  (e) Compute the power consumed by each resistor.

17.10 An ideal 6.0-V supply is connected to three resistors wired in parallel: $R_1 = 12\ \Omega$, $R_2 = 16\ \Omega$, $R_3 = 1.0\ k\Omega$.
  (a) Draw the circuit diagram.
  (b) Calculate the equivalent circuit resistance.
  (c) Find the current through the voltage supply.
  (d) Determine the current through each resistor and the power consumed by each.

17.11 A battery has a measured voltage of 12.6 V when the circuit switch is open. The internal resistance of the battery is 0.25 Ω. The circuit contains resistors of 6.8, 13.7, and 625 Ω connected in parallel.
  (a) Draw the circuit diagram.
  (b) Calculate the equivalent circuit resistance as seen by the battery.
  (c) Find the current flow through the battery when the switch is closed.
  (d) Compute the current through the 6.8-Ω resistor and the power consumed by it.
  (e) Calculate the rate at which heat must be removed from the battery if it is to maintain a constant temperature.

17.12 Given the circuit diagram and the values in Figure 17.17, determine the current through the 15-Ω resistor and voltage drop across it. Find the fraction of the power produced by the battery that is consumed by the 25-Ω resistor.

Figure 17.17

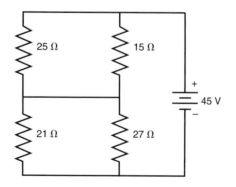

17.13 In Figure 17.18, determine the current and the potential at points S, T, U, and V when $R_1 = 5.0\ \Omega$, $R_2 = 2.6\ \Omega$, $E_G = 120\ V$, $R_G = 0.20\ \Omega$, $E_B = 24\ V$, and $R_B = 0.20\ \Omega$.

17.14 A 155-V generator in Figure 17.19 is charging a 110-V battery and driving a motor. Determine the charging current of the battery and the back-emf of the motor. Assume no internal resistances in the generator, motor, and battery.

17.15 A 125-V generator and a 26-V battery are in parallel with a motor. The current through the motor is 12 A. Determine the back-emf of the motor and current through the battery (see Fig. 17.20).

Figure 17.18

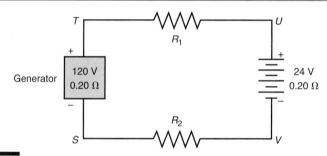

Figure 17.19

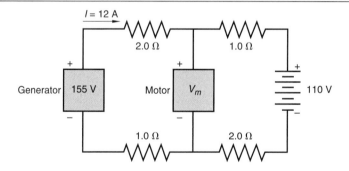

Figure 17.20

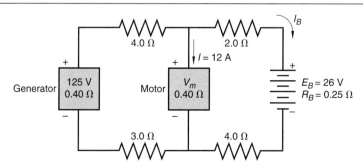

17.16 Figure 17.21 shows a resistance circuit driven by two ideal (no internal resistance) batteries. Determine:

    (*a*) The currents through each resistor

    (*b*) The power delivered to the circuit by the 80-V battery

    (*c*) The voltage across the 24-Ω resistor

    (*d*) The power consumed by the 12-Ω resistor

17.17 An ideal 14-V generator and an ideal 12.6-V battery are connected in the circuit shown in Fig. 17.22.

    (*a*) For R_A = 2.0 Ω and R_B = 3.0 Ω, compute the current through each circuit component (resistors, battery, and generator). Is the battery being charged or discharged?

    (*b*) With R_B = 1.0 Ω, find the current through the battery for 0.50 $\Omega \leq R_A \leq$ 10.0 Ω with ΔR_A = 0.50 Ω. Plot the battery current versus R_A with R_A as the independent variable. Repeat the process with R_B = 5.0 and 10.0 Ω, placing all three curves on the same graph. *Note:* Prepare a computer program and/or work as a team if approved by your instructor.

17.18 For the circuit shown in Figure 17.23:

    (*a*) Calculate the currents I_1, I_2, and I_3 when the switch is placed at A. How much power is consumed by each of the three resistors?

    (*b*) Repeat the problem with the switch placed at B.

Figure 17.21

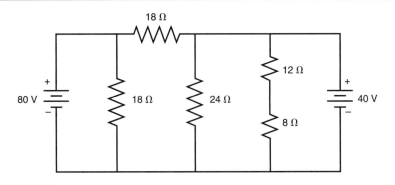

Figure 17.22

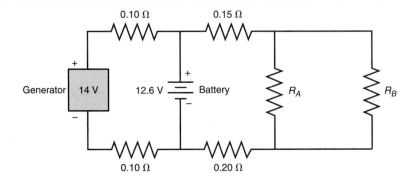

Figure 17.23

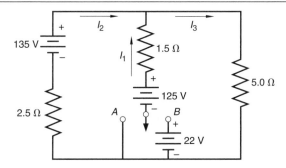

Figure 17.24

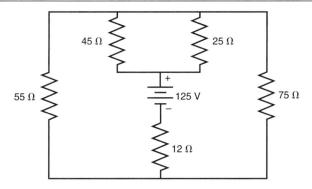

Figure 17.25

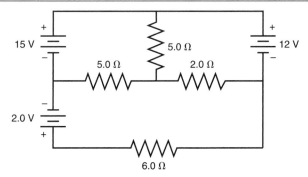

17.19 Compute the current through each resistor and the battery shown in Figure 17.24. Determine the power supplied by the battery and the power consumed by the 25-Ω resistor.

17.20 Determine the current through each component of the circuit in Figure 17.25. Find the power delivered to the 2.0-V battery, the voltage across the 2.0-Ω resistor, and power consumed by the 6.0-Ω resistor.

17.21 With reference to Figure 17.26, calculate the equivalent resistance of the circuit. Also calculate

(a) The current through R_3.

(b) The voltage across R_6.

(c) Repeat parts (*a*) and (*b*) for values of R_4 of 0, 5, 10, 15, 20, 30, 35, 40, 45, and 50 Ω. *Note:* Prepare a computer program and/or work as a team if approved by your instructor.

17.22 With reference to Figure 17.27, the power consumed by R_3 is 20 W. Find the value for R. Repeat this exercise for values of power consumed by R_3 of 2, 4, 6, 8, 10, 12, 14, 16, and 18 W. *Note:* Prepare a computer program and/or work as a team if approved by your instructor.

17.23 With reference to Figure 17.28, the internal resistances of the generator, motor, and battery are 0.25 Ω, 0.75 Ω, and 0.35 Ω respectively. If the back-emf of the motor is measured as 60 V, calculate

(a) Current supplied by the generator.
(b) Current drawn by the motor.
(c) Charging current of the battery.
(d) Power converted to heat in the motor.
(e) If the back-emf of the motor is 50 V, what are the values of (*a*) through (*d*)?

Figure 17.26

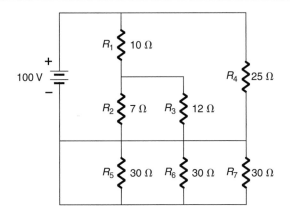

Figure 17.27

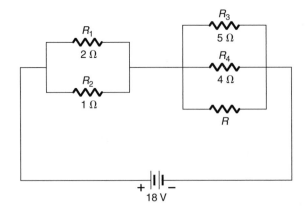

Figure 17.28

407

Problems

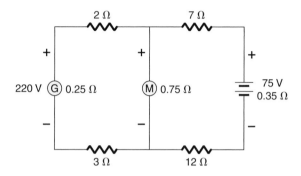

Figure 17.29

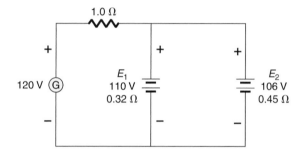

17.24 In the circuit shown (Fig. 17.29) the generator G is used to charge two storage batteries E_1 and E_2. Calculate
   (a) The charging current for both batteries.
   (b) Total power loss in all resistors.
   (c) Power delivered by the generator.
   (d) Efficiency of the charging system.
   (e) How do the values of (a) through (d) change if the line resistor is 10 Ω rather than 1.0 Ω?

17.25 In the circuit of Figure 17.30, calculate (assuming mesh currents)
   (a) The voltage drop V_{BD}.
   (b) Current through R_1.
   (c) Power dissipated in R_4.
   (d) Repeat parts (a) through (c) for battery voltages of 2, 4, 6, 8, 10, and 12 V. Plot the results and interpret the curves. *Note:* Prepare a computer program and/or work as a team if approved by your instructor.

17.26 Find the power delivered by the batteries in the circuit of Fig. 17.31. Find the current through the 1-Ω resistor. What is the voltage across the 0.8-Ω resistor? (Assume mesh currents.)

17.27 In the circuit of Figure 17.32, calculate (assuming mesh currents)
   (a) Current through R_1.
   (b) Voltage across R_2.
   (c) Power dissipated in R_3.
   (d) Power supplied by the batteries.

Figure 17.30

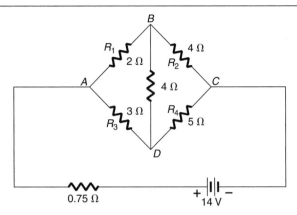

Figure 17.31

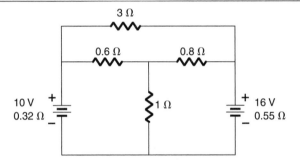

Figure 17.32

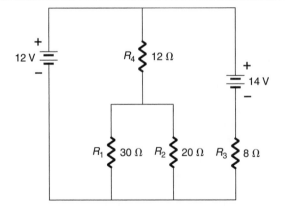

(e) Repeat parts (a) through (d) with the 12-V battery replaced by one of 10 V and then by one of 14 V. Next, with the 12-V battery as shown in the diagram, replace the 14-V battery by one of 12 V and then by one of the 16 V. *Note:* prepare a computer program and/or work as a team if approved by your instructor.

Figure 17.33

409
Problems

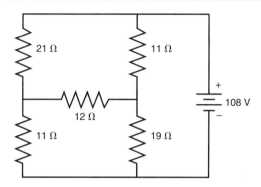

Figure 17.34

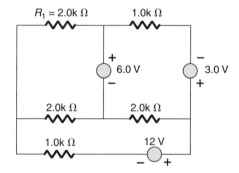

17.28 In the circuit shown in Figure 17.33, calculate (*a*) the current through the 12Ω resistor, (*b*) the voltage across the 19Ω resistor, (*c*) the current supplied by the battery, (*d*) the power consumed by the 21Ω resistor, (*e*) the total power provided to the circuit by the battery, and (*f*) repeat parts (*a*) through (*e*) for batteries of 100V, 102V, 104V, 106V and 110V. *Note:* prepare a computer program and/or work as a team if approved by your instructor.

17.29 For the circuit in Figure 17.34, compute the power supplied to the circuit by each of the batteries and the power consumed by each of the resistors. Repeat for values of R_1 = 3.0kΩ through 10.0kΩ with increments of 1.0kΩ. Plot the power consumed by R_1 as the resistance changes. Discuss the result. *Note:* prepare a computer program and/or work as a team if approved by your instructor.

Part 1 Unit Conversions

Multiply:	By:	To Obtain:
acres	4.356×10^4	ft^2
acres	$4.046\ 9 \times 10^{-1}$	ha
acres	$4.046\ 9 \times 10^3$	m^2
amperes	1	C/s
ampere hours	3.6×10^3	C
angstroms	1×10^{-8}	cm
angstroms	$3.937\ 0 \times 10^{-9}$	in
atmospheres	1.013 3	bars
atmospheres	$2.992\ 1 \times 10^1$	in of Hg
atmospheres	$1.469\ 6 \times 10^1$	lbf/in^2
atmospheres	7.6×10^2	mm of Hg
atmospheres	$1.013\ 3 \times 10^5$	Pa
barrels (petroleum, US)	4.2×10^1	gal (US liquid)
bars	$9.869\ 2 \times 10^{-1}$	atm
bars	$2.953\ 0 \times 10^1$	in of Hg
bars	$1.450\ 4 \times 10^1$	lbf/in^2
bars	1×10^5	Pa
Btu	$7.776\ 5 \times 10^2$	ft · lbf
Btu	$3.927\ 5 \times 10^{-4}$	hp · h
Btu	$1.055\ 1 \times 10^3$	J
Btu	$2.928\ 8 \times 10^{-4}$	kWh
Btu per hour	$2.160\ 1 \times 10^{-1}$	ft · lbf/s
Btu per hour	$3.927\ 5 \times 10^{-4}$	hp
Btu per hour	$2.928\ 8 \times 10^{-1}$	W
Btu per minute	$7.776\ 5 \times 10^2$	ft · lbf/min
Btu per minute	$2.356\ 5 \times 10^{-2}$	hp
Btu per minute	$1.757\ 3 \times 10^{-2}$	kW
bushels (US)	1.244 5	ft^3
bushels (US)	$3.523\ 9 \times 10^1$	L
bushels (US)	$3.523\ 9 \times 10^{-2}$	m^3
candelas	1	lm/sr
candelas per square foot	$3.381\ 6 \times 10^{-3}$	lamberts
centimeters	1×10^8	Å
centimeters	$3.280\ 8 \times 10^{-2}$	ft
centimeters	$3.937\ 0 \times 10^{-1}$	in
centipoises	1×10^{-2}	g/(cm · s)
circular mils	$5.067\ 1 \times 10^{-6}$	cm^2
circular mils	$7.854\ 0 \times 10^{-7}$	in^2
coulombs	1	A · s
cubic centimeters	$6.102\ 4 \times 10^{-2}$	in^3

Multiply:	By:	To Obtain:
cubic centimeters	$3.531\ 5 \times 10^{-5}$	ft³
cubic centimeters	$2.641\ 7 \times 10^{-4}$	gal (US liquid)
cubic centimeters	1×10^{-3}	L
cubic centimeters	$3.381\ 4 \times 10^{-2}$	oz (US fluid)
cubic centimeters per gram	$1.601\ 8 \times 10^{-2}$	ft³/lbm
cubic centimeters per second	$2.118\ 9 \times 10^{-3}$	ft³/min
cubic centimeters per second	$1.585\ 0 \times 10^{-2}$	gal (US liquid)/min
cubic feet	$2.295\ 7 \times 10^{-5}$	acre · ft
cubic feet	$8.035\ 6 \times 10^{-1}$	bushels (US)
cubic feet	$7.480\ 5$	gal (US liquid)
cubic feet	1.728×10^{3}	in³
cubic feet	$2.831\ 7 \times 10^{1}$	L
cubic feet	$2.831\ 7 \times 10^{-2}$	m³
cubic feet per minute	$7.480\ 5$	gal (US liquid)/min
cubic feet per minute	$4.719\ 5 \times 10^{-1}$	L/s
cubic feet per pound-mass	$6.242\ 8 \times 10^{1}$	cm³/g
cubic feet per second	$4.488\ 3 \times 10^{2}$	gal (US liquid)/min
cubic feet per second	$2.831\ 7 \times 10^{1}$	L/s
cubic inches	$4.650\ 3 \times 10^{-4}$	bushels (US)
cubic inches	$1.638\ 7 \times 10^{1}$	cm³
cubic inches	$4.329\ 0 \times 10^{-3}$	gal (US liquid)
cubic inches	$1.638\ 7 \times 10^{-2}$	L
cubic inches	$1.638\ 7 \times 10^{-5}$	m³
cubic inches	$5.541\ 1 \times 10^{-1}$	oz (US fluid)
cubic meters	$8.107\ 1 \times 10^{-4}$	acre · ft
cubic meters	$2.837\ 8 \times 10^{1}$	bushels (US)
cubic meters	$3.531\ 5 \times 10^{1}$	ft³
cubic meters	$2.641\ 7 \times 10^{2}$	gal (US liquid)
cubic meters	1×10^{3}	L
cubic yards	$2.169\ 6 \times 10^{1}$	bushels (US)
cubic yards	$2.019\ 7 \times 10^{2}$	gal (US liquid)
cubic yards	$7.645\ 5 \times 10^{2}$	L
cubic yards	$7.645\ 5 \times 10^{-1}$	m³
dynes	1×10^{-5}	N
dynes per square centimeter	$9.869\ 2 \times 10^{-7}$	atm
dynes per square centimeter	1×10^{-6}	bars
dynes per square centimeter	$1.450\ 4 \times 10^{-5}$	lbf/in²
dyne centimeters	$7.375\ 6 \times 10^{-8}$	ft · lbf
dyne centimeters	1×10^{-7}	N · m
ergs	1	dyne · cm
fathoms	6	ft
feet	3.048×10^{1}	cm
feet	1.2×10^{1}	in
feet	3.048×10^{-4}	km
feet	3.048×10^{-1}	m
feet	$1.893\ 9 \times 10^{-4}$	mi
feet	$6.060\ 6 \times 10^{-2}$	rods
feet per second	$1.097\ 3$	km/h
feet per second	$1.828\ 8 \times 10^{1}$	m/min
feet per second	$6.818\ 2 \times 10^{-1}$	mi/h
feet per second squared	3.048×10^{-1}	m/s²

Multiply:	By:	To Obtain:
foot-candles	1	lm/ft^2
foot-candles	$1.076\ 4 \times 10^1$	lux
foot pounds-force	$1.285\ 9 \times 10^{-3}$	Btu
foot pounds-force	$1.355\ 8 \times 10^7$	dyne · cm
foot pounds-force	$5.050\ 5 \times 10^{-7}$	hp · h
foot pounds-force	$1.355\ 8$	J
foot pounds-force	$3.766\ 2 \times 10^{-7}$	kWh
foot pounds-force	$1.355\ 8$	N · m
foot pounds-force per hour	$2.143\ 2 \times 10^{-5}$	Btu/min
foot pounds-force per hour	$2.259\ 7 \times 10^5$	ergs/min
foot pounds-force per hour	$5.050\ 5 \times 10^{-7}$	hp
foot pounds-force per hour	$3.766\ 2 \times 10^{-7}$	kW
furlongs	6.6×10^2	ft
furlongs	$2.011\ 7 \times 10^2$	m
gallons (US liquid)	$1.336\ 8 \times 10^{-1}$	ft^3
gallons (US liquid)	2.31×10^2	in^3
gallons (US liquid)	$3.785\ 4$	L
gallons (US liquid)	$3.785\ 4 \times 10^{-3}$	m^3
gallons (US liquid)	1.28×10^2	oz (US fluid)
gallons (US liquid)	8	pt (US liquid)
gallons (US liquid)	4	qt (US liquid)
grams	$2.204\ 6 \times 10^{-3}$	lbm
grams per centimeter second	1	poises
grams per cubic centimeter	$6.242\ 8 \times 10^1$	lbm/ft^3
hectares	$2.471\ 1$	acres
hectares	1×10^2	ares
hectares	$1.076\ 4 \times 10^5$	ft^2
hectares	1×10^4	m^2
horsepower	$2.546\ 1 \times 10^3$	Btu/h
horsepower	5.5×10^2	ft · lbf/s
horsepower	$7.457\ 0 \times 10^{-1}$	kW
horsepower	$7.457\ 0 \times 10^2$	W
horsepower hours	$2.546\ 1 \times 10^3$	Btu
horsepower hours	1.98×10^6	ft · lbf
horsepower hours	$2.684\ 5 \times 10^6$	J
horsepower hours	$7.457\ 0 \times 10^{-1}$	kWh
hours	6×10^1	min
hours	3.6×10^3	s
inches	2.54×10^8	Å
inches	2.54	cm
inches	$8.333\ 3 \times 10^{-2}$	ft
inches	1×10^3	mils
inches	$2.777\ 8 \times 10^{-2}$	yd
joules	$9.478\ 2 \times 10^{-4}$	Btu
joules	$7.375\ 6 \times 10^{-1}$	ft · lbf
joules	$3.725\ 1 \times 10^{-7}$	hp · h
joules	$2.777\ 8 \times 10^{-7}$	kWh
joules	1	W · s
joules per second	$5.690\ 7 \times 10^{-2}$	Btu/min
joules per second	1×10^7	ergs/s
joules per second	$7.375\ 6 \times 10^{-1}$	ft · lbf/s

Multiply:	By:	To Obtain:
joules per second	$1.341\ 0 \times 10^{-3}$	hp
joules per second	1	W
kilograms	2.204 6	lbm
kilograms	$6.852\ 2 \times 10^{-2}$	slugs
kilograms	1×10^{-3}	t
kilometers	$3.280\ 8 \times 10^{3}$	ft
kilometers	$6.213\ 7 \times 10^{-1}$	mi
kilometers	$5.399\ 6 \times 10^{-1}$	nmi (nautical mile)
kilometers per hour	$5.468\ 1 \times 10^{1}$	ft/min
kilometers per hour	$9.113\ 4 \times 10^{-1}$	ft/s
kilometers per hour	$5.399\ 6 \times 10^{-1}$	knots
kilometers per hour	$2.777\ 8 \times 10^{-1}$	m/s
kilometers per hour	$6.213\ 7 \times 10^{-1}$	mi/h
kilowatts	$3.414\ 4 \times 10^{3}$	Btu/h
kilowatts	1×10^{10}	ergs/s
kilowatts	$7.375\ 6 \times 10^{2}$	ft · lbf/s
kilowatts	1.341 0	hp
kilowatts	1×10^{3}	J/s
kilowatt hours	$3.414\ 4 \times 10^{3}$	Btu
kilowatt hours	$2.655\ 2 \times 10^{6}$	ft · lbf
kilowatt hours	1.341 0	hp · h
kilowatt hours	3.6×10^{6}	J
knots	1.687 8	ft/s
knots	1.150 8	mi/h
liters	$2.837\ 8 \times 10^{-2}$	bushels (US)
liters	$3.531\ 5 \times 10^{-2}$	ft³
liters	$2.641\ 7 \times 10^{-1}$	gal (US liquid)
liters	$6.102\ 4 \times 10^{1}$	in³
liters per second	2.118 9	ft³/min
liters per second	$1.585\ 0 \times 10^{1}$	gal (US liquid)/min
lumens	$7.957\ 7 \times 10^{-2}$	candle power
lumens per square foot	1	foot-candles
lumens per square meter	$9.290\ 3 \times 10^{-2}$	foot-candles
lux	1	lm/m²
meters	1×10^{10}	Å
meters	3.280 8	ft
meters	$3.937\ 0 \times 10^{1}$	in
meters	$6.213\ 7 \times 10^{-4}$	mi
meters per minute	1.666 7	cm/s
meters per minute	$5.468\ 1 \times 10^{-2}$	ft/s
meters per minute	6×10^{-2}	km/h
meters per minute	$3.239\ 7 \times 10^{-2}$	knots
meters per minute	$3.728\ 2 \times 10^{-2}$	mi/h
microns	1×10^{4}	Å
microns	$3.280\ 8 \times 10^{-6}$	ft
microns	1×10^{-6}	m
miles	5.28×10^{3}	ft
miles	8	furlongs
miles	1.609 3	km
miles	$8.689\ 8 \times 10^{-1}$	nmi (nautical mile)
miles per hour	$4.470\ 4 \times 10^{1}$	cm/s

Multiply:	By:	To Obtain:
miles per hour	8.8×10^1	ft/min
miles per hour	1.466 7	ft/s
miles per hour	1.609 3	km/h
miles per hour	$8.689\ 8 \times 10^{-1}$	knots
miles per hour	$2.682\ 2 \times 10^1$	m/min
nautical miles	1.150 8	mi
newtons	1×10^5	dynes
newtons	$2.248\ 1 \times 10^{-1}$	lbf
newton meters	1×10^7	dyne · cm
newton meters	$7.375\ 6 \times 10^{-1}$	ft · lbf
ounces (US fluid)	$2.957\ 4 \times 10^1$	cm^3
ounces (US fluid)	$7.812\ 5 \times 10^{-3}$	gal (US liquid)
ounces (US fluid)	1.804 7	in^3
ounces (US fluid)	$2.957\ 4 \times 10^{-2}$	L
pascals	$9.869\ 2 \times 10^{-6}$	atm
pascals	$2.088\ 5 \times 10^{-2}$	lbf/ft^2
pascals	$1.450\ 4 \times 10^{-4}$	lbf/in^2
poises	1	g/(cm · s)
pounds-force	4.448 2	N
pounds-mass	$4.535\ 9 \times 10^2$	g
pounds-mass	$4.535\ 9 \times 10^{-1}$	kg
pounds-mass	$3.108\ 1 \times 10^{-2}$	slugs
pounds-mass	$4.535\ 9 \times 10^{-4}$	t
pounds-mass	5×10^{-4}	tons (short)
pounds-force per square foot	$4.725\ 4 \times 10^{-3}$	atm
pounds-force per square foot	$4.788\ 0 \times 10^1$	Pa
pounds-force per square inch	$6.804\ 6 \times 10^{-2}$	atm
pounds-force per square inch	$6.894\ 8 \times 10^{-2}$	bars
pounds-force per square inch	2.036 0	in of Hg
pounds-force per square inch	$5.171\ 5 \times 10^1$	mm of Hg
pounds-force per square inch	$6.894\ 8 \times 10^3$	Pa
pounds-mass per cubic foot	$1.601\ 8 \times 10^{-2}$	g/cm^3
pounds-mass per cubic foot	$1.601\ 8 \times 10^1$	kg/m^3
radians	$5.729\ 6 \times 10^1$	°
radians	$1.591\ 5 \times 10^{-1}$	r (revolutions)
radians per second	9.549 3	r/min
slugs	$1.459\ 4 \times 10^1$	kg
slugs	$3.217\ 4 \times 10^1$	lbm
square centimeters	$1.076\ 4 \times 10^{-3}$	ft^2
square centimeters	$1.550\ 0 \times 10^{-1}$	in^2
square feet	$2.295\ 7 \times 10^{-5}$	acre
square feet	$9.290\ 3 \times 10^2$	cm^2
square feet	$9.290\ 3 \times 10^{-6}$	ha
square feet	$9.290\ 3 \times 10^{-2}$	m^2
square meters	$1.076\ 4 \times 10^1$	ft^2
square meters	$1.550\ 0 \times 10^3$	in^2
square miles	6.4×10^2	acres
square miles	$2.787\ 8 \times 10^7$	ft^2
square miles	$2.590\ 0 \times 10^2$	ha
square miles	2.590 0	km^2
square millimeters	$1.076\ 4 \times 10^{-5}$	ft^2

Multiply:	By:	To Obtain:
square millimeters	$1.550\ 0 \times 10^{-3}$	in^2
stokes	1	cm^2/s
stokes	$1.550\ 0 \times 10^{-1}$	in^2/s
tons (long)	2.24×10^3	lbm
tons (long)	$1.016\ 0$	t
tons (long)	1.12	tons (short)
tons (metric)	$9.017\ 2 \times 10^{-1}$	tons (short)
tons (short)	2×10^3	lbm
watts	$3.414\ 4$	Btu/h
watts	1×10^7	ergs/s
watts	$4.425\ 4 \times 10^1$	ft · lbf/min
watts	$1.341\ 0 \times 10^{-3}$	hp
watts	1	J/s
watt hours	$3.414\ 4$	Btu
watt hours	$2.655\ 2 \times 10^3$	ft · lbf
watt hours	$1.341\ 0 \times 10^{-3}$	hp · h

Part 2 Unit Prefixes

Multiple and submultiple	Prefix	Symbol
$1\,000\,000\,000\,000 = 10^{12}$	tera	T
$1\,000\,000\,000 = 10^{9}$	giga	G
$1\,000\,000 = 10^{6}$	mega	M
$1\,000 = 10^{3}$	kilo	k
$100 = 10^{2}$	hecto	h
$10 = 10$	deka	da
$0.1 = 10^{-1}$	deci	d
$0.01 = 10^{-2}$	centi	c
$0.001 = 10^{-3}$	milli	m
$0.000\,001 = 10^{-6}$	micro	μ
$0.000\,000\,001 = 10^{-9}$	nano	n
$0.000\,000\,000\,001 = 10^{-12}$	pico	p
$0.000\,000\,000\,000\,001 = 10^{-15}$	femto	f
$0.000\,000\,000\,000\,000\,001 = 10^{-18}$	atto	a

Part 3 Physical Constants

Avogadro's number $= 6.022\ 57 \times 10^{23}/\text{mol}$

Density of dry air at 0°C, 1 atm $= 1.293\ \text{kg/m}^3$

Density of water at 3.98°C $= 9.999\ 973 \times 10^2\ \text{kg/m}^3$

Equatorial radius of the earth $= 6\ 378.39\ \text{km} = 3\ 963.34\ \text{mi}$

Gravitational acceleration (standard) at sea level $= 9.806\ 65\ \text{m/s}^2 = 32.174\ \text{ft/s}^2$

Gravitational constant $= 6.672 \times 10^{-11}\ \text{N} \cdot \text{m}^2/\text{kg}^2$

Heat of fusion of water, 0°C $= 3.337\ 5 \times 10^5\ \text{J/kg} = 143.48\ \text{Btu/lbm}$

Heat of vaporization of water, 100°C $= 2.259\ 1 \times 10^6\ \text{J/kg} = 971.19\ \text{Btu/lbm}$

Mass of hydrogen atom $= 1.673\ 39 \times 10^{-27}\ \text{kg}$

Mean density of the earth $= 5.522 \times 10^3\ \text{kg/m}^3 = 344.7\ \text{lbm/ft}^3$

Molar gas constant $= 8.314\ 4\ \text{J/(mol} \cdot \text{K)}$

Planck's constant $= 6.625\ 54 \times 10^{-34}\ \text{J/Hz}$

Polar radius of the earth $= 6\ 356.91\ \text{km} = 3\ 949.99\ \text{mi}$

Velocity of light in a vacuum $= 2.997\ 9 \times 10^8\ \text{m/s}$

Velocity of sound in dry air at 0°C $= 331.36\ \text{m/s} = 1\ 087.1\ \text{ft/s}$

Part 4 Approximate Specific Gravities and Densities

Material	Specific Gravity	Average Density	
		lbm/ft^3	kg/m^3
Gases (0°C and 1 atm)			
Air		0.080 18	1.284
Ammonia		0.048 13	0.771 0
Carbon dioxide		0.123 4	1.977
Carbon monoxide		0.078 06	1.251
Ethane		0.084 69	1.357
Helium		0.011 14	0.178 4
Hydrogen		0.005 611	0.089 88
Methane		0.044 80	0.717 6
Nitrogen		0.078 07	1.251
Oxygen		0.089 21	1.429
Sulfur dioxide		0.182 7	2.927
Liquids (20°C)			
Alcohol, ethyl	0.79	49	790
Alcohol, methyl	0.80	50	800
Benzene	0.88	55	880
Gasoline	0.67	42	670
Heptane	0.68	42	680
Hexane	0.66	41	660
Octane	0.71	44	710
Oil	0.88	55	880
Toluene	0.87	54	870
Water	1.00	62.4	1 000
Metals (20°C)			
Aluminum	2.55−2.80	165	2 640
Brass, cast	8.4−8.7	535	8 570
Bronze	7.4−8.7	510	8 170
Copper, cast	8.9	555	8 900
Gold, cast	19.3	1 210	19 300
Iron, cast	7.04−7.12	440	7 050
Iron, wrought	7.6−7.9	485	7 770
Iron ore	5.2	325	5 210
Lead	11.3	705	11 300
Manganese	7.4	462	7 400
Mercury	13.6	849	13 600
Nickel	8.9	556	8 900
Silver	10.4−10.6	655	10 500
Steel, cold drawn	7.83	489	7 830
Steel, machine	7.80	487	7 800

Material	Specific Gravity	Average Density	
		lbm/ft³	**kg/m³**
Steel, tool	7.70	481	7 700
Tin, cast	7.30	456	7 300
Titanium	4.5	281	4 500
Uranium	18.7	1 170	18 700
Zinc, cast	6.9−7.2	440	7 050

Nonmetallic Solids (20°C)

Material	Specific Gravity	lbm/ft³	kg/m³
Brick, common	1.80	112	1 800
Cedar	0.35	22	350
Clay, damp	1.8−2.6	137	2 200
Coal, bituminous	1.2−1.5	84	1 350
Concrete	2.30	144	2 300
Douglas fir	0.50	31	500
Earth, loose	1.2	75	1 200
Glass, common	2.5−2.8	165	2 650
Gravel, loose	1.4−1.7	97	1 550
Gypsum	2.31	144	2 310
Limestone	2.0−2.9	153	2 450
Mahogany	0.54	34	540
Marble	2.6−2.9	172	2 750
Oak	0.64−0.87	47	750
Paper	0.7−1.2	58	925
Rubber	0.92−0.96	59	940
Salt	0.8−1.2	62	1 000
Sand, loose	1.4−1.7	97	1 550
Sugar	1.61	101	1 610
Sulfur	2.1	131	2 100

Greek Alphabet

Alpha	A	α
Beta	B	β
Gamma	Γ	γ
Delta	Δ	δ
Epsilon	E	ε
Zeta	Z	ζ
Eta	H	η
Theta	Θ	θ
Iota	I	ι
Kappa	K	κ
Lambda	Λ	λ
Mu	M	μ
Nu	N	ν
Xi	Ξ	ξ
Omicron	O	o
Pi	Π	π
Rho	P	ρ
Sigma	Σ	σ
Tau	T	τ
Upsilon	Υ	υ
Phi	Φ	ϕ
Chi	X	χ
Psi	Ψ	ψ
Omega	Ω	ω

Chemical Elements

Element	Symbol	Atomic No.	Atomic Weight
Actinium	Ac	89	
Aluminum	Al	13	26.981 5
Americium	Am	95	
Antimony	Sb	51	121.750
Argon	Ar	18	39.948
Arsenic	As	33	74.921 6
Astatine	At	85	
Barium	Ba	56	137.34
Berkelium	Bk	97	
Beryllium	Be	4	9.012 2
Bismuth	Bi	83	208.980
Boron	B	5	10.811
Bromine	Br	35	79.904
Cadmium	Cd	48	112.40
Calcium	Ca	20	40.08
Californium	Cf	98	
Carbon	C	6	12.011 15
Cerium	Ce	58	140.12
Cesium	Cs	55	132.905
Chlorine	Cl	17	35.453
Chromium	Cr	24	51.996
Cobalt	Co	27	58.933 2
Columbium (see Niobium)			
Copper	Cu	29	63.546
Curium	Cm	96	
Dysprosium	Dy	66	162.50
Einsteinium	Es	99	
Erbium	Er	68	167.26
Europium	Eu	63	151.96
Fermium	Fm	100	
Fluorine	F	9	18.998 4
Francium	Fr	87	
Gadolinium	Gd	64	157.25
Gallium	Ga	31	69.72
Germanium	Ge	32	72.59
Gold	Au	79	196.967
Hafnium	Hf	72	178.49
Helium	He	2	4.002 6
Holmium	Ho	67	164.930
Hydrogen	H	1	1.007 97
Indium	In	49	114.82

Element	Symbol	Atomic No.	Atomic Weight
Iodine	I	53	126.904 4
Iridium	Ir	77	192.2
Iron	Fe	26	55.847
Krypton	Kr	36	83.80
Lanthanum	La	57	138.91
Lead	Pb	82	207.19
Lithium	Li	3	6.939
Lutetium	Lu	71	174.97
Magnesium	Mg	12	24.312
Manganese	Mn	25	54.938 0
Mendelevium	Md	101	
Mercury	Hg	80	200.59
Molybdenum	Mo	42	95.94
Neodymium	Nd	60	144.24
Neon	Ne	10	20.183
Neptunium	Np	93	
Nickel	Ni	28	58.71
Niobium	Nb	41	92.906
Nitrogen	N	7	14.006 7
Nobelium	No	102	
Osmium	Os	76	109.2
Oxygen	O	8	15.999 4
Palladium	Pd	46	106.4
Phosphorus	P	15	30.973 8
Platinum	Pt	78	195.09
Plutonium	Pu	94	
Polonium	Po	84	
Potassium	K	19	39.102
Praseodymium	Pr	59	140.907
Promethium	Pm	61	
Protactinium	Pa	91	
Radium	Ra	88	
Radon	Rn	86	
Rhenium	Re	75	186.2
Rhodium	Rh	45	102.905
Rubidium	Rb	37	85.47
Ruthenium	Ru	44	101.07
Samarium	Sm	62	150.35
Scandium	Sc	21	44.956
Selenium	Se	34	78.96
Silicon	Si	14	28.086
Silver	Ag	47	107.868
Sodium	Na	11	22.989 8
Strontium	Sr	38	87.62
Sulphur	S	16	32.064
Tantalum	Ta	73	180.948
Technetium	Tc	43	
Tellurium	Te	52	127.60
Terbium	Tb	65	158.924
Thallium	Tl	81	204.37
Thorium	Th	90	232.038
Thulium	Tm	69	168.934
Tin	Sn	50	118.69
Titanium	Ti	22	47.90

| Element | Symbol | Atomic | |
		No.	Weight
Tungsten	W	74	183.85
Uranium	U	92	238.03
Vanadium	V	23	50.942
Xenon	Xe	54	131.30
Ytterbium	Yb	70	173.04
Yttrium	Y	39	88.905
Zinc	Zn	30	65.37
Zirconium	Zr	40	91.22

NSPE Code of Ethics for Engineers

Preamble

Engineering is an important and learned profession. As members of this profession, engineers are expected to exhibit the highest standards of honesty and integrity. Engineering has a direct and vital impact on the quality of life for all people. Accordingly, the services provided by engineers require honesty, impartiality, fairness, and equity, and must be dedicated to the protection of the public health, safety, and welfare. Engineers must perform under a standard of professional behavior that requires adherence to the highest principles of ethical conduct.

I. Fundamental Canons

Engineers, in the fulfillment of their professional duties, shall:

1. Hold paramount the safety, health, and welfare of the public.
2. Perform services only in areas of their competence.
3. Issue public statements only in an objective and truthful manner.
4. Act for each employer or client as faithful agents or trustees.
5. Avoid deceptive acts.
6. Conduct themselves honorably, responsibly, ethically, and lawfully so as to enhance the honor, reputation, and usefulness of the profession.

II. Rules of Practice

1. Engineers shall hold paramount the safety, health, and welfare of the public.
    a. If engineers' judgment is overruled under circumstances that endanger life or property, they shall notify their employer or client and such other authority as may be appropriate.
    b. Engineers shall approve only those engineering documents that are in conformity with applicable standards.
    c. Engineers shall not reveal facts, data, or information without the prior consent of the client or employer except as authorized or required by law or this Code.
    d. Engineers shall not permit the use of their name or associate in business ventures with any person or firm that they believe is engaged in fraudulent or dishonest enterprise.

NOTE: In regard to the question of application of the Code to corporations vis-à-vis real persons, business form or type should not negate nor influence conformance of individuals to the Code. The Code deals with professional services, which services must be performed by real persons. Real persons in turn establish and implement policies within business structures. The Code is clearly written to apply to the Engineer, and it is incumbent on members of NSPE to endeavor to live up to its provisions. This applies to all pertinent sections of the Code.

e. Engineers shall not aid or abet the unlawful practice of engineering by a person or firm.

f. Engineers having knowledge of any alleged violation of this Code shall report thereon to appropriate professional bodies and, when relevant, also to public authorities, and cooperate with the proper authorities in furnishing such information or assistance as may be required.

2. Engineers shall perform services only in the areas of their competence.

a. Engineers shall undertake assignments only when qualified by education or experience in the specific technical fields involved.

b. Engineers shall not affix their signatures to any plans or documents dealing with subject matter in which they lack competence, nor to any plan or document not prepared under their direction and control.

c. Engineers may accept assignments and assume responsibility for coordination of an entire project and sign and seal the engineering documents for the entire project, provided that each technical segment is signed and sealed only by the qualified engineers who prepared the segment.

3. Engineers shall issue public statements only in an objective and truthful manner.

a. Engineers shall be objective and truthful in professional reports, statements, or testimony. They shall include all relevant and pertinent information in such reports, statements, or testimony, which should bear the date indicating when it was current.

b. Engineers may express publicly technical opinions that are founded upon knowledge of the facts and competence in the subject matter.

c. Engineers shall issue no statements, criticisms, or arguments on technical matters that are inspired or paid for by interested parties, unless they have prefaced their comments by explicitly identifying the interested parties on whose behalf they are speaking, and by revealing the existence of any interest the engineers may have in the matters.

4. Engineers shall act for each employer or client as faithful agents or trustees.

a. Engineers shall disclose all known or potential conflicts of interest that could influence or appear to influence their judgment or the quality of their services.

b. Engineers shall not accept compensation, financial or otherwise, from more than one party for services on the same project, or for services pertaining to the same project, unless the circumstances are fully disclosed and agreed to by all interested parties.

c. Engineers shall not solicit or accept financial or other valuable consideration, directly or indirectly, from outside agents in connection with the work for which they are responsible.

d. Engineers in public service as members, advisors, or employees of a governmental or quasi-governmental body or department shall not participate in decisions with respect to services solicited or provided by them or their organizations in private or public engineering practice.

e. Engineers shall not solicit or accept a contract from a governmental body on which a principal or officer of their organization serves as a member.

5. Engineers shall avoid deceptive acts.

a. Engineers shall not falsify their qualifications or permit misrepresentation of their or their associates' qualifications. They shall not misrepresent or exaggerate their responsibility in or for the subject matter of prior assignments. Brochures or other presentations incident to the solicitation of employment shall not misrepresent pertinent facts concerning employers, employees, associates, joint venturers, or past accomplishments.

b. Engineers shall not offer, give, solicit, or receive, either directly or indirectly, any contribution to influence the award of a contract by public authority, or which may be reasonably construed by the public as having the effect or intent of influencing the awarding of a contract. They shall not offer any gift or other valuable consideration in order to secure work. They shall not pay a commission, percentage, or brokerage fee in order to secure work, except to a bona fide employee or bona fide established commercial or marketing agencies retained by them.

III. Professional Obligations

1. Engineers shall be guided in all their relations by the highest standards of honesty and integrity.
    a. Engineers shall acknowledge their errors and shall not distort or alter the facts.
    b. Engineers shall advise their clients or employers when they believe a project will not be successful.
    c. Engineers shall not accept outside employment to the detriment of their regular work or interest. Before accepting any outside engineering employment, they will notify their employers.
    d. Engineers shall not attempt to attract an engineer from another employer by false or misleading pretenses.
    e. Engineers shall not promote their own interest at the expense of the dignity and integrity of the profession.
2. Engineers shall at all times strive to serve the public interest.
    a. Engineers shall seek opportunities to participate in civic affairs; career guidance for youths; and work for the advancement of the safety, health, and well-being of their community.
    b. Engineers shall not complete, sign, or seal plans and/or specifications that are not in conformity with applicable engineering standards. If the client or employer insists on such unprofessional conduct, they shall notify the proper authorities and withdraw from further service on the project.
    c. Engineers shall endeavor to extend public knowledge and appreciation of engineering and its achievements.
    d. Engineers shall strive to adhere to the principles of sustainable development[1] in order to protect the environment for future generations.
3. Engineers shall avoid all conduct or practice that deceives the public.
    a. Engineers shall avoid the use of statements containing a material misrepresentation of fact or omitting a material fact.
    b. Consistent with the foregoing, engineers may advertise for recruitment of personnel.
    c. Consistent with the foregoing, engineers may prepare articles for the lay or technical press, but such articles shall not imply credit to the author for work performed by others.

---

1. "Sustainable development" is the challenge of meeting human needs for natural resources, industrial products, energy, food, transportation, shelter, and effective waste management while conserving and protecting environmental quality and the natural resource base essential for future development.

4. Engineers shall not disclose, without consent, confidential information concerning the business affairs or technical processes of any present or former client or employer, or public body on which they serve.

   a. Engineers shall not, without the consent of all interested parties, promote or arrange for new employment or practice in connection with a specific project for which the engineer has gained particular and specialized knowledge.

   b. Engineers shall not, without the consent of all interested parties, participate in or represent an adversary interest in connection with a specific project or proceeding in which the engineer has gained particular specialized knowledge on behalf of a former client or employer.

5. Engineers shall not be influenced in their professional duties by conflicting interests.

   a. Engineers shall not accept financial or other considerations, including free engineering designs, from material or equipment suppliers for specifying their product.

   b. Engineers shall not accept commissions or allowances, directly or indirectly, from contractors or other parties dealing with clients or employers of the engineer in connection with work for which the engineer is responsible.

6. Engineers shall not attempt to obtain employment or advancement or professional engagements by untruthfully criticizing other engineers, or by other improper or questionable methods.

   a. Engineers shall not request, propose, or accept a commission on a contingent basis under circumstances in which their judgment may be compromised.

   b. Engineers in salaried positions shall accept part-time engineering work only to the extent consistent with policies of the employer and in accordance with ethical considerations.

   c. Engineers shall not, without consent, use equipment, supplies, laboratory, or office facilities of an employer to carry on outside private practice.

7. Engineers shall not attempt to injure, maliciously or falsely, directly or indirectly, the professional reputation, prospects, practice, or employment of other engineers. Engineers who believe others are guilty of unethical or illegal practice shall present such information to the proper authority for action.

   a. Engineers in private practice shall not review the work of another engineer for the same client, except with the knowledge of such engineer, or unless the connection of such engineer with the work has been terminated.

   b. Engineers in governmental, industrial, or educational employ are entitled to review and evaluate the work of other engineers when so required by their employment duties.

   c. Engineers in sales or industrial employ are entitled to make engineering comparisons of represented products with products of other suppliers.

8. Engineers shall accept personal responsibility for their professional activities, provided, however, that engineers may seek indemnification for services arising out of their practice for other than gross negligence, where the engineer's interests cannot otherwise be protected.

   a. Engineers shall conform with state registration laws in the practice of engineering.

   b. Engineers shall not use association with a nonengineer, a corporation, or partnership as a "cloak" for unethical acts.

9. Engineers shall give credit for engineering work to those to whom credit is due, and will recognize the proprietary interests of others.

   a. Engineers shall, whenever possible, name the person or persons who may be individually responsible for designs, inventions, writings, or other accomplishments.

b. Engineers using designs supplied by a client recognize that the designs remain the property of the client and may not be duplicated by the engineer for others without express permission.

c. Engineers, before undertaking work for others in connection with which the engineer may make improvements, plans, designs, inventions, or other records that may justify copyrights or patents, should enter into a positive agreement regarding ownership.

d. Engineers' designs, data, records, and notes referring exclusively to an employer's work are the employer's property. The employer should indemnify the engineer for use of the information for any purpose other than the original purpose.

e. Engineers shall continue their professional development throughout their careers and should keep current in their specialty fields by engaging in professional practice, participating in continuing education courses, reading in the technical literature, and attending professional meetings and seminars.

—As Revised January 2006

"By order of the United States District Court for the District of Columbia, former Section 11(c) of the NSPE Code of Ethics prohibiting competitive bidding, and all policy statements, opinions, rulings or other guidelines interpreting its scope, have been rescinded as unlawfully interfering with the legal right of engineers, protected under the antitrust laws, to provide price information to prospective clients; accordingly, nothing contained in the NSPE Code of Ethics, policy statements, opinions, rulings or other guidelines prohibits the submission of price quotations or competitive bids for engineering services at any time or in any amount."

Statement by NSPE Executive Committee

In order to correct misunderstandings which have been indicated in some instances since the issuance of the Supreme Court decision and the entry of the Final Judgment, it is noted that in its decision of April 25, 1978, the Supreme Court of the United States declared: "The Sherman Act does not require competitive bidding."
It is further noted that as made clear in the Supreme Court decision:

1. Engineers and firms may individually refuse to bid for engineering services.
2. Clients are not required to seek bids for engineering services.
3. Federal, state, and local laws governing procedures to procure engineering services are not affected, and remain in full force and effect.
4. State societies and local chapters are free to actively and aggressively seek legislation for professional selection and negotiation procedures by public agencies.
5. State registration board rules of professional conduct, including rules prohibiting competitive bidding for engineering services, are not affected and remain in full force and effect. State registration boards with authority to adopt rules of professional conduct may adopt rules governing procedures to obtain engineering services.
6. As noted by the Supreme Court, "nothing in the judgment prevents NSPE and its members from attempting to influence governmental action . . ."

Flowcharts

E.1 Introduction

Programming languages such as FORTRAN, C+ +, Visual Basic and so forth can be used to custom prepare solutions to certain problems. This approach can be a time-consuming process and it requires much skill and experience to be an effective programmer. If spreadsheets and math calculation packages cannot do the job, you may be forced to program a solution. It is effective to program a solution if the solution is a unique application where commercial software is not likely to be available or if the solution is to be used repetitively without need for changing the programming code.

Custom-solutions may be necessary, but they are likely to be expensive in initial preparation and are not very flexible. They can be difficult to maintain since the original programmer may no longer be available and most engineering programmers are not good at thorough documentation of a program such that someone else readily can understand the process and make needed modifications.

E.2 Flowcharting

An algorithm (solution to a problem) can be described in terms of text-like statements (called pseudocode) or graphically in a form known as a *flowchart*. A flowchart provides a picture of the logic and the steps involved in solving a problem.

As you develop a flowchart, it is advantageous to think in terms of the big picture before focusing on details. For example, when designing a house, an architect must first plan where the kitchen, bathrooms, bedrooms, and other rooms are to be located before specifying where electric, water, and sewer lines should be placed. Similarly, a flowchart is designed by working out large blocks to assure that global logic is satisfied before deciding what detailed procedure should be used within each large block.

A set of graphical symbols is used to describe each step of the flowchart (see Fig. E.1). Although there are many *flowchart symbols* in general use, we will only define a small subset that is generic in nature; that is, each symbol does not denote any particular device or method for performing the operation. For example, a general *input/output* symbol is used that does not suggest that the input or output method is a document, tape, disk storage or display. The symbol only means that communication with the software should occur by using whatever device or method is available and appropriate. However, when coding the input or output process, the programmer will have to be specific about the method or device.

Three general structures are used in algorithm, or flowchart development: the *sequential structure*, the *selection structure*, and the *repetition structure*. The latter two have some variations that will be illustrated in the following paragraphs.

A *sequential structure* defines a series of steps that are performed in order, beginning at the start position and proceeding sequentially from operation to operation until the stop symbol is reached. No decisions are made and no step or series of steps is repeated. A sequential structure is shown in Figure E.2. Dashed lines in this flowchart

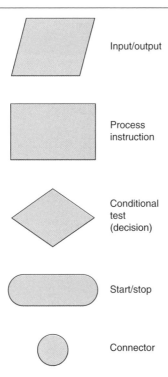

Input/output

Process
instruction

Conditional
test
(decision)

Start/stop

Connector

Flowchart symbols.

and others described later simply mean that repeated symbols have been omitted. A simple example of a sequential structure is shown in Figure E.3 where values are input, a calculation is done and the results are output.

The fundamental *selection structure* is illustrated in Figure E.4. It contains a conditional test symbol that asks a question with a yes/no (true/false) answer or states a condition with two possible outcomes. Thus, based on the outcome of the decision, one of the two sets of operations will be performed. Each branch of the selection structure may contain as many operations as are necessary. There may be one operation or several in a branch, or even no operations in one of the branches. With this structure, flow proceeds through one or the other of the branches and continues on into a later section of the flowchart. Therefore, during a single pass through the structure one of the branches is not used.

Figure E.5 is a simple illustration of a selection structure. Here a variable, x, which has been defined prior to this flowchart segment, is compared to the number 5. If x is greater than 5, the variable, y, is defined as x-squared. Both x and y are then output and control moves to whatever elements are next in the flowchart. If, however, in the decision block, x is not greater than 5, y is defined as x-cubed and then the output occurs. Thus depending on the value of x at this point in the flowchart, y will be defined as either x-squared or x-cubed.

Figure E.2

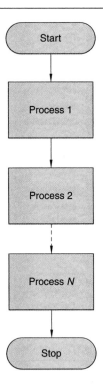

Simple sequence.

Figure E.3

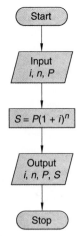

Sequence for future worth calculation.

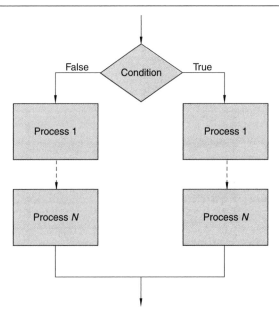

Selection structure.

A somewhat generalized version of the selection structure is shown in Figure E.6. Here the decision or condition at the top of the structure has more than two outcomes, called cases. Each outcome or case can have a unique set of steps to be performed. This structure is useful, for example, in a sorting procedure where one is treating data differently with each of several defined ranges of values.

Figure E.5

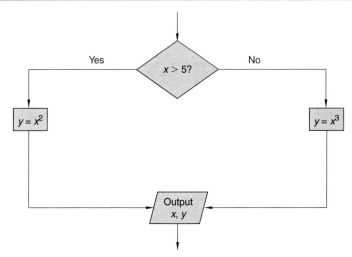

Example of selection structure.

Figure E.6

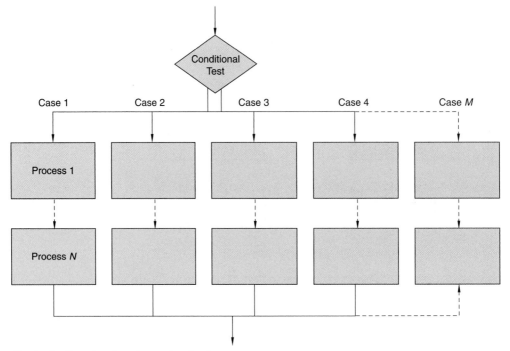

Generalized selection structure (case structure).

Figure E.7

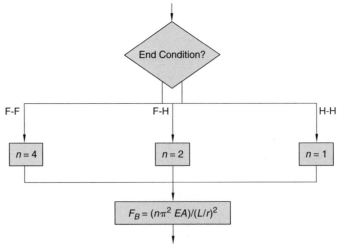

Case structure for slender column.

A computation for the buckling load of a slender column is depicted in Figure E.7. The buckling load depends on the end conditions of the column described as both ends fixed (F-F), one end fixed and one end hinged (F-H) or both ends hinged (H-H). This case structure segment examines the previously defined end condition and sets a variable n to 1, 2 or 4 and then computes the buckling load FB. The buckling load also depends on the modulus of elasticity E, the cross-sectional area A, the length L, and the least moment of gyration r, all of which would have to be input or defined prior to reaching this segment. Output of variables could then occur after the calculation.

The *repetition structure* (looping structure) consists of a step or series of steps that are performed repeatedly until some condition (perhaps a specified number of repetitions) is satisfied, at which time the next step after the loop is executed. Two common repetition structures are presented, one where the conditional test is performed as the last step of the structure and one where the test is the first step of the structure.

The repetition structure where the conditional test is the last step is illustrated in Figure E.8. One or more processes are placed in the forward section of the structure. The conditional test can be reversed; that is, the true and false flow lines can be interchanged depending on the nature of the condition to be tested. There may be no need for the process block in the reverse loop based on the action in the process blocks in the forward section. You may or may not include the reverse-process block as your logic dictates. Frequently, the reverse-loop block performs the action of a counter. Operations such as $X = X + 1$ or $Z = Z + 5$ might appear here. These are not algebraic equations since they are clearly not mathematically correct. They are instructions to replace the current value of X by a new value 1 greater or to replace Z by $Z + 5$. Therefore, they can count the number of times through the loop, as does X, or can increment a variable by a constant, as in the case of Z. Negative increments also are possible so that you can count backward or decrement a variable.

Figure E.8

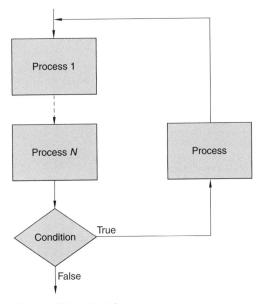

Repetition structure with conditional test last.

Figure E.9

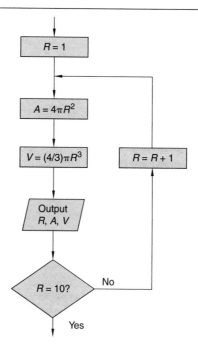

Example of test last.

Figure E.10

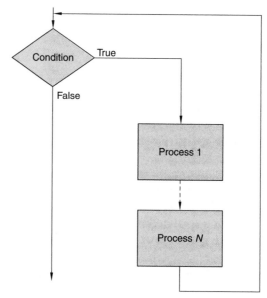

Repetition structure with conditional test first.

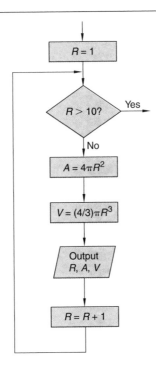

Example of test first.

Figure E.9 is an example of a repetition structure where the test is last. This segment computes and outputs the surface area A and the volume V of spheres ranging from a radius R of 1 to 10 units.

Figure E.10 shows the repetition structure where the conditional test is performed first. Again, the true and false branches can be reversed to match the chosen conditional test. As many process blocks as desired may be used. One of them could be a counter or incrementing block.

The flowchart segment in Figure E.11 does the same thing as the segment in Figure E.9, but is arranged such that the test is first rather than last.

Several selection structures or repetition structures can be combined by a method called *nesting*. In this way one or more loops can be contained within a loop. Similarly, a selection structure can be placed within another selection structure. Figure E.12 shows an example of how a nested loop might appear. The inner loop is performed until condition 2 is satisfied; then control returns to the decision block for condition 1. In each pass through the outer loop the inner loop will be repeated until condition 2 is satisfied. Eventually, condition 1 will be satisfied and flow will pass to the next part of the flowchart.

Examples of various combinations of the structures just discussed can be seen in the problems that follow.

Example problem E.1   Construct a flowchart for calculating the sum of the squares of the even integers from $N1$ to $N2$.

Figure E.12

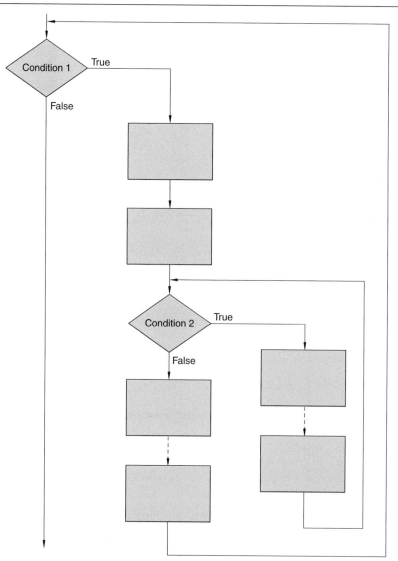

Nested repetition structure (nested loop).

Procedure For purposes of this example, $N1$ and $N2$ will be restricted to even integers only and $N2 > N1$. We could use the sequential structure shown in Figure E.13. This flowchart will result in a variable called SUM as the desired value; SUM is then output. Because of the repetitive nature of the steps, it is far more convenient to use a repetition structure as seen in Figure E.14.

Figure E.13

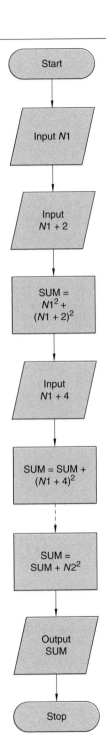

Figure E.14

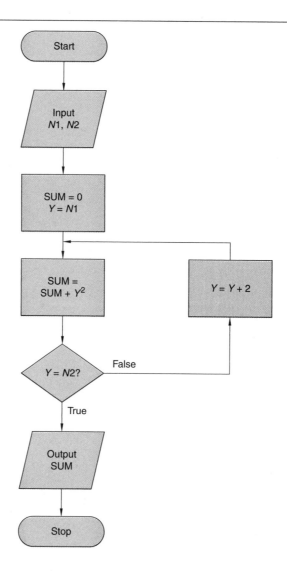

Study Figure E.14 carefully; several important flowcharting concepts are introduced there. First, the flowchart is useful for any pair $N1$, $N2$ as long as each is an even integer and $N2 > N1$. (The flowchart also will work for $N1$ and $N2$ as odd integers, although that is not specified in this problem.) Second, there are two variables, SUM and Y, that will take on numerous values. They must be initialized outside the repetition structure. If the process block containing SUM $= 0$ and $Y = N1$ was inside the loop, SUM and Y would be reset to their initial values each time through the loop and the decision block ($Y = N2$?) could never be satisfied, thereby creating an infinite loop.

The difference between initializing a variable and inputting a variable is important. As a general rule: *Variables that must have initial values but whose values will not change from one use of the flowchart to another (one run of the resulting program to another) should be placed in a process block. Variables that you wish to change from one run to another should be placed in an input block.* This provides you with the necessary flexibility of reusing the flowchart (or program) without having to input the variables that do not change from run to run.

Example problem E.2 Draw a flowchart that will calculate the future sum of a principal (an amount of money) for a given interest rate and the number of interest periods. Allow the user to decide if simple or compound interest is to be used and to compute as many future sums as desired.

Procedure For simple interest $S = P(1 + ni)$, and for compound interest $S = P(1 + i)^n$, where S = future sum, P = principal amount, i = interest rate per period, and n = number of interest periods. One possible flowchart is given in Figure E.15. The user is asked to input $T = 0$ for simple interest and $T = 1$ for compound interest. A decision block checks on P before performing any further calculation. If $P < 0$ (a value not expected to be used), the process terminates. Thus, using a unique value for one of the input variables is one method of terminating the processing. A logical alternative would be to construct a counter and to check to see if a specified number of variable sets has been reached.

The connector symbol has been used to avoid drawing a long flow line in this example. A letter or number (the letter A was used in this case) is placed in the symbols to define uniquely the pair of symbols that should be connected. Connector symbols also are used when a flowchart occupies more than one page and flow lines cannot physically connect portions of the flowchart.

Example problem E.3 The sine of an angle can be approximately calculated from the following series expansion.

$$\sin x \cong \sum_{i=1}^{N}(-1)^{i+1}\left[\frac{x^{2i-1}}{(2i-1)!}\right]$$

$$= x - \frac{x^3}{3!} + \frac{x^5}{5!} - \frac{x^7}{7!} + \cdots + (-1)^{N+1}\left[\frac{x^{2N-1}}{(2N-1)!}\right]$$

where x is the angle in radians. The degree of accuracy is determined by the number of terms in the series that are summed for a given value of x. Prepare a flowchart to calculate the sine of an angle of P degrees and cease the summation when the last term in the series calculated has a magnitude less than 10^{-7}. Of course, an exact answer for the sine of the angle would require the summation of an infinite number of terms. Include the procedure for evaluating a factorial in the flowchart.

Procedure One possible flowchart is shown in Figure E.16. We will discuss several features of this flowchart, after which you should track on paper the first three or four terms of the series expansion to make sure you understand that the flowchart is correctly handling the problem. Note that the general term has been used in the loop and that specific values of each variable are calculated in order to produce the required term each time through the loop:

1. The magnitude of the quantity controlling the number of terms summed is called ERR. It is input so that a magnitude of other than 10^{-7} can be used in a future run.

Figure E.15

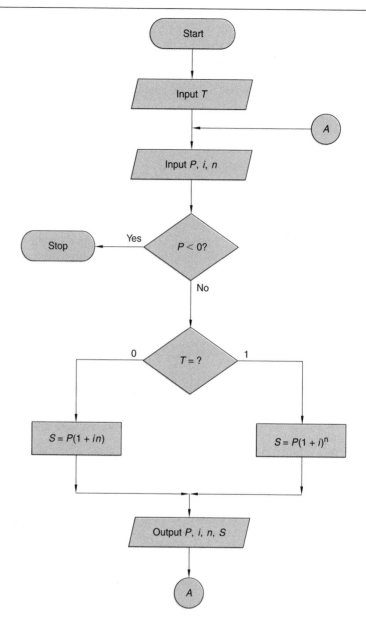

2. i denotes the summation variable and is initialized as 1.
3. M represents $(2i - 1)!$ and is initialized as 1 (its value in the first term of the series).
4. SUM is the accumulated value of the series as each term is added to the previous total. It is initialized as 0.
5. The angle is input in degrees and then immediately converted to radians by multiplying by $\pi/180$.

Figure E.16

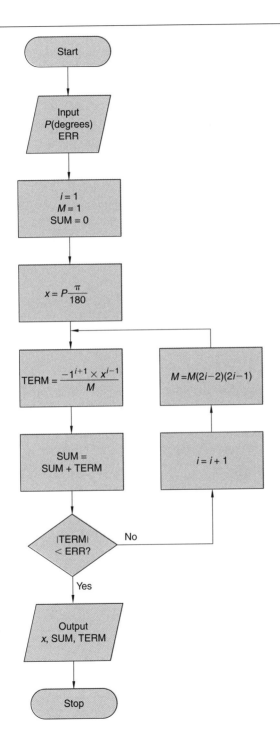

6. TERM is the value of each term beginning with $(-1)^{1+1}\{x^{2(1)-1}/[2(1)-1]!\}$ or simply x. The factor $(-1)^{i+1}$ causes TERM to alternate signs.

7. SUM is equal to $0 + x$ the first time through the procedure.

8. The absolute value of TERM is now checked against the control value ERR to see if computations should cease. Note that the absolute value must be used because of the alternating signs. If a "no" answer is received to this conditional test, the appropriate incrementing of the variable must be undertaken.

9. i is increased by 1. This time through i becomes 2 since it was initialized as 1.

10. M becomes $(1)(2 \times 2 - 2)(2 \times 2 - 1) = 1 \times 2 \times 3 = 3!$

11. Following the flowchart directions, we now return to the evaluation of TERM with the new value of i.

$$\text{TERM} = (-1)^{2+1}x^{2(2)-1}/3! = -x^3/3!$$

12. SUM then becomes $x - x^3/3!$. TERM is again tested against ERR.

13. The process repeats until the magnitude of the last term is less than ERR (10^{-7} in this case). Then the value of x, sin x, and the value of the last term are reported to make certain that the standard of accuracy has been attained.

    You should check several terms for a given value of x and then repeat the process for different angles. You will note that as the size of the angle varies, the number of terms required to achieve the standard of accuracy also varies. For example, when using a value of $P = 5°$, the third term of the series is about $4(10^{-8})$, less than the 10^{-7} requirement. But for $P = 80°$ the third term is approximately $4.4(10^{-2})$. It requires seven terms of the series before the magnitude of the last term becomes less than 10^{-7}.

The preceding examples should give you some insight into the construction of a flowchart as a prelude to writing a computer program or to using other software tools. The mechanisms for calculating, testing, incrementing, looping, and so on vary with the computational device and the programming language or software tool. A flowchart, however, should be valid for all computer systems and software tools because it graphically portrays the steps that must be completed to solve the problem.

Part 1 Selected Algebra Topics

F1.1   Introduction

This appendix includes material on exponents and logarithms, simultaneous equations, and the solution of equations by approximation methods. The material can be used for reference or review. The reader should consult an algebra textbook for more detailed explanations of additional topics for study.

F1.2   Exponents and Radicals

The basic laws of exponents are stated subsequently along with an illustrative example.

Law	Example
$a^m a^n = a^{m+n}$	$x^5 x^{-2} = x^3$
$\dfrac{a^m}{a^n} = a^{m-n} \quad a \neq 0$	$\dfrac{x^5}{x^3} = x^2$
$(a^m) = a^{mn}$	$(x^{-2})^3 = x^{-6}$
$(ab)^m = a^m b^m$	$(xy)^2 = x^2 y^2$
$\left(\dfrac{a}{b}\right)^m = \dfrac{a^m}{b^m} \quad b \neq 0$	$\left(\dfrac{x}{y}\right)^2 = \dfrac{x^2}{y^2}$
$a^{-m} = \dfrac{1}{a^m} \quad a \neq 0$	$x^{-3} = \dfrac{1}{x^3}$
$a^0 = 1 \; a \neq 0$	$2(3x^2)^0 = 2(1) = 2$
$a^1 = a$	$(3x^2)^1 = 3x^2$

These laws are valid for positive and negative integer exponents and for a zero exponent, and can be shown to be valid for rational exponents. Some examples of fractional exponents are illustrated here. Note the use of radical ($\sqrt{}$) notation as an alternative to fractional exponents.

Law	Example
$a^{m/n} = \sqrt[n]{a^m}$	$x^{2/3} = \sqrt[3]{x^2}$
$\dfrac{\sqrt[n]{a}}{\sqrt[n]{b}} = \sqrt[n]{\dfrac{a}{b}} \quad b \neq 0$	$\dfrac{\sqrt[3]{16}}{\sqrt[3]{2}} = \sqrt[3]{8} = 2$
$a^{1/2} = \sqrt[2]{a^1} = \sqrt{a} \quad a \geq 0$	$\sqrt{25} = 5 \quad (\text{not } \pm 5)$

F1.3   Exponential and Power Functions

Functions involving exponents occur in two forms—power and exponential. The power function contains the base as the variable and the exponent is a rational number. An exponential function has a fixed base and variable exponent.

The simplest exponential function is of the form

$$y = b^x \qquad b \geq 0$$

where b is a constant. Note that this function involves a power but is fundamentally different from the power function $y = x^b$.

The inverse of a function is an important concept for the development of logarithmic functions from exponential functions. Consider a function $y = f(x)$. If this function could be solved for x, the result would be expressed as $x = g(y)$. For example, the power function $y = x^2$ has as its inverse $x = \pm\sqrt{y}$. Note that in $y = x^2$, y is a single-valued function of x, whereas the inverse is a double-valued function. For $y = x^2$, x can take on any real value, whereas the inverse $x = \pm\sqrt{y}$ restricts y to only positive values or zero. This result is important in the study and application of logarithmic functions.

F1.4   The Logarithmic Function

The definition of a logarithm may be stated as follows:

A number L is said to be the logarithm of a positive real number N to the base b (where b is real, positive, and different from 1), if L is the exponent to which b must be raised to obtain N.

Symbolically, the logarithm function is expressed as

$$L = \log_b N$$

for which the inverse is

$$N = b^L$$

For instance,

$$\log_2 8 = 3 \qquad \text{since } 8 = 2^3$$
$$\log_{10} 0.01 = -2 \qquad \text{since } 0.01 = 10^{-2}$$
$$\log_5 5 = 1 \qquad \text{since } 5 = 5^1$$
$$\log_b 1 = 0 \qquad \text{since } 1 = b^0$$

Several properties of logarithms and exponential functions can be identified when plotted on a graph.

Example problem F1.1

Plot graphs of $y = \log_2 x$ and $x = 2^y$ that are inverse functions.

Solution

Since $y = \log_2 x$ and $x = 2^y$ are equivalent by definition, they will graph into the same line. Choosing values of y and computing x from $x = 2^y$ yields Figure F1.1.

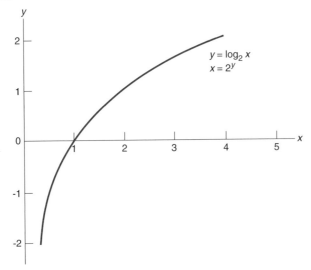

The logarithmic function.

Some properties of logarithms that can be generalized from Figure F1.1 are

1. $\log_b x$ is not defined for negative or zero values of x.
2. $\log_b 1 = 0$.
3. If $x > 1$, then $\log_b x > 0$.
4. If $0 < x < 1$, then $\log_b x < 0$.

Other properties of logarithms that can be proved as a direct consequence of the laws of exponents are, with P and Q being real and positive numbers,

1. $\log_b PQ = \log_b P + \log_b Q$.
2. $\log_b \dfrac{P}{Q} = \log_b P - \log_b Q$.
3. $\log_b (P)^m = m \log_b P$.
4. $\log_b \sqrt[n]{P} = \dfrac{1}{n} \log_b P$.

The base b, as stated in the definition of a logarithm, can be any real number greater than 0 but not equal to 1 since 1 to any power remains 1. When using logarithmic notation, the base is always indicated, with the exception of base 10, in which case the base is frequently omitted. In the expression $y = \log x$, the base is understood to be 10. A somewhat different notation is used for the natural (Naperian) logarithms discussed in Sec. F1.5.

Sometimes it is desirable to change the base of logarithms. The procedure is shown by the following example.

Example problem F1.2

Given that $y = \log_a N$, find $\log_b N$.

Solution

$$y = \log_a N$$

$$N = a^y \qquad \text{(inverse function)}$$

$$\log_b N = y \log_b a \qquad \text{(taking logs to base } b)$$

$$\log_b N = (\log_a N)(\log_b a) \qquad \text{(substitution for } y)$$

$$= \frac{\log_a N}{\log_a b} \qquad \left(\text{since } \log_b a = \frac{1}{\log_a b}\right)$$

F1.5   Natural Logarithms and *e*

In advanced mathematics the base e is usually chosen for logarithms to achieve simpler expressions. Logarithms to the base e are called natural, or Naperian, logarithms. The constant e is defined in the calculus as

$$e = \lim_{n \to 0} (1 + n)^{1/n} = 2.7182818284000\ldots$$

For purposes of calculating e to a desired accuracy, an infinite series is used.

$$e = \sum_{n \to 0}^{\infty} \frac{1}{n!}$$

The required accuracy is obtained by summing sufficient terms. For example,

$$\sum_{n=0}^{6} \frac{1}{n!} = 1 + 1 + \frac{1}{2} + \frac{1}{6} + \frac{1}{24} + \frac{1}{120} + \frac{1}{720}$$

$$= 2.718\,055$$

which is accurate to four significant figures.

Natural logarithms are denoted by the symbol ln, and all the properties defined previously for logarithms apply to natural logarithms. The inverse of $y = \ln x$ is $x = e^y$. The following examples illustrate applications of natural logarithms.

Example problem F1.3

$$\ln 1 = 0 \qquad \text{since } e^0 = 1$$
$$\ln e = 1 \qquad \text{since } e^1 = e$$

Example problem F1.4

Solve for x:

$$2^x = 3^{x-1}$$

Specify your answer to four significant figures.

Taking natural logarithms of both sides of the equation and using a calculator for evaluation of numerical quantities,

$$x \ln 2 = (x - 1)\ln 3$$

$$\frac{x}{x - 1} = \frac{\ln 3}{\ln 2} = 1.585\ 0$$

$$x = 2.709 \text{ (four significant figures)}$$

This problem could have been solved by choosing any base for taking logarithms. However, in general, base e or 10 should be chosen so that a scientific calculator can be used for numerical work.

F1.6   Simultaneous Equations

Several techniques exist for finding the common solution to a set of n algebraic equations in n unknowns. A formal method for solution of a system of linear equations is known as Cramer's rule, which requires a knowledge of determinants.

A second-order determinant is defined and evaluated as

$$\begin{vmatrix} a_1 b_1 \\ a_2 b_2 \end{vmatrix} = a_1 b_2 - a_2 b_1$$

A third-order determinant is defined and evaluated as

$$\begin{vmatrix} a_1 b_1 c_1 \\ a_2 b_2 c_2 \\ a_3 b_3 c_3 \end{vmatrix} = a_1 \begin{vmatrix} b_2 c_2 \\ b_3 c_3 \end{vmatrix} - a_2 \begin{vmatrix} b_1 c_1 \\ b_3 c_3 \end{vmatrix} + a_3 \begin{vmatrix} b_1 c_1 \\ b_2 c_2 \end{vmatrix}$$

where the second-order determinants are evaluated as indicated previously. The procedure may be extended to higher-order determinants.

Cramer's rule for a system of n equations in n unknowns can be stated as follows:

1. Arrange the equations to be solved so that the unknowns x, y, z, and so forth appear in the same order in each equation; if any unknown is missing from an equation, it is to be considered as having a coefficient of zero in that equation.
2. Place all terms that do not involve the unknowns in the right member of each equation.
3. Designate by D the determinant of the coefficients of the unknowns in the same order as they appear in the equations. Designate by D_i the determinant obtained by replacing the elements of the ith column of D by the terms in the right member of the equations.
4. Then if $D \neq 0$ the values of the unknowns x y, z, and so forth, are given by

$$x = \frac{D_1}{D} \qquad y = \frac{D_2}{D} \qquad z = \frac{D_3}{D} \cdots$$

Example problem F1.5

Solve the following system of equations that have already been written in proper form for application of Cramer's rule.

$$3x + y - z = 2$$

$$x - 2y + z = 0$$

$$4x - y + z = 3$$

Solution

$$x = \frac{\begin{vmatrix} 2 & 1 & -1 \\ 0 & -2 & 1 \\ 3 & -1 & 1 \end{vmatrix}}{\begin{vmatrix} 3 & 1 & -1 \\ 1 & -2 & 1 \\ 4 & -1 & 1 \end{vmatrix}} = \frac{2\begin{vmatrix} -2 & 1 \\ -1 & 1 \end{vmatrix} - 1\begin{vmatrix} 0 & 1 \\ 3 & 1 \end{vmatrix} + (-1)\begin{vmatrix} 0 & -2 \\ 3 & -1 \end{vmatrix}}{3\begin{vmatrix} -2 & 1 \\ -1 & 1 \end{vmatrix} - 1\begin{vmatrix} 1 & 1 \\ 4 & 1 \end{vmatrix} + (-1)\begin{vmatrix} 1 & -2 \\ 4 & -1 \end{vmatrix}}$$

$$= \frac{2(-2 + 1) - 1(0 - 3) - 1(0 + 6)}{3(-2 + 1) - 1(1 - 4) - 1(-1 + 8)}$$

$$= \frac{2(-1) - 1(-3) - 1(6)}{3(-1) - 1(-3) - 1(7)}$$

$$= \frac{-5}{-7}$$

$$= \frac{5}{7}$$

The reader may verify the solutions $y = 6/7$ and $z = 1$.

There are several other methods of solution for systems of equations that are illustrated by the following examples.

Example problem F1.6

Solve the system of equations:

$$9x^2 - 16y^2 = 144$$

$$x - 2y = 4$$

Solution

The common solution represents the intersection of a hyperbola and straight line. The method used is substitution. Solving the linear equation for x yields

$$x = 2y + 4$$

Substitution into the second-order equation gives

$$9(2y + 4)^2 - 16y^2 = 144$$

which reduces to

$$20y^2 + 144y = 0$$

Factoring gives

$$4y(5y + 36) = 0$$

which yields

$$y = 0, \frac{-36}{5}$$

Substitution into the linear equation $x = 2y + 4$ gives the corresponding values of x:

$$x = 4, -\frac{52}{5}$$

The solutions thus are the coordinates of intersection of the line and the hyperbola:

$$(4.0), \left(-\frac{52}{5}, -\frac{36}{5}\right)$$

which can be verified by graphical construction.

Example problem F1.7

Solve the system of equations:

$$\text{(a) } 3x - y = 7$$
$$\text{(b) } x + z = 4$$
$$\text{(c) } y - z = -1$$

Solution

Systems of equations similar to these arise frequently in engineering applications. Obviously they can be solved by Cramer's rule. However, a more rapid solution can be obtained directly by elimination.

From Eq. (c),

$$y = z - 1$$

From Eq. (a)

$$y = 7 - 3x$$

From Eq. (b)

$$x = 4 - z$$

Successive substitution yields

$$z - 1 = 7 - 3x$$
$$z - 1 = 7 - 3(4 - z)$$
$$-2z = -4$$
$$z = +2$$

Continued substitution gives

$$y = 1$$
$$x = 2$$

Every system of equations first should be investigated carefully before a method of solution is chosen so that the most direct method, requiring the minimum amount of time, is used.

F1.7 Approximate Solutions

Many equations developed in engineering applications do not lend themselves to direct solution by standard methods. These equations must be solved by approximation methods to the accuracy dictated by the problem conditions. Experience is helpful in choosing the numerical technique for solution.

Example problem F1.8

Find to three significant figures the solution to the equation

$$2 - x = \ln x$$

Solution

One method of solution is graphical. If the equations $y = 2 - x$ and $y = \ln x$ are plotted, the common solution would be the intersection of the two lines. This would not likely give three-significant-figure accuracy, however. A more accurate method requires use of a scientific calculator or computer.

Table F1.1 Solution of $2 - x = \ln x$

x	1	2	1.5	1.6	1.55	1.56	1.557
$2 - x$	1	0	0.500	0.400	0.450	0.440	0.443
$\ln x$	0	0.693	0.405	0.470	0.438	0.445	0.443

Inspection of the equation reveals that the desired solution must lie between 1 and 2. It is then a matter of setting up a routine that will continue to bracket the solution between two increasingly accurate numbers. Table F1.1 shows the intermediate steps and indicates that the solution is $x = 1.56$ to three significant figures.

Computer spreadsheets and solvers or a programmable calculator could be used easily to determine a solution by the method just described. The time available and equipment on hand always will influence the numerical technique to be used.

Part 2 Selected Trigonometry Topics

F2.1 Introduction

This material is intended to be a brief review of concepts from plane trigonometry that are commonly used in engineering calculations. The section deals only with plane trigonometry and furnishes no information about spherical trigonometry. The reader is referred to standard texts in trigonometry for more detailed coverage and analysis.

F2.2 Trigonometric Function Definitions

The trigonometric functions are defined for an angle contained within a right triangle, as shown in Figure F2.1.

$$\text{sine } \theta = \sin \theta = \frac{\text{opposite side}}{\text{hypotenuse}} = \frac{y}{r}$$

$$\text{cosine } \theta = \cos \theta = \frac{\text{adjacent side}}{\text{hypotenuse}} = \frac{x}{r}$$

$$\text{tangent } \theta = \tan \theta = \frac{\text{opposite side}}{\text{adjacent side}} = \frac{y}{x}$$

$$\text{cotangent } \theta = \cot \theta = \frac{\text{adjacent side}}{\text{opposite side}} = \frac{x}{y} = \frac{1}{\tan \theta}$$

$$\text{secant } \theta = \sec \theta = \frac{\text{hypotenuse}}{\text{adjacent side}} = \frac{r}{x} = \frac{1}{\cos \theta}$$

$$\text{cosecant } \theta = \csc \theta = \frac{\text{hypotenuse}}{\text{opposite side}} = \frac{r}{y} = \frac{1}{\sin \theta}$$

The angle θ is by convention measured positive in the counterclockwise direction from the positive x axis.

Figure F2.1

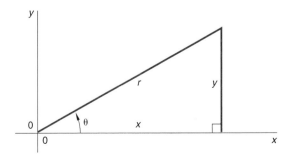

Coordinate definition.

Table F2.1

Quadrant 2	y	Quadrant 1
$x(-)$, $y(+)$		$x(+)$, $y(+)$
sin and csc$(+)$		sin and csc$(+)$
cos and sec$(-)$		cos and sec$(+)$
tan and cot$(-)$		tan and cot$(+)$
		x
Quadrant 3		Quadrant 4
$x(-)$, $y(-)$		$x(+)$, $y(-)$
sin and csc$(-)$		sin and csc$(-)$
cos and sec$(-)$		cos and sec$(+)$
tan and cot$(+)$		tan and cot$(-)$

F2.4   Radians and Degrees

Angles may be measured in either degrees or radians (see Fig. F2.2). By definition

$$1 \text{ degree } (°) = \frac{1}{360} \text{ of the central angle of a circle}$$

1 radian (rad) = angle subtended at center 0 of a circle by an arc equal to the radius

The central angle of a circle is 2π rad or $360°$. Therefore

$$1° = \frac{2\pi}{360°} = \frac{\pi}{180°} = 0.017\ 453\ 29 \cdots \text{ rad}$$

and

$$1 \text{ rad} = \frac{360°}{2\pi} = \frac{180°}{\pi} = 57.295\ 78 \cdots °$$

Figure F2.2

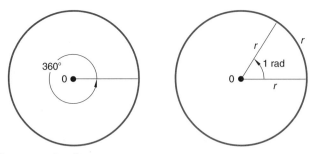

Definition of degrees and radians.

It follows that the conversion of θ in degrees to θ in radians is given by

$$\theta \,(\text{rad}) = \theta \,(°) \,\frac{\pi}{180°}$$

and in like manner

$$\theta \,(°) = \theta \,(\text{rad}) \,\frac{180°}{\pi}$$

F2.5 Plots of Trigonometric Functions

Figure F2.3

$y = \sin\theta$

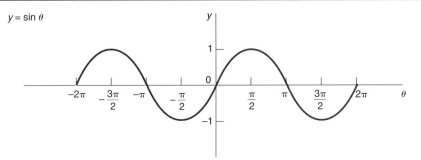

$y = \cos\theta$

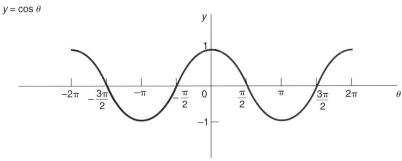

Plots of trigonometric functions.

Figure F2.3 Continued

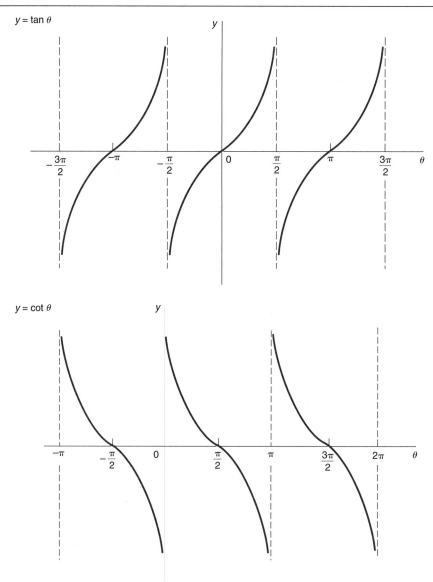

$y = \tan \theta$

$y = \cot \theta$

F2.6 Standard Values of Often-Used Angles

From three basic triangles (see Fig. F2.4), it is possible to compute the values of the trigonometric functions for many standard angles such as 30°, 45°, 60°, 120°, 135°, and so on. We only need to recall that $\sin 30° = \cos 60° = \frac{1}{2}$ and $\tan 45° = 1$ to construct the necessary triangles from which values can be taken to obtain the other functions.

The functions for 0°, 90°, 180°, and so on can be found directly from the function definitions and a simple line sketch. See Table F2.2.

Figure F2.3 Continued

$y = \sec\theta$

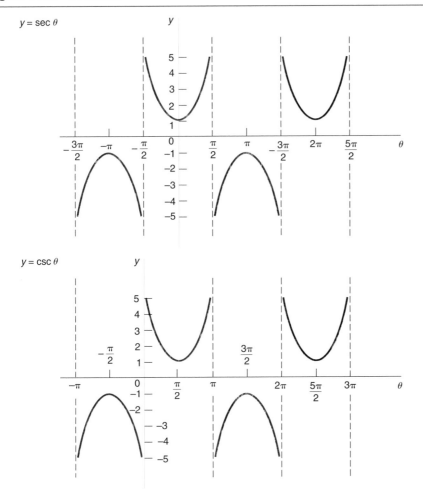

Figure F2.4

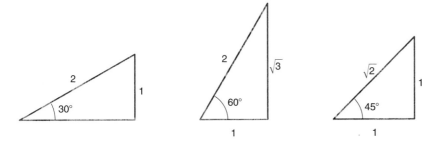

Common triangles.

Table F.2.2 Functions of Common Angles

Angle/Function	0°	30°	45°	60°	90°
sin	0	1/2	$\sqrt{2}/2$	$\sqrt{3}/2$	1
cos	1	$\sqrt{3}/2$	$\sqrt{2}/2$	1/2	0
tan	0	$\sqrt{3}/2$	1	$\sqrt{3}$	∞
cot	∞	$\sqrt{3}$	1	$\sqrt{3}/3$	0
sec	1	$2\sqrt{3}/3$	$\sqrt{2}$	2	∞
csc	∞	2	$\sqrt{2}$	$2\sqrt{3}/3$	1

F2.7   Inverse Trigonometric Functions

Definition
If $y = \sin \theta$, *then θ is an angle whose sine is y.* The symbols ordinarily used to denote an inverse function are

$$\theta = \arcsin y$$

or

$$\theta = \sin^{-1} y$$

Note:

$$\sin^{-1} y \neq \frac{1}{\sin y}$$

This is an exception to the conventional use of exponents.

Inverse functions $\cos^{-1} y$, $\tan^{-1} y$, $\cot^{-1} y$, $\sec^{-1} y$, and $\csc^{-1} y$ are similarly defined. Each of these is a many-valued function of y. The values are grouped into collections called *branches*. One of these branches is defined to be the principal branch, and the values found there are the principal values.

The principal values are as follows:

$$-\frac{\pi}{2} \leq \sin^{-1} y \leq \frac{\pi}{2}$$

$$0 \leq \cos^{-1} y \leq \pi$$

$$-\frac{\pi}{2} < \tan^{-1} y < \frac{\pi}{2}$$

$$0 < \cot^{-1} y < \pi$$

$$0 \leq \sec^{-1} y \leq \pi \qquad \left(\sec^{-1} y \neq \frac{\pi}{2}\right)$$

$$-\frac{\pi}{2} \leq \csc^{-1} y \leq \frac{\pi}{2} \quad (\csc^{-1} y \neq 0)$$

F2.8   Plots of Inverse Trigonometric Functions

All angles are given in radians. Principal branches are shown as solid lines. See Figure F2.5.

Figure F2.5

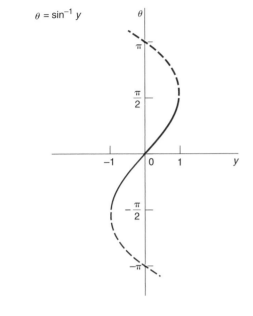

$\theta = \sin^{-1} y$

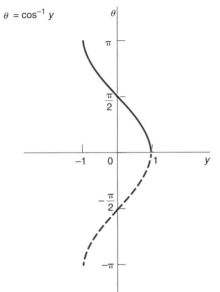

$\theta = \cos^{-1} y$

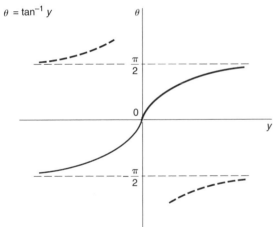

$\theta = \tan^{-1} y$

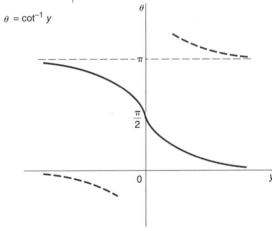

$\theta = \cot^{-1} y$

$\theta = \sec^{-1} y$

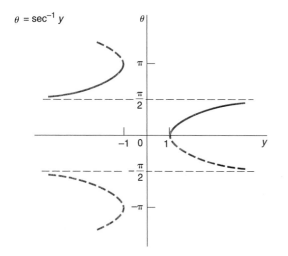

$\theta = \csc^{-1} y$

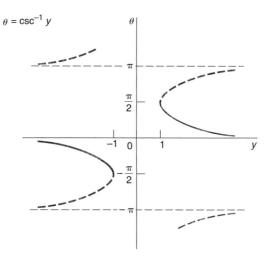

F2.9   Polar-Rectangular Coordinate

See Figure F2.6.

Conversion from polar to rectangular coordinates $(r, \theta) \rightarrow (x, y)$ is given by the following equations:

$$x = r \cos \theta$$

$$y = r \sin \theta$$

Conversion from rectangular to polar coordinates $(x, y) \rightarrow (r, \theta)$ requires the following equations:

$$r = [x^2 + y^2]^{1/2}$$

$$\theta = \tan^{-1}\left(\frac{y}{x}\right)$$

Figure F2.6

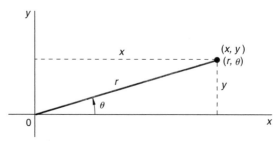

Rectangular and polar coordinate definitions.

The conversion from polar to rectangular coordinates also can be thought of as the determination of the x and y components of a vector (r, θ). Likewise, conversion from rectangular to polar coordinates is the same as finding the resultant vector (r, θ) from its x and y components.

F2.10 Laws of Sines and Cosines

The fundamental definitions of sine, cosine, and so on, apply strictly to right triangles. Solutions needed for oblique triangles then must be accomplished by appropriate constructions that reduce the problem to a series of solutions to right triangles.

Two formulas have been derived for oblique triangles that are much more convenient to use than the construction technique. They are the law of sines and the law of cosines, which apply to any plane triangle. See Figure F2.7.

The *law of sines* states

$$\frac{\sin A}{a} = \frac{\sin B}{b} = \frac{\sin C}{c}$$

The *law of cosines* is

$$a^2 = b^2 + c^2 - 2bc \cos A$$

or

$$b^2 = a^2 + c^2 - 2ac \cos B$$

or

$$c^2 = a^2 + b^2 - 2ab \cos C$$

Application of the law of sines is most convenient in the case where two angles and one side are known and a second side is to be found.

Example problem F2.1

Determine the length of side x for the triangle with base 12.0 m as shown in Figure F2.8.

Solution

The sum of the interior angles must be 180°; therefore

$$\alpha = 180° - 43° - 59° = 78°.$$

Applying the law of sines, we have

$$\frac{\sin 78°}{12.0 \text{ m}} = \frac{\sin 43°}{x}$$

$$x = \frac{\sin 43°}{\sin 78°} \, 12.0 \text{ m}$$

$$= 8.37 \text{ m}$$

The law of cosines is most convenient to use when two sides and the included angle are known for a triangle.

Figure F2.7

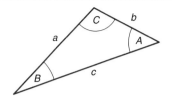

Angle and side designations.

Figure F2.8

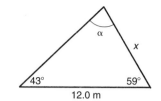

Angle and side designations.

Example problem F2.2

Calculate the length of side y of the triangle shown in Figure F2.9. Its base is 14.0 m.

Solution

Substitute into the law of cosines:

$$y^2 = (14.0 \text{ m})^2 + (7.00 \text{ m})^2 - 2(14.0 \text{ m})(7.00 \text{ m})(\cos 52°)$$

$$= 124.33 \text{ m}^2$$

$$y = 11.2 \text{ m}$$

F2.11   Area of a Triangle

Formulas for the area of a triangle (see Fig. F2.7) in terms of two sides and their included angle are

$$\text{Area} = \frac{1}{2} ab \sin C$$

$$\text{Area} = \frac{1}{2} ac \sin B$$

$$\text{Area} = \frac{1}{2} bc \sin A$$

Figure F2.9

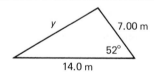

Formulas written in terms of one side and three angles are

$$\text{Area} = \frac{1}{2} a^2 \frac{\sin B \sin C}{\sin A}$$

$$\text{Area} = \frac{1}{2} b^2 \frac{\sin A \sin C}{\sin B}$$

$$\text{Area} = \frac{1}{2} c^2 \frac{\sin A \sin B}{\sin C}$$

The formula for the area in terms of the sides is

$$\text{Area} = [s(s - a)(s - b)(s - c)]^{1/2}$$

where $s = \frac{1}{2} (a + b + c)$.

Example problem F2.3

Determine the areas of the two triangles defined in Sec. F2.10.

Solution

For the 12.0 m base triangle

$$\text{Area} = \frac{1}{2} (12.0 \text{ m})^2 \frac{\sin 59° \sin 43°}{\sin 78°}$$

$$= 43.0 \text{ m}^2$$

For the 14.0 m base triangle

$$\text{Area} = \frac{1}{2} (7.00 \text{ m})(14.0 \text{ m})\sin 52°$$

$$= 38.6 \text{ m}^2$$

F2.12   Series Representation of Trigonometric

Infinite-series representations exist for each of the trigonometric functions. Those for sine, cosine, and tangent are

$$\sin \theta = \theta - \frac{\theta^3}{3!} + \frac{\theta^5}{5!} + \cdots + (-1)^{n-1} \frac{\theta^{2n-1}}{(2n-1)!} + \cdots$$

$$\theta \text{ in radians} \qquad (-\infty < \theta < +\infty)$$

$$\cos \theta = 1 - \frac{\theta^2}{2!} + \frac{\theta^4}{4!} + \cdots + (-1)^{n-1} \frac{\theta^{2n-2}}{(2n-2)!} + \cdots$$

$$\theta \text{ in radians} \qquad (-\infty < \theta < +\infty)$$

$$\tan \theta = \theta + \frac{\theta^3}{3} + \frac{2\theta^5}{15} + \cdots + \frac{2^{2n}(2^{2n} - 1) B_n \theta^{2n-1}}{(2n)!} + \cdots$$

$$\theta \text{ in radians} \qquad \left(-\frac{\pi}{2} < \theta < \frac{\pi}{2}\right)$$

where B_n are the Bernoulli numbers

$$B_1 = \frac{1}{6}$$

$$B_2 = \frac{1}{30}$$

$$B_3 = \frac{1}{42}$$

$$B_4 = \frac{1}{30}$$

$$B_5 = \frac{5}{66}$$

$$B_n = \frac{(2n)!}{2^{2n-1}(\pi)^{2n}}\left(1 + \frac{1}{2^{2n}} + \frac{1}{3^{2n}} + \cdots\right)$$

and where the factorial symbol (!) is defined as

$$n! = n(n = 1)(n - 2) \cdots (3)(2)(1)$$

Trigonometric functions to any accuracy can be calculated from the series if enough terms are used.

For small angles, on the order of 5° or less, it may be sufficient to use only the first term of each series.

$$\sin \theta \cong \theta$$

$$\cos \theta \cong 1$$

$$\tan \theta \cong \theta$$

F2.13 Trigonometric Relationships

Functional Relationships

$$\tan \theta = \frac{\sin \theta}{\cos \theta}$$

$$\sin^2 \theta + \cos^2 \theta = 1$$

$$\sin^2 \theta - \tan^2 \theta = 1$$

$$\csc^2 \theta - \cot^2 \theta = 1$$

$$\sin \theta = \cos(90° - \theta) = \sin(180° - \theta)$$

$$\cos \theta = \sin(90° - \theta) = -\cos(180° - \theta)$$

$$\tan \theta = \cot(90° - \theta) = -\tan(180° - \theta)$$

$$\sin (-\theta) = -\sin \theta$$

$$\cos (-\theta) = \cos \theta$$

$$\tan (-\theta) = -\tan \theta$$

$$\sin (\theta \pm \alpha) = \sin \theta \cos \alpha \pm \cos \theta \sin \alpha$$

$$\cos (\theta \pm \alpha) = \cos \theta \cos \alpha \mp \sin \theta \sin \alpha$$

$$\tan (\theta \pm \alpha) = \frac{\tan \theta \pm \tan \alpha}{1 \mp \tan \theta \tan \alpha}$$

Multiple-Angle Formulas

$$\sin 2\theta = 2 \sin \theta \cos \theta$$

$$\cos 2\theta = \cos^2 \theta - \sin^2 \theta = 2 \cos^2 \theta - 1 = 1 - 2 \sin^2 \theta$$

$$\tan 2\theta = \frac{2 \tan\theta}{1 - \tan^2\theta}$$

$$\sin 3\theta = 3 \sin \theta - 4 \sin^3 \theta$$

$$\cos 3\theta = 4 \cos^3 \theta - 3 \cos \theta$$

$$\tan 3\theta = \frac{3 \tan \theta - \tan^3 \theta}{1 - 3 \tan^2 \theta}$$

$$\sin \frac{\theta}{2} = \pm \sqrt{\frac{1 - \cos \theta}{2}} \qquad \left(\text{sign depends on quadrant of } \frac{\theta}{2}\right)$$

$$\cos \frac{\theta}{2} = \pm \sqrt{\frac{1 + \cos \theta}{2}} \qquad \left(\text{sign depends on quadrant of } \frac{\theta}{2}\right)$$

$$\tan \frac{\theta}{2} = \pm \sqrt{\frac{1 - \cos \theta}{1 + \cos \theta}} = \frac{1 - \cos \theta}{\sin \theta}$$

$$= \frac{\sin \theta}{1 + \cos \theta} = \csc \theta - \cot \theta$$

$$\left(\text{sign depends on quadrant of } \frac{\theta}{2}\right)$$

Sum, Difference, and Product Formulas

$$\sin \theta + \sin \alpha = 2 \sin\left(\frac{\theta + \alpha}{2}\right)\cos\left(\frac{\theta - \alpha}{2}\right)$$

$$\sin \theta - \sin \alpha = 2 \cos\left(\frac{\theta + \alpha}{2}\right)\sin\left(\frac{\theta - \alpha}{2}\right)$$

$$\cos \theta + \cos \alpha = 2 \cos\left(\frac{\theta + \alpha}{2}\right)\cos\left(\frac{\theta - \alpha}{2}\right)$$

$$\cos \theta - \cos \alpha = 2 \sin\left(\frac{\theta + \alpha}{2}\right)\sin\left(\frac{\alpha - \theta}{2}\right)$$

$$\sin \theta \sin \alpha = \frac{1}{2}\left[\cos(\theta - \alpha) - \cos(\theta + \alpha)\right]$$

$$\cos \theta \cos \alpha = \frac{1}{2}\left[\cos(\theta - \alpha) + \cos(\theta + \alpha)\right]$$

$$\sin \theta \cos \alpha = \frac{1}{2}\left[\sin(\theta - \alpha) + \sin(\theta + \alpha)\right]$$

Power Formulas

$$\sin^2 \theta = \frac{1}{2} - \frac{1}{2}\cos 2\theta$$

$$\cos^2 \theta = \frac{1}{2} + \frac{1}{2}\cos 2\theta$$

$$\sin^3 \theta = \frac{3}{4}\sin \theta - \frac{1}{4}\sin 3\theta$$

$$\cos^3 \theta = \frac{3}{4}\cos \theta + \frac{1}{4}\cos 3\theta$$

Answers to Selected Problems

Chapter 4

4.1 $\alpha = 55°$, $B_y = 6.0$ m, $\mathbf{B} = 1.0 \times 10^1$ m

4.3 $XZ = 4.2 \times 10^4$ m

4.6 $\mathbf{R} = 19$ cm $\leftarrow$ horizontal

4.8 Height $= 6.2 \times 10^2$ m

4.11 $AB = 156.8$ m, $\angle ABC = 130.0°$

4.16 1.1 cm

4.20 Heading must be S 12.0°W, actual ground speed $= 587$ km/h

4.22 (*a*) 13 790 more revolutions

   (*b*) 5861 more revolutions

4.25 (*a*) 0 m/s

   (*b*) 11.5 m

   (*c*) 15 m/s downward

   (*d*) 31.8 m/s

   (*e*) 4.77 s

4.27 Shortest height of the tank $= 32.26$ cm

4.30 Refractive index of the glass, $n_b = 1.58$

Chapter 5

5.2 (*c*) V = 0.27 t $-$ 0.4

   (*d*) 0.27 m/s^2

5.4 (*c*) P = 0.71 Q $-$ 17

   (*d*) 9.27 kW

5.6 (*c*) H = 4×10^8 T^3

5.8 (*e*) Linear: R = -27 A + 206

   Exponential: R = 152 e$^{-0.35A}$

   Power: R = 51 A$^{-0.75}$

   (*f*) Power curve

5.12 (*c*) Linear: P = 6432 Y $-$ 1.2648×10^7

   Exponential: P = 5.784×10^{-53} e$^{0.0663Y}$

   (*d*) Exponential

5.14 (*c*) V = 198 e$^{-0.12t}$

5.16 (*c*) Linear: D = $-$ 0.0007T + 0.7824

   Exponential: D = 1.0463 e$^{-0.0017T}$

   Power: D = 215.9 T$^{-0.9974}$

   Power best

5.18 (*c*) C = 5503.8 e$^{-0.0742W}$

   (*d*) 127 counts/second

Chapter 6

6.2 (*b*) 5

   (*f*) exact conversion

   (*h*) 4

6.3 (*c*) 2.64×10^8

6.4 (*c*) 1400 lbm

6.5 (*d*) 168 cm³

6.6 (*a*) 205–215 lb/in²

   (*b*) 82–92 lb/in²

Chapter 7

7.1 (*a*) 1.60×10^2 kW

   (*b*) 390 ha

   (*c*) 1.057×10^3 m³

   (*d*) 64.6 m/s

   (*e*) 9.90×10^2 km/hr

   (*f*) 1.1×10^3 kg

   (*g*) 70.3 kg

   (*h*) 652 m

   (*i*) 1.637×10^4 L

   (*j*) 2.470×10^3 km

7.4 (*a*) 10.8×10^3 J

   (*b*) 1.01×10^5 Pa

   (*c*) 8.850×10^3 m

   (*d*) 48.7 g/cm³

   (*e*) 373K

7.7 (*a*) m = 116 kg

   (*b*) m = 146 kg

7.10 Vol = 1 330 in³

   Mass = 0.048 2 lbm

   #gal = 8.09 gal

7.13 Time = 14 min

7.16 (*a*) H = 43.0×10^3 ft

   (*b*) Dia = 60.8 ft

7.19

CD	cos θ	θ	> OAB	Area	Length	Vol, ft³	Vol, gal	Mass, kg
0	1	0	0	0	25	0	0	0
1	0.8	36.87	73.74	4.09		102.19	764.4	2 111.3
2	0.6	53.13	106.26	11.18		279.56	2 091.3	5 776
3	0.4	66.42	132.84	19.82		495.42	3 706	10 236
4	0.2	78.46	156.93	29.34		733.42	5 486.4	15 153
5	0	90	180	39.27		981.75	7 344	20 284
6				49.2		1 230.07	9 201.6	25 415
7				58.72		1 468.07	10 982	30 332
8				67.36		1 683.94	12 597	34 792
9				74.45		1 861.31	13 924	38 457
10				78.54		1 963.5	14 688	40 569

7.22 Vol = 3.5×10^9 ft³

7.25 $C_{New} = \dfrac{288.8\,\text{gal}}{\text{h} \cdot \text{in}^{2.5}} \dfrac{\text{h}}{3600\,\text{s}} \dfrac{\text{ft}^3}{7.48\,\text{gal}} \dfrac{12^{2.5}\,\text{in}^{2.5}}{\text{ft}^{2.5}} = 5.35\,\text{ft}^{0.5}/\text{s}$

L, in	H, in	Q, gal/h	Q, ft³/s
1	2	817	0.03
2	4	4 621	0.17
3	6	12 733	0.47
4	8	26 139	0.97
5	10	45 663	1.7
6	12	72 031	2.68
7	14	105 898	3.93
8	16	147 866	5.49
9	18	198 494	7.37
10	20	258 311	9.59
11	22	327 811	12.17
12	24	407 470	15.13
13	26	497 738	18.48
14	28	599 049	22.25
15	30	711 820	26.43
16	32	836 454	31.06

Chapter 8

8.2 (*a*) $783.53
   (*b*) $497.18
   (*c*) $327.68
8.4 (*a*) $58.67
   (*b*) $1906.55
   (*c*) $1114.60
   (*d*) $303.60
8.6 (*a*) $1 382 999.74
   (*b*) $1 390 288.11
   (*c*) $1 397 897.40
8.8 (*a*) $58 068.06
   (*b*) 85 675.19
8.10 (*a*) 4.73%
   (*b*) $78 249.10
8.12 $18 052.10 Firm lost money
8.14 $16 166.07
8.16 $1 331.16
8.19 $1142.94
8.21 $3488.21
8.23 $800 000
8.24 $114 022.10, $8087.22
8.26 14.19%
8.28 (*a*) $2014.93
   (*b*) $310.50
8.30 $20.62
8.32 (*a*) $2477.65 (asset)
   (*b*) $1427.22 (asset)
   (*c*) $2825.26 (liability)
   (*d*) $980.15 (liability)
   (*e*) $8702.95 (asset)
   Net worth = $8802.41
8.33 76.1%

8.35 Select option b

8.37 (*a*) $15 473.27

    (*b*) $18 000.00

Chapter 9

9.1 Foam: $4522.29 (net cost)

    Fiberglass: 716.76 (net savings) (select)

9.2 Choose workstation 2

9.3 Choose Sande 10

Chapter 10

10.1 (*a*) 2.53

    (*b*) 2

    (*c*) 2

10.4 (*a*) 78.72

    (*b*) 88

    (*c*) 78

    (*d*) 14.93

10.7 (*a*)

215–219	3
220–224	6
225–229	11
230–234	4
235–239	2
240–244	4

    (*b*)

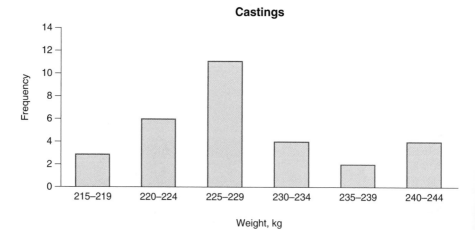

    (*c*) Mean = 228.53

       Mode = 228

       Median = 228

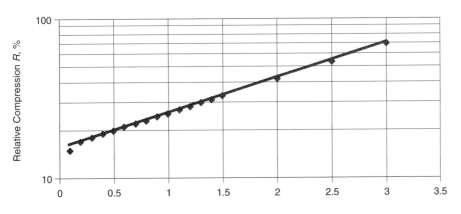

y = 15.293e^{0.5085x}
R² = 0.997 8

Test Results for SILON Q-177

(b) 15.293
(c) $y = 15.293e^{0.5085x}$
(d) r = 0.9989
(e) pressure = 2.72 MPa

10.15

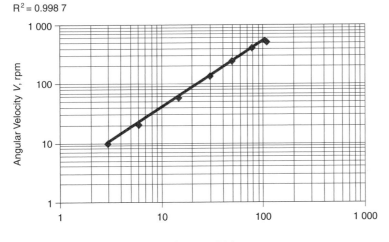

y = 2.859 7x^{1.125 7}
R² = 0.998 7

Conveyor Capacity

(d) r = 0.999 3
(e) C = 6 814 L/s = 6800 L/s

Chapter 11

11.1 (*a*) 15.9%
    (*b*) 0.1%
    (*c*) 95.4%
    (*d*) 50%
11.4 (*a*) 23.3%
    (*b*) 74.5%
    (*c*) 82.8%
11.5 21 190 cans
11.7 (*a*) 1714
    (*b*) 259
    (*c*) 14.29%
    (*d*) 139
11.8 40K±5K p = 0.1587
    35K±15K p = 0.2514

Chapter 12

12.1 $\mathbf{F}_X$ = 156 lbf →
    $\mathbf{F}_y$ = 99.4 lbf ↑
12.7 $\mathbf{R}$ = 42 kN ∠ 34°
12.10 Resultant $\mathbf{T}$ = 85 kN, 36, 77
12.13 $\mathbf{F}_1$ = 1200 N, 5, 2
    $\mathbf{F}_2$ = 1180 N, 1, 2
12.20 $\mathbf{M}_A$ = 4460 ft lbf ⤵
    $\mathbf{M}_B$ = 18 400 ft lbf ⤵
12.22 Unknown force $\mathbf{S}$ = 30 lbf ↑
    Location is 3.2 ft left of B
12.25 $\mathbf{A}$ = 6.3 kN ↑
    $\mathbf{B}$ = 5.4 kN, 130°
12.29 $\mathbf{A}$ = 2.9 kN, 190°
    $\mathbf{B}$ = 2.2 kN ↑
12.31 $\mathbf{A}$ = 360 lbf, 28°
    $\mathbf{B}$ = 420 lbf, 42°
12.34 Cable tension $\mathbf{T}$ = 2090 lbf, 160°
    Reaction at A $\mathbf{A}$ = 2030 lbf, 14°

Chapter 13

13.1 Minimum area for stress: A = 271 mm²
Minimum area for elongation: A = 235 mm²
Therefore: A_{min} = 271 mm²
13.2 Required diameter: d = 2.6 cm
Choose next largest commercial size
13.7 Maximum load for stress: L = 33 kN
Maximum load for deformation: L = 49 kN
Therefore: L_{max} = 33 kN

Chapter 14

14.1 A = 1260 kg
B = 872 kg
14.3 106 lb
14.6 291 kg
14.10 Product = 0.3 t/h; alcohol = 80%; water = 14%; inert = 6%
14.13 Total input to evaporator = 19 t/h
Feed rate to crystallizer/filter = 17 t/h
Water removed by evaporator = 2.0 t/h
14.16 Dirty solvent produced = 12.7 t
Percent solvent in discard = 91%
Percent toxic removed in discard = 92.6%
14.19 40% concentrate req'd = 35 lb; 98% concentrate req'd = 215 lb
14.23 44 lbm
14.25 Feed rate = 99 t/h
Refuse composition: coal = 10%; ash = 45%; water = 13%
14.28 Flow rate of processed livers = 2000 lbm/h
Percent oil in processed livers = 1.0%
Percent ether in processed livers = 48%
Percent inert material in processed livers = 51%

Chapter 15

No solutions

Chapter 16

16.2 V = 59.0 mph
16.4 d = 660 ft
16.8 V = 91 mph
16.11 V = 54.7 ft/s
16.16 $u_2 - u_1$ = 4.08 kJ/kg
16.19 The efficiency of the invention is 42%, which is greater than the Carnot
Efficiency of 35.4%; therefore it is not a reasonable claim.
16.20 T_H = 319°C
16.23 (*a*) 70 hp
(*b*) 61.3 kW
16.26 # homes = 38 900
16.28 (*a*) 97.6%
(*b*) 3.432 kWh
(*c*) $0.34

17.2 1.9×10^5 C

17.4 I = 1.07 A, R = 7.03 Ω

17.6 P = 50 W, R = 3.2 Ω

17.8 (*b*) R_E = 52 Ω

　　(*c*) I = 1.8 A

　　(*d*) E_1 = 22 V, E_2 = 27 V, E_3 = 46 V

17.10 (*b*) R_E = 6.81 Ω

　　(*c*) I = 0.881 A

　　(*d*) I_1 = 0.50 A, P_1 = 3.00 W

　　I_2 = 0.375 A, P_2 = 2.25 W

　　I_3 = 6.0×10^{-3} A, P_3 = 0.036 W

17.12 Current through 15-Ω resistor = 1.3 A

　　Voltage across 15-Ω resistor = 2.0×10^2 V

　　Fraction of power consumed by 25-Ω resistor = 17%

17.14 Charging current = 3 A

　　Back-emf = 1.2×10^2 V

17.16 (*a*) I_{18H} = 2.22 A, I_{18V} = 4.44 A, I_{24} = 1.67 A, I_{12} = 2.00 A, I_8 = 2.00 A

　　(*b*) 533 W

　　(*c*) 40.0 V

　　(*d*) 48 W

17.18 (*a*) I_1 = 11.1 A, I_2 = 10.6 A, I_3 = 21.7 A, $P_{2.5}$ = 283 W, $P_{1.5}$ = 183 W, $P_{5.0}$ = 2.35 kW

　　(*b*) I_1 = 18.0 A, I_2 = 6.00 A, I_3 = 24.0 A, $P_{2.5}$ = 90.0 W, $P_{1.5}$ = 486 W, $P_{5.0}$ = 2.88 kW

17.20 I_{15Bat} = 1.17 A, I_{12Bat} = 0.83 A, $I_{5.0V}$ = 2.00 A, $I_{2.0}$ = 1.00 A, $I_{5.0H}$ = 1.00A, $I_{6.0}$ = 0.17 A

　　Power to 2.0-V battery = 0.33 W

　　Voltage across 2.0-Ω resistor = 2.00 V

　　Power consumed by 6.0-Ω resistor = 0.17 W

17.22 R = 1.33 Ω

17.24 (*a*) Current for E_1 = 0.55 A, Current for E_2 = 9.3 A

　　(*b*) 1.4×10^2 W

　　(*c*) 1.2 kW

　　(*d*) 89%

17.26 Power delivered by 10-V battery = 15 W

　　Power delivered by 16-V battery = 1.0×10^2 W

　　Current through 1-Ω resistor = 8.0 A

　　Voltage across 0.8-Ω resistor = 4.4 V

Credits

Chapter 1
Figure 1.1: Brand X Pictures
Figure 1.2: Royalty-Free/CORBIS
Figure 1.3: © Corbis
Figure 1.4: Kim Steele/Getty Images
Figure 1.5: Creatas Images/Jupiterimages
Figure 1.6: © Royalty Free Corbis
Figure 1.7: © Tom Grill/Corbis
Figure 1.8: © Business & Technology/Corbis RF
Figure 1.9: Keith Brofsky/Getty Images
Figure 1.10: © Business & Technology/Corbis RF
Figure 1.11: © Manufacturing & Industry/Corbis RF
Figure 1.12: Liquidlibrary/Dynamic; Graphics/Jupiterimages
Figure 1.13: © Business Perspectives/Corbis
Figure 1.14: © Business Perspectives/Corbis
Figure 1.15: BananaStock/PictureQuest
Figure 1.17: Royalty-Free/CORBIS
Figure 1.18: © Vol. 45/Corbis
Figure 1.19: © Digital Vision/PunchStock
Figure 1.20: © Brand X Pictures/PunchStock
Figure 1.21: © Energy Issues/Corbis
Figure 1.22: © Business & Technology/Corbis RF
Figure 1.23: © Brand X Pictures/PunchStock

Chapter 2
Figure 2.1: BananaStock/PictureQuest
Figure 2.3: © Image Source/PunchStock

Chapter 3
Figure 3.1: © Brand X Pictures/PunchStock
Figure 3.2: © Creatas/PunchStock
Figure 3.3: © Vol. 66/PhotoDisc/Getty Images
Figure 3.8: © Energy Issues/Corbis
Figure 3.9: © Energy Issues/Corbis
Figure 3.11: Royalty-Free/CORBIS
Figure 3.12: © Beathan/Corbis

Chapter 6
ta6.1: Courtesy of Boeing Company

Chapter 10
Figure 10.1: © Vol. 25/PhotoDisc/Getty Images

Chapter 12
Figure 12.1: Imagestate Media (John Foxx)

Chapter 13
Figure 13.5: Author photo

Chapter 14
Figure 14.1: © Royalty Free Corbis

Chapter 15
Figure 15.10: © Royalty Free Corbis
Figure 15.12: © Royalty Free Corbis

Index

477